44 Springer Series in Solid-State Sciences

Edited by Ekkehart Kröner

Springer Series in Solid-State Sciences

Editors: M. Cardona P. Fulde H.-J. Queisser

Volume 40 **Semiconductor Physics** - An Introduction By K. Seeger

Volume 41 **The LMTO Method** By H.L. Skriver

Volume 42 **Crystal Optics with Spatial Dispersion and the Theory of Excitations**
By V.M. Agranovich and V.L. Ginzburg

Volume 43 **Resonant Nonlinear Interactions of Light with Matter**
By V.S. Butylkin, A.E. Kaplan, Yu.G. Khronopulo, and E.I. Yakubovich

Volume 44 **Elastic Media with Microstructure II** Three-Dimensional Models
By I.A. Kunin

Volume 45 **Electronic Properties of Doped Semiconductors**
By B.I. Shklovsky and A.L. Efros

Volume 46 **Topological Disorder in Condensed Matter**
Editors: F. Yonezawa and T. Ninomiya

Volumes 1 - 39 are listed on the back inside cover

Isaak A. Kunin

Elastic Media with Microstructure II

Three-Dimensional Models

With 20 Figures

Springer-Verlag
Berlin Heidelberg New York Tokyo 1983

Professor *Isaak A. Kunin*

Department of Mechanical Engineering, University of Houston
Houston, TX 77004, USA

Guest Editor:

Professor Dr. Ekkehart Kröner

Institut für Physik, Universität Stuttgart
D-7000 Stuttgart 80, Fed. Rep. of Germany

Title of the original Russian edition: *Teoriya uprugikh sred s mikrostrukturoi*

ISBN 3-540-12078-5 Springer-Verlag Berlin Heidelberg New York Tokyo
ISBN 0-387-12078-5 Springer-Verlag New York Heidelberg Berlin Tokyo

Library of Congress Cataloging in Publication Data. (Revised for volume 2). Kunin, I. A. (Isaak Abramovich), 1924- Elastic media with microstructure. (Springer series in solid-state sciences ;) Rev. and updated translation of: Teoriia uprugikh sred s mikrostrukturoĭ. Bibliography: v. 1, p. Includes index. Contents: 1. One-dimensional models. - v. 2. Three-dimensional models. 1. Elasticity. I. Title. II. Series.
QA931.K8913 1982 531'.3823 81-18268

Offset printing: Beltz Offsetdruck 6944 Hemsbach. Bookbinding: J. Schäffer OHG, 6718 Grünstadt
2153/3130-543210

Preface

Crystals and polycrystals, composites and polymers, grids and multibar systems can be considered as examples of media with microstructure. A characteristic feature of all such models is the existence of scale parameters which are connected with microgeometry or long-range interacting forces. As a result the corresponding theory must essentially be a nonlocal one.

This treatment provides a systematic investigation of the effects of microstructure, inner degrees of freedom and nonlocality in elastic media. The propagation of linear and nonlinear waves in dispersive media, static, deterministic and stochastic problems, and the theory of local defects and dislocations are considered in detail. Especial attention is paid to approximate models and limiting transitions to classical elasticity.

The book forms the second part of a revised and updated edition of the author's monograph published under the same title in Russian in 1975. The first part (Vol. 26 of Springer Series in Solid-State Sciences) presents a self-contained theory of one-dimensional models. The theory of three-dimensional models is considered in this volume.

I would like to thank E. Kröner and A. Seeger for supporting the idea of an English edition of my original Russian book. I am also grateful to E. Borie, H. Lotsch and H. Zorski who read the manuscript and offered many suggestions.

Houston, Texas
January, 1983

Isaak A. Kunin

Contents

1. Introduction

In recent years, new physical and mathematical models of material media, which can be considered far-reaching generalizations of classical theories of elasticity, plasticity, and ideal and viscous liquids, have been developed intensively. Such models have appeared for a number of reasons. Primary among them are the use of new construction materials in extreme conditions and the intensification of technological processes. The increasing tendency toward rapprochement of mechanics with physics is closely connected with these factors. The internal logic of the development of continuum mechanics as a science is also important.

This treatment is devoted to the study of models of elastic media with microstructure and to the development of the nonlocal theory of elasticity. Starting from such models as a crystal lattice and simple discrete mechanical systems, we develop the theory and its applications in a systematic way.

The Cosserat continuum was historically one of the first models of elastic media which could not be described within the scope of classical elasticity [1.1] However, the memoirs of *E.* and *F. Cosserat* (1909) remained unnoticed for a long time, and only around 1960 did the generalized models of the Cosserat continuum start to be developed intensively. They are known as oriented media, asymmetric, multipolar, micromorphic, couple-stress, etc., theories. For short we shall call them couple-stress theories. Essential contributions to the development of couple-stress theories were made, for example, by Aero, Eringen, Green, Grioli, Günther, Herrmann, Koiter, Kuvshinsky, Mindlin, Naghdi, Nowacki, Palmov, Rivlin, Sternberg, Toupin, and Wozniak; their fundamental works are listed in the Bibliography. The Bibliography is far from being complete. A survey of works before 1960 can be found in the fundamental treatment of *Truesdell* and *Toupin* [1.2]; later ones are quoted in papers by *Wozniak* [1.3], *Savin* and *Nemish* [1.4], *Iliushin* and *Lomakin* [1.5], as well as in monographs by *Misicu* [1.6] and *Nowacki* [1.7].

From the very beginning of the development of the generalized Cosserat models, attention was turned to their connections with the continuum theory of dislocations. In 1967 a symposium was organized by the International Union of Theoretical and Applied Mechanics, which had great significance in summing up the ten-year period of development [1.8]. In the symposium a new trend closely connected with the theory of the crystal lattice was also presented which contained the above-indicated models as a long-wave approximation, namely, a nonlocal theory of elasticity. The nonlocal theory of elasticity was also developed in works of Edelen, Eringen, Green, Kröner, Kunin, Laws and others, given in the Bibliography. A rather complete listing on media with microstructure is contained in [1.9]. It is worth mentioning here that the very term "nonlocal elasticy" seems to have been introduced by *Kröner* in 1963 [1.10], and the first monograph on the subject was published by the present author in 1975 [1.11].

We start out with a brief classification of the theories of elastic media with microstructure. Explicit or implicit nonlocality is the characteristic feature of all such theories. The latter, in its turn, displays itself in that the theories contain parameters which have the dimension of length. These scale parameters can have different physcial meanings: a distance between particles in discrete structures, the dimension of a grain or a cell, a characteristic radius of correlation or action-at-a-distance forces, etc. However, we shall always assume that the scale parameters are small in comparison with dimensions of the body.

One has to distinguish the cases of strong and weak nonlocality. If the "resolving power" of the model has the order of the scale parameter, i.e., if, in the corresponding theory, it is physically acceptable to consider wavelengths comparable with the scale parameter, then we shall call the theory nonlocal or strongly nonlocal (when intending to emphasize this). In such models, one can consider elements of the medium of the order of the scale parameter, but, as a rule, distances much smaller than the parameter have no physcial meaning. The equations of motion of a consistently nonlocal theory necessarily contain integral, integrodifferential, or finite-difference operators in the spatial variables. In nonlocal models, the velocity of wave propagation depends on wavelength; therefore, the term "medium with spatial dispersion" is also used frequently.

Let us emphasize that nonlocality or spatial dispersion can have different origins. They can be caused by a microstructure of the medium (in particular, by the discreteness of the micromodel) or by approximate consideration of such parameters as thickness of a rod or plate. One can speak therefore about the physical or geometrical nature of nonlocality. In the latter case, the nonlocal model is, as a rule, one- or two-dimensional and serves as an effective approximate description of a local three-dimensional medium.

If the scale parameter is small in comparison with the wavelengths considered, but the effects of nonlocality cannot be neglected completely, then a transition is possible to approximate models, for which integral and finite-difference operators are replaced by differential operators with small parameters attached to their highest derivatives. In such a case, one can speak about the model of the medium with weak spatial dispersion. The corresponding theory will be called weakly nonlocal. All above-mentioned couple-stress theories belong to this type, although they are usually constructed on a purely phenomenological basis.

Finally, the consideration of sufficiently long waves (zeroth long-wave approximation) leads to a transition to a local theory in the limit, already containing no scale parameters. This property of locality, i.e., the possibility of considering "infinitesimally small" elements of the medium, is inherent in all the classical models of the mechanics of continuous media.

Let us return to nonlocal models. They can be divided into two classes: discrete and continuous. Discrete structure of a medium could be taken into account in the usual way, for example, as is done in the theory of the crystal lattice. However, the apparatus of discrete mathematics is most cumbersome; therefore, we shall also use the mathematical model of quasicontinuum for an adequate description of the discrete medium. Its essence is an interpolation of functions of discrete argument by a special class of analytical functions in such a way that the condition of one-to-one correspondence between quasicontinuum and the discrete medium is fulfilled. The advantages of such an approach consist in an ability to describe discrete and continuous media within the scope of a unified formalism and, in particular, to generalize correctly such concepts of continuum mechanics as strain and stress. It is to be emphasized that the model of quasicontinuum is applicable not only to crystal lattices but also to macrosystems.

We shall also distinguish media of simple and complex structures. In the first case, the displacement vector is the only kinematic variable and it determines a state of the medium completely. Body forces are the corresponding force variable. To describe a medium of complex structure, a set of microrotations and microdeformations of different orders characterizing the internal degrees of freedom and the corresponding force micromoments is additionally introduced.

The difference between media of simple and complex structures, generally speaking, is conserved in the approximation of weak nonlocality, but this is displayed only for high enough frequencies of the order of the natural frequencies of the internal degrees of freedom. At low frequencies, the internal degrees of freedom can be excluded from the equations of motion so that they will contribute to the effective characteristics of the medium only. The difference between the quasicontinuum and the continuous medium completely disappears in the approximation of weak nonlocality.

In the zeroth long-wave approximation, at not very high frequencies, a complete identification of different models of a medium with microstructure takes place: all of them are equivalent to the classical model of elastic continua which was obtained on the basis of general phenomenological postulates. Only effective elastic moduli "know" about the structure of the initial micromodel, but this information cannot, of course, be derived from them. It follows that an explicit consideration of microstructure effects and, in particular, of the internal degrees of freedom is possible only with the simultaneous consideration of nonlocality, i.e., a consistent theory of elastic medium with microstructure must necessarily be nonlocal.

Schematically, the connections between different theories are shown in Fig. 1.1.

Our main purpose is the investigation of the effects of microstructure and nonlocality. In addition, we wish to elucidate the domain of applicability of different theories of media with microstructure. Such theories are considerably more complex than the usual theory of elasticity, although they reduce to this in

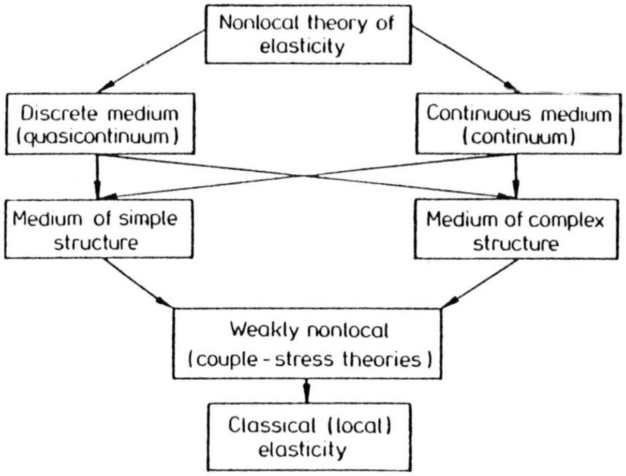

Fig. 1.1. Illustrating the interconnections of the various theories

certain limits; their application is, as a rule, reasonable only when they describe qualitatively new effects which are not derivable from the local theory.

We find it advisable to consider a number of simple models of media with microstructure in order to acquire some nonlocal intuition before proceeding to generalizations.

For these reasons, the treatment is divided into two parts. One-dimensional models, for which cumbersome tensor algebra is not needed, are studied in the first volume [1.12]. At the same time, one can trace a number of distinctions of the nonlocal theory from the local one already in one-dimensional models. These distinctions have both physical and methodological character. Particularly, one has to analyze critically the possibility of using in nonlocal theories such habitual notions as stress, strain, energy, density and flux.

In this volume, three-dimensional models of media with microstructure are considered, the main attention being payed to specific three-dimensional effects. The second and third chapters are devoted to the general theory of media with simple and complex structures. Local defects in media with microstructure as well as in elasticity are considered in Chapter 4. Internal stress and dislocations in the nonlocal elasticity are considered in Chapters 5 and 6. Chapter 7 (written by S.K. Kanaun) is devoted to random fields of inhomogeneities.

The author has tried to avoid complex mathematical methods in the first part. It is assumed that the reader is acquainted with the Fourier transform and has some skill in working with δ-functions (though it is easy to acquire it in the process of reading the book).

The necessity of simultaneous use of space-time and frequency-wave representations is one of the peculiarities of the formalism used in the nonlocal theory. This way of thinking is quite habitual for physicists, but could be, to some extent, unusual for engineers. Because of this, the rate of presenting the material is slow in the beginning and then speeds up gradually. For the same purpose, a number of results are presented in the form of problems, which are considered a part of the text and they are referred to in the later text.

In conclusion, let us note that a number of important questions are omitted. In particular, this is related to the thermodynamics of the nonlocal models, which was contained in the original plan of the book. Unfortunately, the author has not succeeded in representing this problem in a sufficiently simple and physically motivated form. An axiomatic approach to the nonlocal thermodynamics was developed by *Edelen* [1.13].

2. Medium of Simple Structure

In this chapter, we consider in detail the three-dimensional elastic medium of simple structure. To describe, in the same language, discrete and continuum non-local models, we introduce the three-dimensional quasicontinuum. Then, starting from crystal dynamics, we obtain the general nonlocal equations of motion for the medium of simple structure. We investigate the consequences of such physical requirements as energy conservation, finite action-at-a-distance, stability and invariance of the energy with respect to rigid body motions. In particular, the last requirement leads to the existence of a symmetric stress tensor. In contrast to the local elasticity, the stress and energy density are not defined uniquely in nonlocal elasticity.

The second part of the chapter is devoted to homogeneous media. Wave propagation (including surface waves) in dispersive media and the corresponding physical phemomena are examined. Approximate theories that can describe partially these effects are discussed. As examples of nonlocal media, a cubic lattice and isotropic models are considered.

2.1 Quasicontinuum

Let us suppose that in a three-dimensional Euclidean space E_3 with points x, a triple of noncoplanar vectors $e_\alpha (\alpha = 1, 2, 3)$ with a common origin at the point x_0 is given. The set of points in E_3, which is obtained through all displacements of the origin by vectors[1] $n^\alpha e_\alpha (n$ and α being aribtrary integers) forms a (simple) lattice with an elementary cell of the form of parallelepiped, constructed on e_α. Points of this lattice are also called knots.

It is convenient to introduce oblique lattice coordinates x^α with the origin at the point x_0, basis vectors e_α and a metric tensor $g_{\alpha\beta}$, which is equal to the scalar product of the basis vectors: $g_{\alpha\beta} = (e_\alpha, e_\beta)$. Then, to knots of a lattice there correspond combinations of vectors $n = n^\alpha e_\alpha$ with integer components n^α.

Let $u(n)$ be a scalar or tensor, generally a complexvalued function, which is given at knots and which increases no faster than a power of $|n|$ as $|n| \to \infty$. Then analogously to the one-dimensional case [1.12], the function $u(n)$ can be interpolated by a generalized analytic function $u(x)$, which is uniquely deter-

---

[1]Here and in the following material by identical superscripts and subscripts we contemplate summation.

mined by some natural conditions. In order to avoid repetition, we shall concentrate on the special features of the three-dimensional case in the following.

Together with the interpolating function $u(x)$ let us also consider its Fourier transform:

$$u(k) = \int u(x)\mathrm{e}^{-ikx}\,dx\,. \tag{2.1.1}$$

Here, kx is, strictly speaking, a linear function, which is conveniently written in the form of a scalar product. As is well-known, the space E_3 of all linear functions of x is a three-dimensional vector space (k-space), where the basis e^β is introduced; it is reciprocal to the basis e_α:

$$e^\beta e_\alpha = e^\beta(e_\alpha) = \delta^\beta_\alpha\,. \tag{2.1.2}$$

Then $k = k_\beta e^\beta$ and $kx = k_\alpha x^\alpha$. Note that the last expression is also understood as an ordinary scalar product (k, x), if between the spaces E_3 (with fixed origin) and E_3' an identification is established with the help of the metric tensor $g_{\alpha\beta}$, i.e., the operation of raising and lowering of indices. For our purpose such an identification is not convenient and we shall consider separately the "physical" x-space E_3 and the dual k-space E_3'.

Let us construct a parallelepiped $\mathrm{B}\{-\pi \le k_\beta \le \pi\}$, in the k-space, whose edges are parallel to vectors of the reciprocal basis e^β and set

$$\delta_\mathrm{B}(x) \stackrel{\mathrm{def}}{=} \frac{1}{(2\pi)^3}\int_\mathrm{B} \mathrm{e}^{ixk}\,dk = \frac{1}{\pi^3 v_0}\prod_{\beta=1}^{3}\frac{\sin(xe^\beta)}{xe^\beta}\,, \tag{2.1.3}$$

where v_0 is the volume of an elementary cell constructed on the vectors e_α. It is obvious that $\delta_\mathrm{B}(0) = v_0^{-1}$ and $\delta_\mathrm{B}(n) = 0$ for all other knots.

Let us assume that $u(n)$ decreases sufficiently rapidly as $|n| \to \infty$. Then in B one can define a function $u(k)$, such that the coefficients of its Fourier series are equal to $v_0 u(n)$, i.e.

$$u(k) = v_0 \sum_n u(n)\mathrm{e}^{ink}, \qquad k \in \mathrm{B}\,. \tag{2.1.4}$$

Defining $u(k)$ to be zero outside B, and taking into account (2.1.3) we find for the inverse Fourier transform $u(x)$

$$u(x) = v_0 \sum_n u(n)\delta_\mathrm{B}(x - n)\,. \tag{2.1.5}$$

It is possible to prove that the function $u(x)$ can be continued analytically into the complex plane as an entire function of exponential type $\le \pi$.

In view of the properties of $\delta_\mathrm{B}(x)$, it is easy to see that $u(x)$, defined by (2.1.5), is in fact the required interpolating function. The one-to-one correspondence

$u(n) \leftrightarrow u(x) \leftrightarrow u(k)$ is well guaranteed by the condition of truncating the Fourier-transform $u(k)$ of the function $u(x)$ and the uniqueness of the expansion (2.1.4).

Let us proceed to the general case of the space $N'(\mathrm{B})$, whose elements are functions $u(n)$, which increase no faster than some power of $|n|$ as $|n| \to \infty$. The space $N'(\mathrm{B})$ can also be regarded as a space of generalized functions on the space $N(\mathrm{B})$, the latter being the space of rapidly decreasing functions.

Recall that in the one-dimensional case [1.12], when interpolating functions $u(n) \in N'(\mathrm{B})$, it turned out to be necessary to identify the ends of the segment B, transforming it into a circle. Analogously, let us identify the opposite faces of the present parallelepiped B or, in other words, let us transform it into a three-dimensional torus.

The series (2.1.4) now converges only in the sense of the generalized functions [2.1], and $u(k)$ belongs to the space of the generalized functions $K'(\mathrm{B})$, defined on the space $K(\mathrm{B})$ of infinitely differentiable test functions with supports in B.

Let us denote the inverse Fourier transforms of the spaces $K(\mathrm{B})$ and $K'(\mathrm{B})$ by $X(\mathrm{B})$ and $X'(\mathrm{B})$, respectively. The inverse Fourier transform $u(x) \in X'(\mathrm{B})$ of a function $u(k) \in K'(\mathrm{B})$ is defined, as usual, with the help of the Parseval equality

$$\int \overline{u(x)}\varphi(x)dx = \frac{1}{(2\pi)^3} \int \overline{u(k)}\varphi(k)dk \,, \tag{2.1.6}$$

where $\varphi(x) \in X(\mathrm{B})$ and $\varphi(k) \in K(\mathrm{B})$ are connected with each other by the usual Fourier transform.

It can be proved [2.1] that $u(x) \in X'(\mathrm{B})$ are regular generalized functions (i.e. corresponding to ordinary functions), which increase no faster than a power of $|x|$ as $|x| \to \infty$ and which can be analytically continued to a complex region as entire functions of the exponential type $\leq \pi$. The proof of the fact that the function $u(x)$ has values $u(n)$ at the knots, can be performed analogously to the one-dimensional case [1.12].

The uniqueness of interpolation is ensured by the identification of the opposite faces of the parallelepiped B. This excludes from $X'(\mathrm{B})$ functions of the type $P(x)\sin[\pi(xe)^\beta]$, [$P(x)$ being a polynomial], which vanish at all the knots.

Thus we have:

Theorem. The formulae (2.1.4, 6) fix linear isomorphisms between the spaces

$$N'(\mathrm{B}) \leftrightarrow X'(\mathrm{B}) \leftrightarrow K'(\mathrm{B}) \,, \tag{2.1.7}$$

such that $u(x) \in X'(\mathrm{B})$ interpolates $u(n) \in N'(\mathrm{B})$.

We may now consider $u(n)$, $u(x)$ and $u(k)$ as representations of one and the same function in different functional bases. Of course, to this end it is also necessary to know the corresponding representations of operations over these functions.

A function $q(n)$, which decreases rapidly, as $|n| \to \infty$ (in particular, a function with finite support), determines the linear functional

$$\langle q|u\rangle \overset{\text{def}}{=} \sum_n \overline{q(n)}u(n) = \int \overline{q(x)}u(x)dx = \frac{1}{(2\pi)^3}\int \overline{q(k)}u(k)dk\,, \tag{2.1.8}$$

which is invariant with respect to the $n-$, $x-$, and k-representations, if the kernels $q(n)$, $q(x)$ and $q(k)$ are connected by the relations

$$q(k) = \sum_n q(n)e^{-ink}, \qquad k \in \mathrm{B}\,,$$

$$q(x) = \sum_n q(n)\delta_{\mathrm{B}}(x-n)\,. \tag{2.1.9}$$

Problem 2.1.1. Prove the above proposition.

Note that, as distinct from (2.1.4, 5), in the correspondence $q(n) \leftrightarrow q(x) \leftrightarrow q(x)$ the factor v_0 is absent. This is connected with the fact that in the further consideration the quantities $q(n)$ will have the meaning of forces acting at the knots of a lattice. Thus $q(x)$ can be interpreted as the corresponding density of body forces, $q(k)$ as its Fourier transform and the functional $\langle q|u\rangle$ as the work, done by the forces q on the displacement u.

As an important example, consider the functional

$$\langle \delta|u\rangle = u(n)\,|_{n=0} = u(x)\,|_{x=0} = \int_{\mathrm{B}} u(k)dk \tag{2.1.10}$$

with the kernels

$$\delta(n) \leftrightarrow \delta_{\mathrm{B}}(x) \leftrightarrow 1(k)\,. \tag{2.1.11}$$

Here $\delta(n) = 1$ at $n = 0$, and it is equal to zero at all other knots. Evidently, $\delta_{\mathrm{B}}(x)$ plays in $X'(\mathrm{B})$ the role of an ordinary δ-function but, as distinct from the latter $\delta_{\mathrm{B}}(x)$ is not singular. In what follows, we shall use the notation $\delta(x)$ instead of $\delta_{\mathrm{B}}(x)$ when this does not lead to any misunderstandings.

Problem 2.1.2. Show that the following identities are generated by the functional with the kernel $1(x)$:

$$v_0 \sum u(n) = \int u(x)dx = u(k)\,|_{k=0}\,. \tag{2.1.12}$$

Analogously to the one-dimensional case [1.12] we define the invariant form

$$\langle u|\Phi|w\rangle \stackrel{\text{def}}{=} \sum_{nn'} \overline{u(n)}\Phi(n, n')w(n') = \iint \overline{u(x)}\Phi(x, x')w(x')dx\, dx'$$

$$= \frac{1}{(2\pi)^3} \iint \overline{u(k)}\Phi(k, k')w(k')dk\, dk' = \overline{\langle w|\Phi^+|u\rangle} \,. \tag{2.1.13}$$

Here

$$\Phi(k, k') = \frac{1}{(2\pi)^3} \sum_{nn'} \Phi(n, n')e^{-i(kn-k'n')}$$

$$= \frac{1}{(2\pi)^3} \iint \Phi(x, x')e^{-i(kx-k'x')}dx\, dx' \,, \tag{2.1.14}$$

$$\Phi^+(k, k') = \overline{\Phi(k', k)}, \qquad \Phi^+(x, x') = \overline{\Phi(x', x)} \,. \tag{2.1.15}$$

One can establish the correspondences between the discrete and continuous operations of multiplication, convolution, differentiation and so on, distinctions compared with the one-dimensional case being small.

Thus, the functions $u(n) \in N'(\mathrm{B})$ constitute a ring with respect to the operation of multiplication. At the same time, the ordinary product of two functions from $X'(\mathrm{B})$, generally speaking, does not belong to $X'(\mathrm{B})$. Nevertheless, it is possible to introduce in $X'(\mathrm{B})$ such an operation of multiplication, which corresponds to the multiplication of the inverse transforms in $N'(\mathrm{B})$ and whose result is still contained in $X'(\mathrm{B})$.

In fact, on the torus B we can define an operation of displacement and, hence, an integral convolution

$$u(k) * w(k) = \int_{\mathrm{B}} u(k - k')w(k')dk', \tag{2.1.16}$$

with respect to which $K'(\mathrm{B})$ is a ring.

Problem 2.1.3. Verify the correspondence

$$f(n) = u(n)w(n) \leftrightarrow f(k) = u(k) * w(k) \,. \tag{2.1.17}$$

This leads to the natural definition of the product, in $X'(\mathrm{B})$:

$$f(x) = u(x) \cdot w(x) \leftrightarrow f(k) = u(k) * w(k) \,. \tag{2.1.18}$$

Problem 2.1.4. Show that $f(x)$ coincides with $f(n)$ at all knotes.

In particular, the following formulae are valid:

$$u(x)\delta(x) = u(0)\delta(x), \quad \delta(x)\delta(x) = v_0\delta(x) \,. \tag{2.1.19}$$

Here, we observe from a new point of view, the expedience of transforming the parrallelepiped B into a torus, since it is impossible to find a reasonable definition of a convolution on a parallelepiped such that the support of $u(k)*w(k)$ is always contained in the parallelepiped.

Note that if the supports of $u(k)$, $w(k) \in K'(\mathrm{B})$ are contained in a parallelepied similar to B, but twice smaller, then, the convolution in (2.1.16) coincides with the ordinary one and hence the above-defined multiplication also coincides with the ordinary one. Of course, this will also happen if one of the factors is a polynomial (in the x-representation), since its support (in the k-representation) is concentrated at the point $k = 0$.

The algorithm considered above for passing from functions of discrete argument to functions of continuous argument may be clearly interpreted as spanning some analytic structure over the lattice in such a way that one-to-one correspondence between the structures is conserved. The Euclidean space in which only functions from $X'(\mathrm{B})$ are allowed, is conveniently interpreted as a quasicontinuum, its geometry being isomorphic to that of a lattice. When the size of elementary cell of the lattice tends to zero (this is equivalent to the consideration of functions which change only slightly over distances of the order of the cell size), the region B extends over all k-space and the quasicontinuum is transformed into a continuum with the ordinary geometry.

The idea of quasicontinuum can be naturally generalized by taking, instead of the parallelepiped B, another manifold. The isotropic continuum will be of considerable importance, the corresponding region B being a sphere. The replacement of the parallelepiped by a sphere in k-space was first carried out by Debye in order to calculate the heat capacity of a crystal and was further widely used for qualitative estimates of physical properties of crystals (without any connection with the idea of the quasicontinuum). The radius κ of the sphere is usually determined by the condition of equality of the volumes of the sphere and the parallelepiped. The exact correspondence with the lattice structure is lost in this case, but the corresponding quasicontinuum model, which will be called the Debye model, describes qualitatively correctly the presence of a discrete structure with the characteristic scale parameter $a \simeq \pi/\kappa$.

The Debye model can be somewhat improved by identifying the opposite points of the boundary sphere analogously to identifying opposite faces of the parallelepiped. In this case, the manifold B, will be homeomorphic to a three-dimensional projective space.

As another example we can mention a layered medium, in which the characteristic scale parameter is to be introduced in one direction only. Accordingly, in the k-space, transform is to be carried out with respect to one coordinate only.

In the general case, the quasicontinuum of class B may be defined as the set of two objects: an Euclidean space and a space of admissible functions $X'(\mathrm{B})$ given on that space, B being a fixed manifold.

2.2 Equations of Motion

The model for a medium with simple structure will be based on Born's model of a simple lattice with the basis e_α in the harmonic approximation. For the elastic energy Φ we have, analogously to the case of a one-dimensional chain [1.12],

$$\Phi = \frac{1}{2} \sum_{nn'} u_\alpha(n)\, \Phi^{\alpha\beta}(n, n') u_\beta(n') , \tag{2.2.1}$$

where $\Phi^{\alpha\beta}(n, n')$ is a given tensor of the force constants, and $u_\alpha(n)$ is the displacement of the particles from the equilibrium position, i.e. it is assumed that initial external forces are absent.

This model may be interpreted, if necessary, as a system of pointwise particles situated at the knots of the lattice and interacting by means of linear elastic bonds of the most gêneral nature.

The Lagrangian of this system has the form

$$\begin{aligned} L = &\frac{1}{2} g^{\alpha\beta} \sum_n m(n) \dot{u}_\alpha(n) \dot{u}_\beta(n) \\ &- \frac{1}{2} \sum_{nn'} u_\alpha(n) \Phi^{\alpha\beta}(n, n') u_\beta(n') - \sum_n q^\alpha(n) u_\alpha(n) , \end{aligned} \tag{2.2.2}$$

where $m(n)$ is the mass of a particle at the point n, and $q^\alpha(n)$ is the external force acting on the particle. The dependence of the field parameters on time is not indicated explicitly.

In the usual way we find the equations of motion

$$m(n) \ddot{u}^\alpha(n) + \sum_{n'} \Phi^{\alpha\beta}(n, n') u_\beta(n') = q^\alpha(n) . \tag{2.2.3}$$

Using the above-described algorithm, let us proceed to a quasicontinuum representation. For the elastic energy Φ we have, cf. (2.1.13)

$$\begin{aligned} \Phi &= \frac{1}{2} \langle u_\alpha | \Phi^{\alpha\beta} | u_\beta \rangle = \frac{1}{2} \iint u_\alpha(x) \Phi^{\alpha\beta}(x, x')\, u_\beta(x') dx\, dx' \\ &= \frac{1}{(2\pi)^3} \iint \overline{u_\alpha(k)} \Phi^{\alpha\beta}(k, k') u_\beta(k') dk\, dk' , \end{aligned} \tag{2.2.4}$$

where $\Phi^{\alpha\beta}(x, x')$ and $\Phi^{\alpha\beta}(k, k')$ are expressed in terms of the force constants

$$\begin{aligned} \Phi^{\alpha\beta}(k, k') &= \frac{1}{(2\pi)^3} \sum_{nn'} \Phi^{\alpha\beta}(n, n') \mathrm{e}^{-\mathrm{i}(kn - k'n')} \\ &= \frac{1}{(2\pi)^3} \iint \Phi^{\alpha\beta}(x, x') \mathrm{e}^{-\mathrm{i}(kx - k'x')}\, dx\, dx' . \end{aligned} \tag{2.2.5}$$

Introducing the mass density $\rho(x)$ and force density $q(x)$

$$\rho(x) = \sum_n m(n)\delta(x - n), \; q(x) = \sum_n q(n)\delta(x - n), \tag{2.2.6}$$

we write the Lagrangian (2.2.2) in the form

$$L = \frac{1}{2}\langle \dot{u}_\alpha | \rho g^{\alpha\beta} | \dot{u}_\beta \rangle - \frac{1}{2}\langle u_\alpha | \Phi^{\alpha\beta} | u_\beta \rangle + \langle q^\alpha | u_\alpha \rangle, \tag{2.2.7}$$

where the operator ρ has the kernel $\rho(x)\delta(x - x')$.

If we do not assume the existence of a discrete micromodel, then the Lagrangian in the form or (2.2.7) corresponds to the most general nonlocal theory of an elastic medium with simple structure. In this case, the admissible functions are not restricted by truncation of their Fourier transforms, i.e., all the field quantities are considered to be given on an ordinary manifold. The quasi-continuum (discrete) model is, in fact, a particular case of the general theory.

The equations of motion, which correspond to the Lagrangian (2.2.7) have the form

$$\rho(x)\ddot{u}^\alpha(x) + \int \Phi^{\alpha\beta}(x, x')u_\beta(x')dx' = q^\alpha(x), \tag{2.2.8}$$

$$(2\pi)^{-3}\rho(k) * \ddot{u}^\alpha(k) + \int \Phi^{\alpha\beta}(k, k')u_\beta(k')dk' = q^\alpha(k) \tag{2.2.9}$$

in the x- and k-representations, respectively. In the case of the quasicontinuum, the n-representation (2.2.3) is to be added.

The equations of motion are conveniently considered in the ω-representation. In operator form, they have the form cf. (2.1.37),

$$(-\rho\omega^2\delta^{\alpha\beta} + \Phi^{\alpha\beta})u_\beta = q^\alpha. \tag{2.2.10}$$

Obviously, the physical content of the equations of motion is determined, first of all, by properties of the operator of the elastic energy $\Phi^{\alpha\beta}$; the next section will be devoted to its consideration.

2.3 Elastic Energy Operator

The properties of the operator $\Phi^{\alpha\beta}(x, x')$ are in many aspects analogous to the properties of the operator $\Phi(x, x')$, which were considered in details in [Ref. 1.12, Sect. 2.4]. We shall therefore concentrate on features specific to the three-dimensional case.

Hermiticity. From (2.2.4), it follows that

$$\Phi^{\alpha\beta}(x, x') = \Phi^{\beta\alpha}(x', x), \; \Phi^{\alpha\beta}(k, k') = \overline{\Phi^{\beta\alpha}(k', k)}. \tag{2.3.1}$$

Also, since $\Phi^{\alpha\beta}(x, x')$ is real

$$\overline{\Phi^{\alpha\beta}(k, k')} = \Phi^{\alpha\beta}(-k, -k'). \tag{2.3.2}$$

Finiteness of action-at-a-distance: In what follows it is assumed that $\Phi^{\alpha\beta}(x, x') = 0$ if $|x - x'| > l$ for any value of x. In other words, it is assumed that there exists a characteristic radius of interaction, l. An argument analogous to that discussed in [Ref. 1.12, Sect. 2.4] implies that $\Phi^{\alpha\beta}(k, k')$ is an entire function of exponential type $\leq l$ with respect to the argument $k + k'$.

If, moreover, the medium is bounded by a characteristic dimension L, then $\Phi^{\alpha\beta}(k, k')$ is also an entire function of exponential type $\leq L$ of the argument $k - k'$.

Invariance of Φ with respect to translation and rotation, as will be shown in the next section, is equivalent to the conditions[2]

$$\int \Phi^{\alpha\beta}(x, x')dx = 0, \; \int x^{[\lambda}\Phi^{\alpha]\beta}(x, x')dx = 0 \tag{2.3.3}$$

or, in the quasicontinuum (lattice) terminology,

$$\sum_n \Phi^{\alpha\beta}(n, n') = 0, \; \sum_n n^{[\lambda}\Phi^{\alpha]\beta}(n, n') = 0. \tag{2.3.4}$$

From this follows the possibility of representing $\Phi^{\alpha\beta}(k, k')$ in the form

$$\Phi^{\alpha\beta}(k, k') = k_\lambda k'_\mu c^{\lambda\alpha\mu\beta}(k, k'), \tag{2.3.5}$$

where $c^{\lambda\alpha\mu\beta}(k, k')$ is an analytic function of the same type as $\Phi^{\alpha\beta}(k, k')$; without loss of generality it can be considered to be symmetric in the indices inside the first and second pairs and to be Hermitian with respect to permutation of the pairs, i.e.

$$c^{\lambda\alpha\mu\beta}(k, k') = c^{\alpha\lambda\mu\beta}(k, k') = \overline{c^{\mu\beta\lambda\alpha}(k', k)}. \tag{2.3.6}$$

Local approximation: If the field parameters vary slowly over distances of the order of l, then

$$c^{\lambda\alpha\mu\beta}(x, x') = c_0^{\lambda\alpha\mu\beta}(x)\delta(x - x'), \tag{2.3.7}$$

where $c_0^{\lambda\alpha\mu\beta}(x)$ is a tensor of the elastic moduli of the corresponding local model of the zero-order approximation (with respect to l).

Stability: This requirement is equivalent to positive definiteness of the elastic energy Φ for all admissible displacements, excluding translation and rigid body rotation, for which $\Phi = 0$.

---

[2]Here and subsequently indices which are contained in brackets (parentheses) are assumed to be subject to the operation of alternation (symmetrization).

Media with binary interaction: One of the possible specific properties is the following one. The forces appearing due to a change of the distance between points x and x', are proportional to this change and are directed along the line, which connects these two points. It is easy to show that such a medium can be described by an operator of elastic bonds $\Psi^{\alpha\beta}$ with the kernel

$$\Psi^{\alpha\beta}(x, x') = \frac{(x^\alpha - x'^\alpha)(x^\beta - x'^\beta)}{|x - x'|^2} \Psi(x, x'), \tag{2.3.8}$$

where $\Psi(x, x') = \Psi(x', x)$ is a scalar characteristic of the elastic bond which connects the points x and x', such that $\Psi(x, x') = 0$ when $|x - x'| > L$.

For the energy operator Φ we may write

$$\Phi^{\alpha\beta}(x - x') = \psi^{\alpha\beta}(x)\delta(x - x') - \Psi^{\alpha\beta}(x, x'), \tag{2.3.9}$$

$$\psi^{\alpha\beta}(x) = \int \Psi^{\alpha\beta}(x, x')dx'. \tag{2.3.10}$$

The conditions of invariance with respect to rotation are here satisfied automatically.

Homogeneous media: In this case, which is important for applications

$$\Phi^{\alpha\beta}(x, x') = \Phi^{\alpha\beta}(x - x'). \tag{2.3.11}$$

From (2.2.5) we find

$$\Phi^{\alpha\beta}(k, k') = \Phi^{\alpha\beta}(k)\delta(k - k'), \tag{2.3.12}$$

$$\Phi^{\alpha\beta}(k) = \int \Phi^{\alpha\beta}(x)e^{-ikx}\, dx = \overline{\Phi^{\beta\alpha}(k)}. \tag{2.3.13}$$

In case of the quasicontinuum we also have

$$\Phi^{\alpha\beta}(n, n') = \Phi^{\alpha\beta}(n - n'), \tag{2.3.14}$$

$$\Phi^{\alpha\beta}(k) = \frac{1}{v_0} \sum_n \Phi^{\alpha\beta}(n)e^{-ikn}. \tag{2.3.15}$$

The finiteness of action-at-a-distance is now equivalent to the functions $\Phi^{\alpha\beta}(x)$ or $\Phi^{\alpha\beta}(n)$ having finite support, from which it follows that $\Phi^{\alpha\beta}(k)$ is an entire function of exponential type $\leq L$.

The conditions (2.3.3, 4) now take the form

$$\int \Phi^{\alpha\beta}(x)dx = 0, \int x^{[\lambda}\Phi^{\alpha]\beta}(x)dx = 0, \tag{2.3.16}$$

$$\sum_n \Phi^{\alpha\beta}(n) = 0, \sum_n n^{[\lambda}\Phi^{\alpha]\beta}(n) = 0\,, \tag{2.3.17}$$

and to (2.3.5, 6) there correspond

$$\Phi^{\alpha\beta}(k) = k_\lambda k_\mu c^{\lambda\alpha\mu\beta}(k)\,, \tag{2.3.18}$$

$$c^{\lambda\alpha\mu\beta}(k) = c^{\alpha\lambda\mu\beta}(k) = \overline{c^{\mu\beta\lambda\alpha}(k)} = c^{\mu\beta\lambda\alpha}(-k)\,. \tag{2.3.19}$$

The static Green's tensor $G_{\alpha\beta}$ is defined as an operator which is inverse with respect to the energy operator $\Phi^{\alpha\beta}$ and hence

$$\begin{aligned} \int \Phi^{\alpha\lambda}(x, x'')G_{\lambda\beta}(x'', x')dx'' &= \delta^\alpha_\beta\delta(x - x')\,, \\ \int \Phi^{\alpha\lambda}(k, k'')G_{\lambda\beta}(k'', k')dk'' &= \delta^\alpha_\beta\delta(k - k')\,, \end{aligned} \tag{2.3.20}$$

$G_{\alpha\beta}(x, x')$ satisfying the corresponding boundary conditions as $x, x' \to \infty$.

For a homogeneous medium,

$$\Phi^{\alpha\lambda}(k)G_{\lambda\beta}(k) = \delta^\alpha_\beta\,, \tag{2.3.21}$$

i.e. in the k representation, the kernel of the Green's tensor can be obtained by the purely algebraic operation of inverting the matrix $\Phi^{\alpha\beta}(k)$.

It is easy to show that from the definition of the Green's tensor, the elastic energy can be represented in the static case, as

$$\Phi = \frac{1}{2}\langle u_\alpha|\Phi^{\alpha\beta}|u_\beta\rangle = \langle q^\alpha|G_{\alpha\beta}|q^\beta\rangle\,, \tag{2.3.22}$$

where the q^α's are body forces. Conversely, this relation can be taken as a definition of the static Green's tensor.

Problem 2.3.1. Prove the equivalence of the two definitions of $G_{\alpha\beta}$.

It is obvious that the kernels $G_{\alpha\beta}(x, x')$, $G_{\alpha\beta}(k, k')$ possess the symmetry properties (2.3.1, 2). For the quasicontinuum, due to the condition $k \in \mathrm{B}$, the kernel $G_{\alpha\beta}(x, x')$ is an entire function of the corresponding arguments.

The dynamic Green's tensor $G_{\alpha\beta}$ is defined as an operator inverse to the operator

$$L^{\alpha\beta} \stackrel{\text{def}}{=} \rho g^{\alpha\beta}\,\partial_t^2 + \Phi^{\alpha\beta}\,. \tag{2.3.23}$$

We leave it to the reader to write the analogs of the equations (2.3.20) for the kernels $G_{\alpha\beta}(x, x', t)$, $G_{\alpha\beta}(x, x', \omega)$ and $G_{\alpha\beta}(k, k', \omega)$.

2.4 Symmetric Stress Tensor and Energy Density

In this section, we introduce a symmetric stress tensor and the corresponding elastic energy density. In order to achieve this, the relations (7.3.5 and 6) have to be proved.

We need three auxiliary propositions.

Lemma 1. Let $A(k)$ be an analytic tensor function (indices are omitted) and $A(0) = 0$. Then

$$A(k) = k_\lambda A^\lambda(k)\,, \tag{2.4.1}$$

where the analytic function $A^\lambda(k)$ is uniquely determined by the condition that coefficients of its Taylor series

$$A^\lambda(k) = \sum_{n=0}^{\infty} A^{\lambda\lambda_1\ldots\lambda_n} k_{\lambda_1} \ldots k_{\lambda_n} \tag{2.4.2}$$

are symmetric with respect to the indices $\lambda, \lambda_1, \ldots, \lambda_n$.

The proof is obvious: it is sufficient to compare the Taylor series for $A(k)$ and $A^\lambda(k)$.

Lemma 2. Let $A^{\lambda\alpha\mu}$ be a tensor, symmetric with respect to the indices λ, μ. Then it can be associated with the tensor

$$A^{\{\lambda\alpha\mu\}} \overset{\text{def}}{=} A^{\lambda\alpha\mu} + A^{\alpha\lambda\mu} - A^{\lambda\mu\alpha}\,, \tag{2.4.3}$$

which is symmetric with respect to the indices λ, α; the one-to-one correspondence:

$$A^{\lambda\alpha\mu} = \frac{1}{2}\left(A^{\{\lambda\alpha\mu\}} + A^{\{\mu\alpha\lambda\}}\right). \tag{2.4.4}$$

holds.

Lemma 3. If an analytic function $A^\alpha(k)$ satisfies the conditions

$$A^\alpha(0) = 0,\ \partial^{[\lambda}A^{\alpha]}(0) = 0\ (\partial^\lambda = \partial/\partial k_\lambda)\,, \tag{2.4.5}$$

then it admits the representation

$$A^\alpha(k) = k_\lambda B^{\lambda\alpha}(k)\,, \tag{2.4.6}$$

where the analytic function $B^{\lambda\alpha}(k)$ is symmetric with respect to the indices λ, α.

In order to prove this let us write the function $A^\alpha(k)$ in the form

$$A^\alpha(k) = k_\lambda A_0^{\lambda\alpha} + k_\lambda k_\mu A^{\lambda\alpha\mu}(k)\,. \tag{2.4.7}$$

The constant tensor $A_0^{\lambda\alpha}$ appearing here is symmetric because of the second of the conditions (2.4.5); the tensor function $A^{\lambda\alpha\mu}(k)$ is symmetric with respect to the indices $\lambda\mu$ and, according to Lemma 1, is defined uniquely. This function can be replaced by a tensor function $A^{\{\lambda\alpha\mu\}}(k)$ from Lemma 2, which is symmetric with respect to the indices λ, α, since the factor $k_\lambda k_\mu$ provides the operation of symmetrization needed in the identity (2.4.4). As a result, we find

$$B^{\lambda\alpha}(k) = A_0^{\lambda\alpha} + k_\mu A^{\{\lambda\alpha\mu\}}(k) = B^{\alpha\lambda}(k) . \tag{2.4.8}$$

This completes the proof.

Let us now prove the relations (2.3.5). The main physical assumption that the elastic medium is situated in a homogeneous isotropic Euclidean space and is equivalent to requiring the invariance of all properties of the medium with respect to translation and rigid rotation of the whole medium. As it is known from Lie group theory, it is sufficient to restrict ourselves to invariance with respect to infinitesimal transformations.

Let the state of the medium be described by a displacement field u_α. In the linear theory the principle of superposition is always valid and to an additional infinitesimal displacement, there corresponds the transformation $u_\alpha \to u_\alpha + u_\alpha^0$, where, in the case of translation,

$$u_\alpha^0(x) = a_\alpha = \text{const}, \; u_\alpha^0(k) = (2\pi)^3 a_\alpha \delta(k) , \tag{2.4.9}$$

and for an infinitesimal rigid body rotation

$$\begin{aligned} u_\alpha^0(x) &= a_{\alpha\lambda} x^\lambda, \; a_{\alpha\lambda} = -a_{\lambda\alpha} = \text{const} , \\ u_\alpha^0(k) &= -(2\pi)^3 i a_{\alpha\lambda} \partial^\lambda \delta(k) . \end{aligned} \tag{2.4.10}$$

The requirement concerning invariance of the energy Φ with respect to the transformation $u_\alpha \to u_\alpha + u_\alpha^0$, (2.2.4) and the Hermiticity of the operator $\Phi^{\alpha\beta}$ lead to the following equation

$$2\langle u_\alpha^0 | \Phi^{\alpha\beta} | u_\beta \rangle + \langle u_\alpha^0 | \Phi^{\alpha\beta} | u_\beta^0 \rangle = 0 , \tag{2.4.11}$$

which is to be satisfied for any u_α^0 and u_β. From this one can obtain conditions which must be satisfied by the kernel of the operator $\Phi_{\alpha\beta}$.

First, let us make one remark. In the case of an infinite medium, the expression for the energy (2.2.4) and hence the second term in (2.4.11) are not a priori defined on displacements u_α^0 of the form (2.4.9) or (2.4.10). Consequently let us consider next a bounded medium and then perform the transition to the limit.

Taking into account the arbitrariness of u_β, we conclude that (2.4.11) is equivalent to the condition

$$\int u_\alpha^0(x) \Phi^{\alpha\beta}(x, x') dx = 0 \tag{2.4.12}$$

or in the k representation

$$\int u_\alpha^0(k)\Phi^{\alpha\beta}(k, k')dk = 0\,. \tag{2.4.13}$$

Using (2.4.9), a_α being arbitrary, gives two equivalent conditions for $\Phi^{\alpha\beta}$

$$\int \Phi^{\alpha\beta}(x, x')dx = 0,\ \Phi^{\alpha\beta}(k, k')\,|_{k=0} = 0\,. \tag{2.4.14}$$

Due to the assumption about the boundedness of the medium the function $\Phi^{\alpha\beta}(x, x')$ has finite support and hence the function $\Phi^{\alpha\beta}(k, k')$ is analytic with respect to the pair of points k, k'. Taking into account also the conditions of Hermiticity (2.3.1) we find that the relations (2.4.14) are equivalent to the possibility of representing $\Phi^{\alpha\beta}(k, k')$ in the form

$$\Phi^{\alpha\beta}(k, k') = k_\lambda k'_\mu \Phi^{\lambda\alpha\mu\beta}(k, k')\,, \tag{2.4.15}$$

where, according to Lemma 1, the analytic function $\Phi^{\lambda\alpha\mu\beta}(k, k')$ is determined uniquely.

The substitution of (2.4.10) into (2.4.13), $a_{\alpha\lambda}$ being an arbitrary antisymmetric tensor, gives in the x-representation

$$\int x^{[\lambda}\Phi^{\alpha]\beta}(x, x')dx = 0\,. \tag{2.4.16}$$

In the k representation, taking into account (2.4.15) and the identity

$$k_\lambda \partial^\mu \delta(k) = -\,\delta_\lambda^\mu \delta(k)\,, \tag{2.4.17}$$

we have

$$k'_\mu \Phi^{[\lambda\alpha]\mu\beta}(0, k') = 0\,, \tag{2.4.18}$$

and, due to the Hermitian symmetry of $\Phi^{\alpha\beta}(k, k')$,

$$k_\lambda \Phi^{\lambda\alpha[\mu\beta]}(k, 0) = 0\,. \tag{2.4.19}$$

We now apply Lemma 3 to the function $\Phi^{\alpha\beta}(k, k')$ for each argument k and k'. This leads to the required result:

$$\Phi^{\alpha\beta}(k, k') = k_\lambda k'_\mu \Phi^{\lambda\alpha\mu\beta}(k, k') = k_\lambda k'_\mu c^{\lambda\alpha\mu\beta}(k, k')\,. \tag{2.4.20}$$

Here, the tensor $c^{\lambda\alpha\mu\beta}(k, k')$ satisfies the symmetry conditions (2.3.6) and is connected with $\Phi^{\lambda\alpha\mu\beta}(k, k')$ by

$$\begin{aligned} c^{\lambda\alpha\mu\beta}(k, k') &= \Phi^{\lambda\alpha\mu\beta}(0, 0) + k'_\tau \Phi^{\lambda\alpha\{\mu\beta\tau\}}(0, k') \\ &\quad + k_\sigma \Phi^{\{\lambda\alpha\sigma\}\mu\beta}(k, 0) + k_\sigma k'_\tau \Phi^{\{\lambda\alpha\sigma\}\{\mu\beta\tau\}}(k, k')\,. \end{aligned} \tag{2.4.21}$$

Problem 2.4.1. Check that $c^{\lambda\alpha\mu\beta}(k, k')$ satisfies the conditions (2.3.6).

It was assumed above that the medium is finite. However, the derived result about the representability of $\Phi^{\alpha\beta}(k, k')$ in the form of (2.3.5) with the tensor $c^{\lambda\alpha\mu\beta}(k, k')$ possessing the symmetry (2.3.6) depends neither upon the shape nor upon the size of the elastic body. Hence this remains valid also for the limiting case of an infinite medium. According to the same reasoning, if the second term in (2.4.11), vanishes for an abritrary finite medium, it can be considered to be equal to zero also in the limiting case of an infinite medium (natural regularization).

Let us now introduce the strain tensor

$$\varepsilon_{\mu\beta}(x) \stackrel{\text{def}}{=} \partial_{(\mu}u_{\beta)}(x), \ \varepsilon_{\mu\beta}(k) = ik_{(\mu}u_{\beta)}(k) \tag{2.4.22}$$

and the symmetric tensor

$$\sigma^{\lambda\alpha}(x) \stackrel{\text{def}}{=} \int c^{\lambda\alpha\mu\beta}(x, x')\varepsilon_{\mu\beta}(x')dx' \tag{2.4.23}$$

or in the k-representation

$$\sigma^{\lambda\alpha}(k) = \int c^{\lambda\alpha\mu\beta}(k, k')\, \varepsilon_{\mu\beta}(k')dk' \,. \tag{2.4.24}$$

The equations of motion (2.8) take the form

$$\rho(x)\ddot{u}^{\alpha}(x) - \partial_{\lambda}\sigma^{\lambda\alpha}(x) = q^{\alpha}(x) \tag{2.4.25}$$

or, in direct notations,

$$\rho\ddot{u} - \operatorname{div}\sigma = q \,. \tag{2.4.26}$$

The expression for elastic energy can be represented in the form

$$2\Phi = \langle u_{\alpha}|\Phi^{\alpha\beta}|u_{\beta}\rangle = \langle \varepsilon_{\lambda\alpha}|c^{\lambda\alpha\mu\beta}|\varepsilon_{\mu\beta}\rangle \tag{2.4.27}$$

or

$$\Phi = \int \varphi(x)dx, \qquad \varphi(x) = \frac{1}{2}\sigma^{\lambda\alpha}(x)\varepsilon_{\lambda\alpha}(x) \,. \tag{2.4.28}$$

From the above expressions it follows that $\sigma^{\lambda\alpha}(x)$ and $\varphi(x)$ may be interpreted as the stress tensor and the elastic energy density, respectively, both quantities being invariant with respect to translation and rotation, as they should be according to their physical meaning.

The expressions (2.4.23) and (2.4.24) which can be rewritten in operator form as

$$\sigma = C\varepsilon , \qquad (2.4.29)$$

can be interpreted naturally as Hooke's law in operator form with C playing the role of the operator of the elastic moduli. In essence, we have proved the following

Theorem. In the nonlocal theory of a linear elastic medium of simple structure with finite action-at-a-distance, it is always possible to introduce a symmetric stress tensor and an energy density, which can be expressed in terms of stress and strain in the usual way.

At first glance, it may seem that we have proved even more, since the expression obtained above determines the stress tensor and the energy density uniquely. However, this conclusion is of a formal nature, since the derivation itself contains an element of arbitrariness. From reasoning analogous to that described in [Ref. 1.12, Chap. 2], it follows that the stress and energy density in a nonlocal theory cannot be determined uniquely from a physical point of view, if only the operator of the elastic energy $\Phi^{\alpha\beta}(x, x')$ is given; information on the structure of the elastic bonds is also needed. At the same time, in the three-dimensional case additional difficulties arise as compared with the one-dimensional one. The first difficulty is connected with the nonuniqueness of tensor representations of the type (2.4.20). The second and more important one is due to the fact that, generally speaking, even the effective bonds do not allow representation in the form of a binary interaction. As a consequence, the stress tensor (2.4.23) only approximately determines the force which acts on a small area, and the larger the latter is (in comparison with the characteristic length l), the more accurate is the relation considered.

Nevertheless, the conclusion which we have drawn is of principle significance. This is because it is a widespread opinion that as distinct from the ordinary theory of elasticity, a nonsymmetric stress tensor is necessarily associated with a weakly nonlocal theory of media of simple structure. As an example, we mention the so-called couple-stress theory of elasticity with constrained rotation. The corresponding proofs cannot be considered as entirely correct.

This is connected with the fact that in the above-mentioned theories the stress tensor is introduced in a formal manner, by analogy with the local theory of elasticity and without due consideration of specific features of the nonlocal model. In particular, no attention is paid to the nonuniqueness of tensor representations of the type (2.4.20) and, as a rule, the natural principle of correspondence is not taken into account: in the long-wave length approximation, the stress tensor must coincide with the classical symmetric stress tensor.

In conclusion, let us note that, as in the one-dimensional model, one can also introduce two-point measures of deformation, stress and energy density, which have, however, a clear physical meaning only in the presence of binary interaction.

2.5 Homogeneous Media

Let us consider an infinite homogeneous medium with the difference operator $\Phi^{\alpha\beta}(x - x')$ and the constant density ρ. In the k-representation

$$\Phi^{\alpha\beta}(k, k') = \Phi^{\alpha\beta}(k)\delta(k - k'), \tag{2.5.1}$$

where $\Phi^{\alpha\beta}(k)$ is defined by (2.3.13) or (2.3.15).

In view of the essentially non-analytic nature of $\Phi^{\alpha\beta}(k, k')$ a direct generalization of the results of the last section to the case of a homogeneous medium is impossible; exact limiting transitions are necessary, but these are connected with some difficulties. Therefore we prefer a more direct approach.

Let us begin with the properties which $\Phi^{\alpha\beta}(k)$ possesses due to the invariance of the energy with respect to translation and rotation.

Problem 2.5.1. Show that a necessary condition for this invariance is the representability of $\Phi^{\alpha\beta}(k)$ in the form

$$\Phi^{\alpha\beta}(k) = k_\lambda k_\mu \Phi^{\lambda\alpha\mu\beta}(k). \tag{2.5.2}$$

[*Hint*: Expand $\Phi^{\alpha\beta}(k)$ in a series, write (2.4.12) and (2.4.16) in the k-representation and take into account the condition of Hermiticity.]

However, the condition (2.5.2) is not sufficient. In order to obtain an additional condition, let us take into account that due to (2.4.10), this can be imposed only on $\Phi^{\lambda\alpha\mu\beta}$, i.e. on the constants of the zeroth long-wavelength approximation. But in this limiting case there must exist a uniquely determined elastic energy density $\varphi_0(x)$, which is invariant with respect to translation and rotation, and also makes sense for displacements which increase linearly as $|x| \to \infty$ (as distinct from the energy Φ).

To the operator of the zeroth-order approximation

$$\Phi_0^{\alpha\beta}(k) = k_\lambda k_\mu \Phi_0^{\lambda\alpha\mu\beta} \tag{2.5.3}$$

corresponds the Lagrangian density

$$\varphi_0'(x) = \frac{1}{2}\,\partial_\lambda u_\alpha(x)\Phi_0^{\lambda\alpha\mu\beta}\partial_\mu u_\beta(x). \tag{2.5.4}$$

The energy density $\varphi_0(x)$ differs from $\varphi_0'(x)$ by divergence terms, which must be chosen in accordance with the conditions of invariance of $\varphi_0(x)$ with respect to translation and rotation. Since $\varphi_0'(x)$, which is determined by (2.5.4), is already invariant with respect to translation, the divergence part must also possess this property.

Problem 2.5.2. Verify that the most general expression for the divergence part, which is quadratic with respect to the displacement and invariant with respect to translation, has the form

$$\varphi_0(x) - \varphi_0'(x) = \frac{1}{2}\,\partial_\lambda[u_\alpha(x)b^{\lambda\alpha\mu\beta}\partial_\mu u_\beta(x)]. \tag{2.5.5}$$

where $b^{\lambda\alpha\mu\beta} = -b^{\mu\alpha\lambda\beta}$ is a constant tensor.

Problem 2.5.3. Show that without loss of generality it is possible to replace $b^{\lambda\alpha\mu\beta}$ by a tensor, which is antisymmetric with respect to $\lambda\mu$ and $\alpha\beta$ and is symmetric with respect to permutations of these pairs.

Thus, $\varphi_0(x)$ can be represented in the form

$$\varphi_0(x) = \frac{1}{2}\,\partial_\lambda u_\alpha(x)(\Phi_0^{\lambda\alpha\mu\beta} + b^{\lambda\alpha\mu\beta})\partial_\mu u_\beta(x)\,, \tag{2.5.6}$$

where $\Phi_0^{\lambda\alpha\mu\beta}$ and $b^{\lambda\alpha\mu\beta}$ possess the above-described symmetries. The conditions of invariance with respect to rotation yield

$$(\Phi_0^{\lambda\alpha\mu\beta} + b^{\lambda\alpha\mu\beta})_{|\mu\beta|} = 0\,. \tag{2.5.7}$$

Regarding (2.5.7) as an equation for $b^{\lambda\alpha\mu\beta}$, let us find conditions for its solvability. For this purpose it is necessary to employ a simple but somewhat cumbersome procedure: to symmetrize (2.5.7) with respect to $\lambda\mu$ and $\alpha\beta$ separately, then to take the difference of these expressions and, finally, to symmetrize the difference with respect to $\lambda\,\alpha$. The final result will contain only two components of the tensor $\Phi_0^{\lambda\alpha\mu\beta}$ with permuted pairs of indices,

$$\Phi_0^{\lambda\alpha\mu\beta} = \Phi_0^{\alpha\lambda\beta\mu}\,. \tag{2.5.8}$$

This is, in fact, the desired additional condition which ensures the invariance of $\varphi_0(x)$ with respect to rotation.

Problem 2.5.4. Show that if the condition (2.5.8) is satisfied, (2.5.7) has only one solution:

$$b^{\lambda\alpha\mu\beta} = \Phi_0^{\lambda\alpha\mu\beta} - \Phi_0^{\lambda\mu\alpha\beta}\,, \tag{2.5.9}$$

which possesses the necessary symmetry.

The substitution of (2.5.9) in to (2.5.6) leads to the final expression for the elastic energy density

$$\varphi_0(x) = \frac{1}{2}\,\partial_\lambda u_\alpha(x) c_0^{\lambda\alpha\mu\beta} \partial_\mu u_\beta(x)\,, \tag{2.5.10}$$

where

$$c_0^{\lambda\alpha\mu\beta} = \Phi_0^{\{\lambda\alpha\mu\}\beta} = \Phi_0^{\lambda\alpha\mu\beta} + \Phi_0^{\alpha\lambda\mu\beta} - \Phi_0^{\lambda\mu\alpha\beta} \tag{2.5.11}$$

is the tensor of the elastic constants in zeroth approximation. As desired, this tensor possesses the symmetry with respect to the first and second pairs of indices as well as with respect to permutation of the pairs, i.e.

$$c_0^{\lambda\alpha\mu\beta} = c_0^{\alpha\lambda\mu\beta} = c_0^{\mu\beta\lambda\alpha}\,. \tag{2.5.12}$$

It can easily be seen that (5.3) now admits another form:

$$\Phi_0^{\alpha\beta}(k) = k_\lambda k_\mu \Phi_0^{\lambda\alpha\mu\beta} = k_\lambda k_\mu c_0^{\lambda\alpha\mu\beta}\,. \tag{2.5.13}$$

Such equalities reflect features specific to the three-dimensional case. For a one-dimensional medium $\Phi(k)$ uniquely determined the coefficient of k^2, whereas the tensor coefficient of $k_\lambda k_\mu$ is given only when its symmetry is specified. The physical meaning of the tensor of elastic constants $c_0^{\lambda\alpha\mu\beta}$ is clear. The tensor $\Phi_0^{\lambda\alpha\mu\beta}$ is convenient since it is directly connected with the force constants. In fact, from (2.5.2) and (2.3.13) we have

$$\Phi_0^{\lambda\alpha\mu\beta} = [\partial^\lambda \partial^\mu \Phi^{\alpha\beta}(k)]_{k=0} = -\int x^\lambda x^\mu \Phi^{\alpha\beta}(x) dx \tag{2.5.14}$$

or in the n-representation, taking into account (2.3.15),

$$\Phi_0^{\lambda\alpha\mu\beta} = -\frac{1}{v_0}\sum_n n^\lambda n^\mu \Phi^{\alpha\beta}(n)\,. \tag{2.5.15}$$

These expressions enable us to express explicitly $c_0^{\lambda\alpha\mu\beta}$ in terms of the force constants and to interpret (2.5.8) as restrictions on the force constants, due to the requirements of invariance of the energy with respect to rotation.

Thus, we have completed the first part of the problem. Now it is necessary to show that (2.5.2) admits a representation in the form

$$\Phi^{\alpha\beta}(k) = k_\lambda k_\mu \Phi^{\lambda\alpha\mu\beta}(k) = k_\lambda k_\mu c^{\lambda\alpha\mu\beta}(k)\,, \tag{2.5.16}$$

where the operator of the elastic moduli $c^{\lambda\alpha\mu\beta}(k)$ possesses the symmetry (2.3.19), and to find a connection between $\Phi^{\lambda\alpha\mu\beta}(k)$ and $c^{\lambda\alpha\mu\beta}(k)$.

Let us now write down (2.2.4) for the elastic energy Φ in the form

$$2\Phi = \langle k_\lambda u_\alpha | \Phi^{\lambda\alpha\mu\beta} | k_\mu u_\beta \rangle\,. \tag{2.5.17}$$

It is easily seen that in this expression it is possible to replace $\Phi^{\lambda\alpha\mu\beta}(k)$ by the tensor

$$a_1^{\lambda\alpha\mu\beta}(k) \overset{\text{def}}{=} \Phi^{\{\lambda\alpha\mu\}\beta}(k)\,, \tag{2.5.18}$$

which is symmetric with respect to the indices $\lambda\alpha$ because of the symmetry properties of $\Phi^{\lambda\alpha\mu\beta}(k)$ (Lemma 2). This enables us to write (2.5.17) in the form

$$2\Phi = \langle \varepsilon_{\lambda\alpha} | a_1^{\lambda\alpha\mu\beta} | k_\mu u_\beta \rangle\,, \tag{2.5.19}$$

where $\varepsilon_{\lambda\alpha}(k)$ is the strain tensor (2.4.22).

Let

$$a_1^{\lambda\alpha\mu\beta}(k) = a_1^{\lambda\alpha(\mu\beta)}(k) + a_1^{\lambda\alpha\,[\mu\beta]}(k)\,. \tag{2.5.20}$$

Taking into account (2.5.11) and (2.5.18), we have

$$a_1^{\lambda\alpha\mu\beta}(0) = a_1^{\lambda\alpha\{\mu\beta\}}(0) = c_0^{\lambda\alpha\mu\beta}\,, \tag{2.5.21}$$

from which it follows that

$$a_1^{\lambda\alpha\,[\mu\beta]}(0) = 0\,, \tag{2.5.22}$$

or

$$a_1^{\lambda\alpha\,[\mu\beta]}(k) = k_\tau a_1^{\lambda\alpha\mu\beta\tau}(k)\,, \tag{2.5.23}$$

where $a_1^{\lambda\alpha\mu\beta\tau}(k)$ is a tensor, antisymmetric with respect to $\mu\beta$.

The expression for Φ may be now rewritten in the form

$$2\Phi = \langle \varepsilon_{\lambda\alpha} | a_1^{\lambda\alpha(\mu\beta)} | \varepsilon_{\mu\beta} \rangle + \langle \varepsilon_{\lambda\alpha} | a_1^{\lambda\alpha\mu\beta\tau} | k_\tau \Omega_{\mu\beta} \rangle\,, \tag{2.5.24}$$

where

$$\Omega_{\mu\beta}(x) \overset{\text{def}}{=} \partial_{[\mu} u_{\beta]}(x), \qquad \Omega_{\mu\beta}(k) = \mathrm{i}k_{[\mu} u_{\beta]}(k) \tag{2.5.25}$$

is an antisymmetric rotation tensor.

The first term on the right-hand side of (2.5.24) does not change if we replace $a_1^{\lambda\alpha(\mu\beta)}$ by the tensor

$$a_2^{\lambda\alpha\mu\beta} = \frac{1}{2}\left(a_1^{\lambda\alpha(\mu\beta)} + \overline{a_1^{(\mu\beta)\lambda\alpha}}\right), \tag{2.5.26}$$

which has the required symmetry.

Taking into account the identity

$$k_\tau \Omega_{\mu\beta}(k) = 2k_{[\mu}\varepsilon_{\beta]\tau}(k)\,, \tag{2.5.27}$$

let us represent the second term in (2.5.24) in the form

$$\langle \varepsilon_{\lambda\alpha} | a_3^{\lambda\alpha\mu\beta} | \varepsilon_{\mu\beta} \rangle \,,$$

where

$$a_3^{\lambda\alpha\mu\beta} = \frac{1}{2} k_\tau (a_1^{\lambda\alpha\tau\beta\mu} + a_1^{\lambda\alpha\tau\mu\beta} + \overline{a_1^{\beta\mu\tau\alpha\lambda}} + \overline{a_1^{\mu\beta\tau\lambda\alpha}}) \tag{2.5.28}$$

also has the required symmetry.

Finally, combining both terms, we obtain

$$2\Phi = \langle \varepsilon_{\lambda\alpha} | c^{\lambda\alpha\mu\beta} | \varepsilon_{\mu\beta} \rangle \,, \tag{2.5.29}$$

where the operator of the elastic moduli

$$c^{\lambda\alpha\mu\beta}(k) = a_2^{\lambda\alpha\mu\beta}(k) + a_3^{\lambda\alpha\mu\beta}(k) = S\Phi^{\lambda\alpha\mu\beta}(k) \tag{2.5.30}$$

satisfies the symmetry conditions (2.3.19) and is connected with $\Phi^{\lambda\alpha\mu\beta}(k)$ by the above-described expressions, which implicitly determine the operator S.

Hooke's law in the operator form (2.4.23, 24) takes now the form

$$\sigma^{\lambda\alpha}(x) = \int c^{\lambda\alpha\mu\beta}(x - x')\, \varepsilon_{\mu\beta}(x') dx' \,, \tag{2.5.31}$$

$$\sigma^{\lambda\alpha}(k) = c^{\lambda\alpha\mu\beta}(k) \varepsilon_{\mu\beta}\,(k) \,. \tag{2.5.32}$$

Problem 2.5.5. Show that the real (imaginary) part of $c^{\lambda\alpha\mu\beta}(k)$ is an even (odd) function of k and hence, for an isotropic medium and for a medium with central symmetry, $c^{\lambda\alpha\mu\beta}(k)$ is a real function.

The equation of motion (2.2.8) for a homogeneous medium has the form

$$\rho u^\alpha(x) + \int \Phi^{\alpha\beta}(x - x') u_\beta(x')\, dx' = q^\alpha(x) \tag{2.5.33}$$

or in the (k, ω)-representation:

$$\rho\omega^2 u^\alpha(k) - \Phi^{\alpha\beta}(k) u_\beta(k) = -q^\alpha(k) \,. \tag{2.5.34}$$

Taking into acount (2.5.16), the equations of motion may be also written in the following form

$$\rho u^\alpha(x) - \partial_\lambda \int c^{\lambda\alpha\mu\beta}(x - x') \partial_\mu u_\beta(x')\, dx' = q^\alpha(x) \,, \tag{2.5.35}$$

$$\rho\omega^2 u^\alpha(k) - k_\lambda k_\mu c^{\lambda\alpha\mu\beta}(k) u_\beta(k) = -q^\alpha(k) \,. \tag{2.5.36}$$

For an infinite medium, free vibrations have a special significance. By free vibrations we mean solutions of the equations of motion with $q = 0$, which correspond to propagating waves. It is clear that nontrivial solutions are possible only when ω and k are connected by the dispersion equation

$$\det[\rho\omega^2 g^{\alpha\beta} - \Phi^{\alpha\beta}(k)] = 0\,, \tag{2.5.37}$$

the three roots of which

$$\omega_i = \omega_i(k),\ i = 1,\ 2,\ 3 \tag{2.5.38}$$

give the dependence of the frequencies on the wave vector for the corresponding mode of free vibrations. At the same time, according to (2.5.2) all the modes pass through the points $\omega = 0$, $k = 0$.

To real values of ω and k, correspond nondecaying waves of the type

$$v_\alpha(x, t) = u_\alpha(k)\mathrm{e}^{\mathrm{i}(\omega t - kx)}\,, \tag{2.5.39}$$

which propagate with the group velocity

$$v_i^\lambda(k) \overset{\text{def}}{=} \frac{\partial\omega_i(k)}{\partial k_\lambda}\,. \tag{2.5.40}$$

Generally speaking, the group velocity depends upon the modulus of the wave vector k, i.e. spatial dispersion takes place. The absence of nonzero real roots of the equation $\omega_i(k) = 0$ is a necessary condition of stability of the medium. In fact, such roots would correspond to static displacements different from translation and rotation, which do not contribute to the elastic energy; this is not possible for a stable elastic medium.

However, as in the one-dimensional case, it is possible to have complex roots of the function $\omega_i(k)$, which are caused by the action-at-a-distance. To these correspond exponentially increasing (or decreasing) displacements $u_\alpha(x)$, which cannot be considered to be admissible functions for an infinite medium. At the same time such solutions of the static equation should be taken into account in boundary-value problems, where they can make a considerable contribution to the elastic energy by means of the forces which act in the boundary region.

As a rule, for high frequencies only complex values of the wave vector can satisfy (2.5.37). Solutions having such frequencies must decay, while propagating, but this is not connected with dissipation of energy, since for such waves the energy flux is equal to zero.

2.6 Approximate Models

As in the case of a one-dimensional medium [1.12], when we consider fields which change slowly enough over distances of the order of the characteristic scale parameter l, we can perform a transition to approximate models by replacing the integral operators by differential ones. As we shall see, in this situation, there are possibly some qualitative distinctions between the three-dimensional medium and the one-dimensional one.

We start with the simpler case of a homogeneous medium with an operator of elastic moduli $c^{\lambda\alpha\mu\beta}(k)$ which, as usual, will be assumed to be an analytic function, i.e.

$$c^{\lambda\alpha\mu\beta}(k) = \sum_{n=0}^{\infty} c_n^{\lambda\alpha\mu\beta\tau_1\ldots\tau_n}(\mathrm{i}k_{\tau_1}) \ldots (\mathrm{i}k_{\tau_n}), \tag{2.6.1}$$

where the c_n's are real constant tensors, which can be explicitly expressed in terms of force constants of the micro-model.

For example, in the zeroth approximation, taking into account that

$$\Phi_0^{\lambda\mu\alpha\beta} = [\partial^\lambda\partial^\mu\Phi^{\alpha\beta}(k)]_{k=0} = -\int x^\lambda x^\mu \Phi^{\alpha\beta}(x)\, dx - = \frac{1}{v_0}\sum_n n^\lambda n^\mu \Phi^{\alpha\beta}(n), \tag{2.6.2}$$

(the last expression is for a quasicontinuum) from (2.5.4), we have

$$c_0^{\lambda\alpha\mu\beta} = \int x^{\{\lambda} x^\mu \Phi^{\alpha\}\beta}(x)\, dx = \frac{1}{v_0}\sum_n n^{\{\lambda} n^\mu \Phi^{\alpha\}\beta}(n), \tag{2.6.3}$$

which coincides with the expression for the elastic moduli, known in the theory of a crystal lattice [2.2].

Thus, in the zeroth approximation we obtain the equations of the classical theory of elasticity

$$\rho u^\alpha(x) - \partial_\lambda c_0^{\lambda\alpha\mu\beta}\partial_\mu u_\beta(x) = q^\alpha(x) \tag{2.6.4}$$

with the elastic-constants tensor $c_0^{\lambda\alpha\mu\beta}$. Note that this approximation is universal in the sense that it is physically correct for an infinite medium, as well as for boundary-value problems.

As we saw in the examples of one-dimensional models, when constructing successive approximations, it is usually necessary to distinguish the cases of infinite and finite media. In this section, we consider infinite media only. As distinct form the classical theory of elasticity, for nonlocal and weakly nonlocal theories this case has a fundamental significance.

Recall that in models of media with weak dispersion, which correspond to the long-wavelength approximation, only the fields $u(x)$ whose Fourier transforms $u(k)$ are localized in the region of small (as compared with l^{-1}) wave numbers are strictly speaking, to be considered. This enables one to restrict oneself to a finite number of terms in (2.6.1), or, equivalently in the expansion of the functions of bounded support $c^{\lambda\alpha\mu\beta}(x)$ in a multipole series.

To a first approximation

$$c^{\mu\alpha\mu\beta}(k) = c_0^{\lambda\alpha\mu\beta} + \mathrm{i}c_1^{\lambda\alpha\mu\beta\tau} k_\tau\,. \tag{2.6.5}$$

The operator Hooke's law (2.5.20) takes the form

$$\sigma^{\lambda\alpha}(x) = c_0^{\lambda\alpha\mu\beta}\varepsilon_{\mu\beta}(x) + c_1^{\lambda\alpha\mu\beta\tau}\partial_\tau\varepsilon_{\mu\beta}(x)\,, \tag{2.6.6}$$

and the equations of motion (2.5.35) are differential equations of third order with respect to space derivatives.

It is known [2.2] that such models in crystal optics describe the specific phenomenon of a rotation of the polarization plane of a propogating electromagnetic wave. Media for which this effect takes place are called gyrotropic. An anologous phenomenon is possible also for an elastic medium, when $c_1^{\lambda\alpha\mu\beta\tau} \neq 0$. Unfortunately, as far as the author knows, no experimental investigations of rotating of the polarization plane of elastic waves in a gyrotropic medium have been carried out up to now. At the same time, such an investigation must be of interest, because gyrotropy is one of not very many qualitatively new phenomena which arise already in taking into account weak spatial dispersion. This is connected with the fact that, despite the small value of the second term in (2.6.5), the effect accumulates over a sufficiently large path and hence the plane of polarization rotates through a finite angle.

For media with central symmetry, as well as for the isotropic medium, we have

$$\Phi^{\alpha\beta}(x) = \Phi^{\alpha\beta}(-x) = \Phi^{\beta\alpha}(x)\,, \tag{2.6.7}$$

and, as can be easily shown (Problem 2.5.5), $c_1^{\lambda\alpha\mu\beta\tau} = 0$, i.e. such media are not gyrotropic. For such media the next representation after the zeroth one is the second approximation

$$c^{\lambda\alpha\mu\beta}(k) = c_0^{\lambda\alpha\mu\beta} - c_2^{\lambda\alpha\mu\beta\tau_1\tau_2}k_{\tau_1}k_{\tau_2}\,, \tag{2.6.8}$$

to which corresponds Hooke's law

$$\sigma^{\lambda\alpha}(x) = (c_0^{\lambda\alpha\mu\beta} + c_2^{\lambda\alpha\mu\beta\tau_1\tau_2}\partial_{\tau_1}\partial_{\tau_2})\varepsilon_{\mu\beta}(x)\,. \tag{2.6.9}$$

It is convenient to set

$$c_2 = l^2 c_2'\,, \tag{2.6.10}$$

where c_2' has the dimension of ordinary elastic moduli and l is the characteristic scale parameter. The latter can also be considered as a small parameter with respect to which the expansion is carried out. The equations of motion (2.5.35) are now written in the form

$$\rho u^\alpha(x) - \partial_\lambda(c_0^{\lambda\alpha\mu\beta} + l^2 c_2'^{\lambda\alpha\mu\beta\tau_1\tau_2}\partial_{\tau_1}\partial_{\tau_2})\partial_\mu u_\beta(x) = q^\alpha(x)\,. \tag{2.6.11}$$

For an inhomogeneous medium, the elastic moduli c_0, c_1, c_2 will be functions of the point x, and it is natural to regard them as changing slowly over distances of the order of l. The connection between the elastic moduli and $\Phi^{\alpha\beta}(x, x')$ or $c^{\lambda\alpha\mu\beta}(x, x')$ can be found, as in [Ref. 1.12, Sect. 2.4].

Thus the equations for the medium with weak dispersion belong to the well studied class of equations with a small coefficient of higher derivatives. It is necessary to take this coefficient into account when constructing solutions as well as when determining the field of applicability. In particular, such equations are also connected with phenomena of the boundary-layer type.

In an obvious way, the equations for higher-order approximations can be written down, but they have no significance, since they do not describe qualitatively new phenomena and may provide at most a small correction.

Equation (2.6.11) of a medium of simple structure with weak spatial dispersion does not differ in essence from the equations of the so-called couple-stress theory of elasticity with constrained rotation, which was developed by a number of authors [2.3–5]. A superficial distinction is only in the form in which the equations of motion are written down, since in the couple-stress theory of elasticity a nonsymmetric stress tensor usually appears. Such a distinction is connected with the different definitions of the stress tensor and has mainly a methodological character: the equations of motion expressed in terms of displacements coincide.

However, the above-accepted physical interpretation of the approximate model (2.6.11) differs fundamentally from the usual interpretation of the couple-stress theory of elasticity. The latter is introduced purely phenomenologically in analogy with the classical theory of elasticity, and the parameter l is not regarded as small.[3] As a result, typical problems, which are often solved in connection with couple-stress theories, concern calculations of stress concentration on inhomogeneties having dimensions of the order of or even less than l. It is natural that values of the stress concentration can differ strongly from values obtained on the basis of the classical theory of elasticity, and this creates the illusion of a theory being significantly improved. The preceeding analysis shows that these results should rather be considered as a brave extrapolation of the approximate model (2.6.11) in a field in which, generally speaking, only the strongly nonlocal model is applicable.

2.7 Cubic Lattice

As an example, let us consider a simple cubic lattice with interaction between nearest neighbors. It is evident in that case that one cannot restrict oneself

[3]Note that when deriving the equations of the couple-stress theory of elasticity from a consideration of equillibrium of the volume element, the existence of a scale parameter, i.e. a nonlocal nature of the model, is not taken into account. At the same time, propositions concerning the forces of short-range action and infinitesimal volume elements, which are typical for such considerations, are justifiable only for the consistent local classical theory of elasticity, where there is no scale parameter.

to the binary interaction (with the bond of the type of a spring) since such a lattice would be unstable; it would be possible to fold up such a lattice without changing the distances between neighboring knots. Therefore, for a lattice with nearest-neighbor interaction, it is necessary to introduce bonds of the rod type. In the simplest case, each of these rods would be characterized by two parameters: the longitudinal and the transverse stiffnesses A_i and B_i, respectively ($i = 1, 2, 3$).

Taking into account the above reasoning let us give, in the lattice coordinate system, the matrix of force constants in the form (see also [2.6]):

n	$1, 0, 0$ $-1, 0, 0$	$0, 1, 0$ $0, -1, 0$	$0, 0, 1$ $0, 0, -1$
$-v_0^{-1}\Phi^{\alpha\beta}(n)$	$\begin{matrix} A_1 & 0 & 0 \\ 0 & B_1 & 0 \\ 0 & 0 & B_1 \end{matrix}$	$\begin{matrix} B_2 & 0 & 0 \\ 0 & A_2 & 0 \\ 0 & 0 & B_2 \end{matrix}$	$\begin{matrix} B_3 & 0 & 0 \\ 0 & B_3 & 0 \\ 0 & 0 & A_3 \end{matrix}$

(2.7.1)

The first condition from (2.3.17) gives (here and subsequently only nonzero components are presented and there is no summation over identical indices):

$$v_0^{-1}\Phi^{\alpha\alpha}(0) = 2(A_\alpha + B_\beta + B_\gamma), \ \alpha \neq \beta \neq \gamma \,. \tag{2.7.2}$$

From (2.3.15) it follows that the matrix $\Phi^{\alpha\beta}(k)$ is diagonal. For its components in the crystallographic system of coordinates we find ($\alpha \neq \beta \neq \gamma$)

$$\begin{aligned} \Phi^{\alpha\alpha}(k) &= A_\alpha k_\alpha^2 s^2(k_\alpha) + B_\beta k_\beta^2 s^2(k_\beta) + B_\gamma k_\gamma^2 s(k_\gamma)\,, \\ s(k_\alpha) &\overset{\text{def}}{=} \frac{2}{k_\alpha} \sin\frac{k_\alpha}{2} (-\pi \leq k_\alpha \leq \pi)\,. \end{aligned} \tag{2.7.3}$$

From these expressions, according to (2.5.2) we easily find the tensor $\Phi^{\lambda\alpha\mu\beta}(k)$, which is symmetric with respect to the indices λ, μ

$$\Phi^{\alpha\alpha\alpha\alpha}(k) = A_\alpha s^2(k_\alpha), \ \Phi^{\alpha\beta\alpha\beta}(k) = B_\alpha s^2(k_\alpha)\,. \tag{2.7.4}$$

The substitution of components of the tensor $\Phi^{\lambda\alpha\mu\beta}(0)$ into (2.5.8) shows that the conditions of invariance with respect to rotation are not fulfilled. This should not cause a surprise, since for real rod systems the elastic energy depends not only on displacements of the ends of the rods but also on the angles of their rotation. Consideration of such a system in the scope of the model of simple structure is equivalent to postulating that the angles of rotation of the rods' ends are equal to zero. This means that masses located at the knots are displaced in slots which prevent their rotation. However, such a construction is not, in general invariant with respect to rotation.

For an adequate description of the system of rods a model of a medium with complex structure is needed, as we have seen in [Ref. 1.12, Chap. 3], when considering one-dimensional models. The corresponding generalization to the three-dimensional case will be carried out in the following chapter, and here we make use of the incidental fact that for rods with equal transverse stiffness, i.e. for $B_\alpha = B_\beta$, the conditions (2.5.8), of invariance with respect to rotation, are satisfied automatically. Hence this model admits a formal consideration in terms of the medium with simple structure, but one should not expect all results to be valid for a real-rod system.

Thus, assuming that the conditions $B_\alpha = B_\beta = B$ are satisfied let us construct the expression for the operator of elastic moduli $c^{\lambda\alpha\mu\beta}(k)$, following the procedure which was considered in Sect. 2.5.

To this end, first of all, let us find the tensor $a_1^{\lambda\alpha\mu\beta}$, which is defined by (2.5.18). Its nonzero components have the form

$$\begin{aligned} a_1^{\alpha\alpha\alpha\alpha} &= A_\alpha s^2(k_\alpha),\ a_1^{\alpha\alpha\beta\beta} = -\ B s^2(k_\alpha)\,, \\ a_1^{\beta\alpha\alpha\beta} &= a_1^{\alpha\beta\alpha\beta} = B s^2(k_\alpha)\,. \end{aligned} \tag{2.7.5}$$

For the components (2.5.20) which are symmetric and scewsymmetric with respect to the second pair of indices we obtain

$$\begin{aligned} a_1^{\alpha\beta(\alpha\beta)} &= \frac{1}{2} B[s^2(k_\alpha) + s^2(k_\alpha)]\,, \\ a_1^{\alpha\beta[\alpha\beta]} &= \frac{1}{2} B[s^2(k_\alpha) - s^2(k_\beta)]\,. \end{aligned} \tag{2.7.6}$$

According to (2.5.26) for the components $a_2^{\lambda\alpha\mu\beta}$ we have

$$\begin{aligned} a_2^{\alpha\alpha\alpha\alpha} &= A_\alpha s^2(k_\alpha), \qquad a_2^{\alpha\alpha\beta\beta} = -\frac{1}{2} B[s^2(k_\alpha) + s^2(k_\beta)]\,, \\ a_2^{\alpha\beta\alpha\beta} &= \frac{1}{2} B[s^2(k_\alpha) + s^2(k_\beta)]\,. \end{aligned} \tag{2.7.7}$$

The pentavalent tensor $a_1^{\lambda\alpha\mu\beta\tau}$ of fifth rank, which is defined by (2.5.23), has only the following nonzero components:

$$a_1^{\alpha\beta\alpha\beta\alpha} = \frac{1}{2} B \frac{s^2(k_\alpha)}{k_\alpha}\,. \tag{2.7.8}$$

Substitution in (2.5.28) gives

$$\begin{aligned} a_3^{\alpha\beta\alpha\beta} &= \frac{1}{2} B[s^2(k_\alpha) + s^2(k_\beta) - 2]\,, \\ a_3^{\alpha\alpha\alpha\beta} &= \frac{1}{2} B k_\beta \frac{1 - s^2(k_\alpha)}{k_\alpha}\,. \end{aligned} \tag{2.7.9}$$

Finally, according to (2.5.30), for nontrivial components of the elastic moduli operator $c^{\lambda\alpha\mu\beta}(k)$ we may write down the expressions

$$
\begin{aligned}
c^{\alpha\alpha\alpha\alpha} &= A_\alpha s^2(k_\alpha)\,, \\
c^{\alpha\alpha\alpha\beta} &= \frac{1}{2}\,Bk_\beta \frac{1 - s^2(k_\alpha)}{k_\alpha}\,, \\
c^{\alpha\beta\alpha\beta} &= B[s^2(k_\alpha) + s^2(k_\beta) - 1]\,, \\
c^{\alpha\alpha\beta\beta} &= -\frac{1}{2}\,B[s^2(k_\alpha) + s^2(k_\beta)]\,.
\end{aligned}
\tag{2.7.10}
$$

Note that all the computations here were carried out in the lattice coordinate system and therefore the components of the tensor $c^{\lambda\alpha\mu\beta}$ have the dimension [Nm^{-6}]. In order to obtain the usual dimension, i.e. [Nm^{-2}], it is necessary to consider the mixed components $c^{\lambda\alpha}_{\cdot\cdot\mu\beta}$. Since $g_{\alpha\beta} \sim l^2$, where l is the length of a rod, equal to the parameter of the cubic lattice, we have

$$
c^{\lambda\alpha}_{\cdot\cdot\mu\beta} = g_{\mu\tau} g_{\beta\sigma} c^{\lambda\alpha\tau\sigma} \doteq l^4 c^{\lambda\alpha\mu\beta}\,, \tag{2.7.11}
$$

where the last equality is valid only in the lattice coordinate system.

For the elastic-constants tensor in the zeroth approximation, we find

$$
c_{0\cdot\cdot\mu\beta}^{\lambda\alpha} \doteq \lambda_0 \delta^{\lambda\alpha} \delta_{\mu\beta} + \mu_0(\delta^\lambda_\mu \delta^\alpha_\beta + \delta^\lambda_\beta \delta^\alpha_\mu) + \sum_i \gamma_i \delta^{i\lambda} \delta^{i\alpha} \delta_{i\mu} \delta_{i\beta}\,, \tag{2.7.12}
$$

where

$$
\mu_0 = -\lambda_0 = l^4 B,\ \gamma_i = l^4(A_i - B)\,. \tag{2.7.13}
$$

Note that the different signs of the Lamé constants λ_0 and μ_0 additionally indicate the artificial nature of the model and, in particular, the fact that the model does not correspond to a system of rods. In fact, in the latter, a side thrust should not result from tension along a lattice axis, and this is possible only if $\lambda_0 = 0$. In the next chapter we shall see that in a correct consideration of rod systems in terms of media of complex structure, such contradictions do not appear.

Problem 2.7.1. Write down the elastic constants in the second approximation.

Problem 2.7.2. Taking into account (2.7.3), derive an expression for the dispersive modes $\omega_j(k)$ ($j = 1, 2, 3$).

2.8 Isotropic Homogeneous Medium

As a start, let us define our notation in the case of isotropic media. The position vectors of the physical and Fourier spaces will be denoted by $\mathbf{r}$ and $\mathbf{k}$, respectively. Their moduli are denoted by r and k, i.e. $r = |\mathbf{r}|$, $k = |\mathbf{k}|$. Note also that in the three-dimensional case, a function $\varphi(r)$ and its Fourier transform $\varphi(k)$ are connected by the following relations

$$\varphi(k) = \frac{2\pi i}{k} \int_{-\infty}^{\infty} r\varphi(r) e^{-ikr}\, dr\,, \tag{2.8.1}$$

$$\varphi(r) = -\frac{i}{4\pi^2 r} \int_{-\infty}^{\infty} k\varphi(k) e^{irk}\, dk\,, \tag{2.8.2}$$

in the presence of poles, the appropriate regularization of the integrals is necessary.

Obviously, for an infinite isotropic medium, in cartesian coordinates, the operator $\Phi^{\alpha\beta}(\mathbf{k})$ may be constructed only by means of combinations of the tensor $\delta^{\lambda\mu}$ and the components of the vector **k**. The most general expression which satisfies the conditions (2.3.13,18), has the form

$$\Phi^{\alpha\beta}(\mathbf{k}) = \mu(k) k^2 \delta^{\alpha\beta} + [\lambda(k) + \mu(k)] k^\alpha k^\beta\,, \tag{2.8.3}$$

where $\lambda(k)$, $\mu(k)$ are real scalar functions. When the action-at-a-distance is finite, λ and μ are analytic functions of the vector **k**, and their expansions with respect to **k** can contain only terms with k^2 (as a consequence of isotropy), i.e. $\lambda(k)$ and $\mu(k)$ are even analytic functions of k.

It can be easily seen that (2.8.3) for the operator $\Phi^{\alpha\beta}(\mathbf{k})$ can be associated with the corresponding operator of the elastic moduli

$$c^{\lambda\alpha\mu\beta}(\mathbf{k}) = \lambda(k)\delta^{\lambda\alpha}\delta^{\mu\beta} + \mu(k)(\delta^{\alpha\beta}\delta^{\lambda\mu} + \delta^{\lambda\beta}\delta^{\alpha\mu})\,, \tag{2.8.4}$$

which has the same structure as the tensor of elastic moduli in the local theory of elasticity. However, as distinct from the local theory, this representation for $c^{\lambda\alpha\mu\beta}(\mathbf{k})$ is not unique. In fact, without destroying the symmetry, one can add to (2.8.4) terms of the form

$$\alpha(k)\delta^{\lambda\alpha} k^\mu k^\beta + \delta^{\mu\beta} k^\lambda k^\alpha + \beta(k) k^\lambda k^\alpha k^\mu k^\beta\,,$$

which would only change the connection between the scalar coefficients of the tensors $\Phi^{\alpha\beta}(\mathbf{k})$ and $c^{\lambda\alpha\mu\beta}(\mathbf{k})$. Thus, we have here one more example of nonuniqueness of the operator of elastic moduli (and hence the stress tensor) in a nonlocal theory.

The equations of motion (2.5.33) in direct notation have the form

$$\rho u - M\,\Delta u - (\Lambda + M)\,\mathrm{grad}\,\mathrm{div}\, u = q\,, \tag{2.8.5}$$

which is identical with the equation for an isotropic medium in the local theory, but with substitution of the Lamé constrants λ_0, μ_0 by scalar operators Λ, M

with the kernels $\lambda(r)$, $\mu(r)$. The latter are inverse Fourier transforms of $\lambda(k)$, $\mu(k)$. Note that the nonuniqueness of the operator of elastic moduli $c^{\lambda\alpha\mu\beta}$ does not influence the equations of motion whatsoever: these are fully determined by the elastic-energy operator $\Phi^{\alpha\beta}$.

In order to obtain solutions of the equations of motion it is sufficient to construct the Green's tensor $G(\mathbf{r}, t)$. Formally this can be acheived by replacing the Lamé constants in the Green's tensor of the local theory of elasticity by the corresponding operators. For example, for the static Green's tensor we have

$$G_{\alpha\beta}(\mathbf{k}) = \frac{1}{k^2\mu(k)}\left(\delta^{\alpha\beta} - \frac{\lambda(k) + \mu(k)}{\lambda(k) + 2\mu(k)}\frac{k_\alpha k_\beta}{k^2}\right). \tag{2.8.6}$$

However, this representation is not very suitable for the investigation of specific nonlocal phenomena. Consequently, we shall consider another approach.

An arbitrary vector function $u_\beta(\mathbf{k})$ can be decomposed into its longitudinal (l) and transverse (t) components, as a result of the projection operators

$$\pi^{\alpha\beta}(\mathbf{k}) \stackrel{\text{def}}{=} \frac{k^\alpha k^\beta}{k^2}, \quad \vartheta^{\alpha\beta}(\mathbf{k}) \stackrel{\text{def}}{=} \delta^{\alpha\beta} - \frac{k^\alpha k^\beta}{k^2}. \tag{2.8.7}$$

In the $\mathbf{r}$-representation this corresponds to the decomposition of the vector field $u_\beta(\mathbf{r})$ into potential and rotational parts.

Problem 2.8.1. Check that in the $\mathbf{r}$-representation

$$\pi = -\frac{1}{4\pi r} * \text{grad div}, \qquad \vartheta = \frac{1}{4\pi r} * \text{rot rot}. \tag{2.8.8}$$

Note that $\Phi^{\alpha\beta}(\mathbf{k})$ can be represented in the form

$$\Phi^{\alpha\beta}(\mathbf{k}) = \Phi_l(k)\pi^{\alpha\beta}(\mathbf{k}) + \Phi_t(k)\vartheta^{\alpha\beta}(\mathbf{k}), \tag{2.8.9}$$

where

$$\Phi_l(k) \stackrel{\text{def}}{=} k^2[\lambda(k) + 2\mu(k)], \; \Phi_t(k) \stackrel{\text{def}}{=} k^2\mu(k). \tag{2.8.10}$$

Applying the operators π and ϑ to the equations of motion we find that these equations are decomposed into independent equations with scalar operators

$$[-\rho\omega^2 + \Phi_j(k)]\, u_j^\alpha(\mathbf{k}, \omega) = q_j^\alpha(\mathbf{k}, \omega), j = \text{l, t} \tag{2.8.11}$$

with respect to the longitudinal and transverse components of the displacements and forces.

The dispersion equation (2.5.37) is also decomposed into an equation for the longitudinal mode

$$\omega^2 = \omega_l^2(k) \stackrel{\text{def}}{=} \rho^{-1}\Phi_l(k) \tag{2.8.12}$$

and a doubly degenerate equation for the transverse mode

$$\omega^2 = \omega_t^2(k) \stackrel{\text{def}}{=} \rho^{-1}\Phi_t(k)\,. \tag{2.8.13}$$

To the decomposition (2.8.9) of the operator $\Phi^{\alpha\beta}(k)$ into two orthogonal components corresponds the representation of the Green's tensor $G_{\alpha\beta}(\mathbf{k}, \omega)$ in the form

$$G_{\alpha\beta}(\mathbf{k}, \omega) = \pi_{\alpha\beta}(\mathbf{k})G^l(k, \omega) + \vartheta_{\alpha\beta}(\mathbf{k})G^t(k, \omega)\,, \tag{2.8.14}$$

where

$$G^j(k, \omega) \stackrel{\text{def}}{=} \frac{1}{-\rho\omega^2 + \Phi_j(k)} = \frac{1}{\rho[\omega_j^2(k) - \omega^2]}\,, \quad j = l, t\,. \tag{2.8.15}$$

If the action-at-a-distance is finite, the analytic properties of the entire functions $\Phi_j(k, \omega) = -\rho\omega^2 + \Phi_j(k)$ are identical to the properties of the function $\Phi(k, \omega)$ for a one-dimensional model. This allows us to use the method developed in [Ref. 1.12, Sect. 2.8, 9] for constructing the meromorphic functions $G^j(k, \omega)$, but in order find their inverse Fourier transforms $G^j(r, \omega)$, it is necessary to apply the three-dimensional inversion formula (2.8.2).

Analogous to (I.2.8.16) we have

$$\Phi_j(k, \omega) = -\rho\omega^2 \prod_{m=0}^{\infty}\left(1 - \frac{k^2}{k_{j,2m}^2(\omega)}\right), \tag{2.8.16}$$

where, as $\omega \to 0$,

$$k_{l,0}^2(\omega) \to \frac{\rho\omega^2}{\lambda_0 + 2\mu_0}\,, \quad k_{t,0}^2(\omega) \to \frac{\rho\omega^2}{\mu_0}\,. \tag{2.8.17}$$

The inverse functions $G^j(k, \omega) = \Phi_j^{-1}(k, \omega)$ can be represented in the form ([Ref. 1.12., Eq. (2.9.17)])

$$G^j(k, \omega) = 2\sum_{m=0}^{\infty} \frac{k_{j,2m}(\omega)}{\Phi_j'(k_{j,2m}(\omega))} \frac{1}{k^2 - k_{j,2m}^2(\omega)}\,, \tag{2.8.18}$$

and hence, in view of (2.8.2), we obtain

$$G^j(r, \omega) = \frac{1}{4\pi r} \sum_{m=0}^{\infty} \frac{k_{j,2m}(\omega)}{\Phi_j'(k_{j,2m}(\omega))} \exp\left[ik_{j,2m}(\omega)\, r\right]. \tag{2.8.19}$$

Thus for the isotropic medium, the structure of the solutions is determined by the distribution of roots of the functions $\Phi_j(k, \omega)$ as was also true in the one-dimensional case.

When proceeding to models with weak nonlocality, we used the long-wavelength approximation. However, for isotropic media, in a number of cases, it may be more suitable to use an approximation with respect to the first roots (Sect. 2.10). In this approximation, in the static case ($\omega = 0$)

$$G^{l}_{\alpha\beta}(\mathbf{r}) = \frac{1}{8\pi(\lambda_0 + 2\mu_0)}\,\partial_\alpha\partial_\beta\left(r + 4\,\mathrm{Re}\left\{\frac{\kappa_l^2}{\kappa_l^4 - |\kappa_l|^4}\cdot\frac{1 - e^{i\mathscr{H}_t r}}{r}\right\}\right),$$

$$G^{t}_{\alpha\beta}(\mathbf{r}) = \frac{\delta_{\alpha\beta}}{4\pi\mu_0 r}\left(1 + 2\,\mathrm{Re}\,\frac{\kappa_t^2 e^{i\mathscr{H}_t r}}{\kappa_t^2 - \bar{\mathscr{H}}_t^2}\right) \qquad (2.8.20)$$

$$-\frac{1}{8\pi\mu_0 r}\,\partial_\alpha\partial_\beta\left(r + 4\,\mathrm{Re}\left\{\frac{\kappa_t^2}{\kappa_t^4 - |\kappa_t|^4}\cdot\frac{1 - e^{i\mathscr{H}_t r}}{r}\right\}\right),$$

where $\kappa_j = k_{j,2}$ and $\mathrm{Im}\{\kappa_j\} > 0$.

From the above expressions it follows that together with the classical asymptotic behavior $\sim r^{-1}$ there are also terms $\sim r^{-3}$ and terms which decay exponentially. Note that for one-dimensional media, the asymptotic, behavior contained only terms $\sim |x|^{-1}$ and exponential terms.

When passing to the local theory, it can be assumed that $\mathrm{Im}\{\kappa_j\} \to \infty$, and the Green's tensor coincides with the usual one.

2.9 Debye Quasicontinuum

In the preceding section, we considered the isotropic elastic continuum with nonlocal interaction. In order to obtain a correct qualitative description of the discrete structure, and to conserve isotropy simultaneously one can use the Debye quasicontinuum model. All the formulae (2.8.1–15) remain valid, but the vector $\mathbf{k}$ has to satisfy the additional condition $k \leq \kappa$ where $\kappa \simeq \pi/a$ is the Debye radius and a is the parameter of the discrete structure.

The role of the δ-function is now played by $\delta_\kappa(\mathbf{r})$, whose Fourier transform is equal to 1 when $k \leq \kappa$ and equal to zero when $k > \kappa$. Applying (2.8.2), we obtain

$$\delta_\kappa(\mathbf{r}) = \frac{\kappa}{2\pi^2 r^2}\left(\frac{\sin\kappa r}{\kappa r} - \cos\kappa r\right). \qquad (2.9.1)$$

We shall also need the Green's functions of the Laplacian $g(r)$ and bi-Laplacian $h(r)$, i.e. the inverse Fourier transforms of $-k^2$ and k^4, which satisfy the condition

$$\Delta^2 h(r) = \Delta g(r) = \delta_\kappa(\mathbf{r})\,. \qquad (2.9.2)$$

Simple computations yield

$$g(r) = -\frac{1}{2\pi^2 r}\,\mathrm{Si}(\kappa r), \tag{2.9.3}$$

$$h(r) = -\frac{1}{4\pi^2\kappa}\left(\kappa r\,\mathrm{Si}(\kappa r) + \frac{\sin\kappa r}{\kappa r} + \cos\kappa r\right), \tag{2.9.4}$$

where Si (x) is the integral sine.

When $\kappa \to \infty$ $(a \to 0)$ or when $r \to \infty$, all these expressions are transformed into the ordinary ones.

The Green's tensor in the **r**-representation can be constructed, as in the previous section, with the help of the expansion with respect to the roots. However, in the case of a quasicontinuum there is another possibility which is connected with a suitable approximation of the functions $\omega_j(k)$ over the segment $0 \leq k \leq \kappa$.

Various approximations for $\omega_j(k)$ which are used in the theory of specific heat [2.7], are known. Thus in the simplest Debye model it is assumed that

$$\omega_j(k) = v_j k, \quad j = l, t, \tag{2.9.5}$$

where v_l, v_t are the logitudinal and transverse velocities of sound in the corresponding local model of the elastic continuum. It is evident that this model does not take into account the spatial dispersion.

In a more realistic Born-Karman model, a dispersion law is accepted, which is analogous to the case of a chain with interaction between nearest neighbors

$$\omega_j(k) = \frac{2\kappa v_j}{\pi}\sin\frac{\pi k}{2\kappa}. \tag{2.9.6}$$

The natural conditions

$$\frac{d\omega_j(0)}{dk} = v_j, \quad \frac{d\omega_j(\kappa)}{dk} = 0 \tag{2.9.7}$$

are then fulfilled.

For our purposes the polynomial approximation of the form

$$\omega_j^2(k) = \frac{v_j^2 k^2}{P_j(k^2)} \tag{2.9.8}$$

is more convenient, where $P_j(k^2)$ is an appropriate polynomial, which satisfies the conditions (2.9.7.) The simplest polynomial is

$$P_j(k^2) = 1 + \gamma_j\left(\frac{k}{\kappa}\right)^2 + \left(\frac{k}{\kappa}\right)^4; \tag{2.9.9}$$

it depends on the parameter γ_j. The corresponding family of dispersion curves is shown in Fig. 2.1. The straight line corresponds to the Debye model where $\omega_j^0 = v_j\kappa$ is the Debye frequency. The dotted curve corresponds to the Born-Karman model and, as is clear from the picture, practically it coincides with the curve for $\gamma_j = 0.5$.

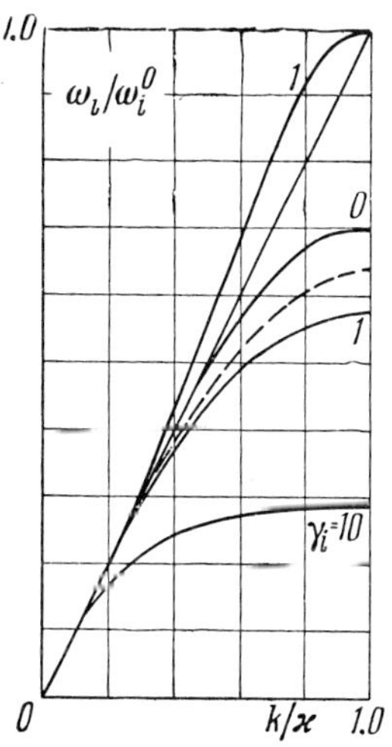

Fig. 2.1. Dispersion curves

The parameter γ_j may be connected with the ratio of the boundary frequency $\omega_j(\kappa)$ to the Debye frequency ω_j^0, i.e.,

$$\frac{\omega_j(\kappa)}{\omega_j^0} = \frac{1}{\sqrt{2 + \gamma_j}}, \tag{2.9.10}$$

from which the condition $-2 < \gamma_j < \infty$ follows.

The approximation (2.9.8) allows us to construct an explicit expression for the static Green's tensor. From (2.8.7, 14 and 15) with $\omega = 0$, we have

$$\begin{aligned} G^l_{\alpha\beta}(k) &= \frac{k_\alpha k_\beta}{\rho k^2 \omega_l^2(k)}, \\ G^t_{\alpha\beta}(k) &= \left(\delta_{\alpha\beta} - \frac{k_\alpha k_\beta}{k^2}\right)\frac{1}{\rho\omega_t^2(k)}. \end{aligned} \tag{2.9.11}$$

Taking into account (2.9.2, 8) we finally obtain

$$G^l_{\alpha\beta}(\mathbf{r}) = -\frac{1}{\rho v_l^2}\partial_\alpha\partial_\beta h_l(r), \tag{2.9.12}$$

$$G^t_{\alpha\beta}(\mathbf{r}) = \frac{1}{\rho v_t^2}[\partial_\alpha\partial_\beta h_t(r) - g_t(r)\delta_{\alpha\beta}], \tag{2.9.13}$$

where

$$h_j(r) \stackrel{\text{def}}{=} P_j(-\Delta)h(r), \; g_j(r) = P_j(-\Delta)g(r)\,, \tag{2.9.14}$$

and $h(r)$ and $g(r) = \Delta h(r)$ are determined by (2.9.3, 4).

In particular, for the model (2.9.9)

$$h_j(r) = h(r) - \gamma_j \kappa^{-2}\, g(r) + \kappa^{-4}\delta_\kappa(r)\,, \tag{2.9.15}$$

and for the Debye model (2.9.5) we have $P_j(-\Delta) = 1$.

An important property of the constructed Green's tensors is the absence of a singularity at $r = 0$, since $G_{\alpha\beta}(r)$ is an even analytic function. This corresponds to a physical meaning of the nonlocal theory for a medium with an elementary unit of length, which models the discrete medium and formally is an effect of truncating the Fourier spectrum. As a result, all the divergencies connected with point sources of force, which are inherent to the model of an elastic continuum are eliminated.

Let us consider for example the concentrated force

$$q^\alpha(\mathbf{r}) = Q^\alpha \delta(\mathbf{r})\,, \tag{2.9.16}$$

where $Q^\alpha = \text{const}$, and $\delta(\mathbf{r})$ can also be replaced by $\delta_\kappa(\mathbf{r})$ since due to the truncation of the Fourier spectrum of the Green's tensor, the final result is the same. This reasoning enables us, in what follows, to make no distinction between the "true" δ-function and the δ-function with the truncated Fourier spectrum, when we consider the quasicontinuum model.

For the energy of the elastic field of a concentrated force, taking into account (2.3.16), we have

$$\Phi = \frac{1}{2}\, Q^\alpha Q^\beta G_{\alpha\beta}(0) = \frac{Q^2\kappa}{12\pi^2\rho}\left(\frac{\alpha_l}{v_l^2} + \frac{2\alpha_t}{v_t^2}\right), \tag{2.9.17}$$

where the α_j are dimensionless constants, which depend on the form of the functions $\omega_j(k)$,

$$\alpha_j = \frac{v_j^2}{\kappa}\int_0^\kappa \frac{k^2}{\omega_j^2(k)}\, dk\,. \tag{2.9.18}$$

For the Debye model and for the model (2.9.9) we find

$$\alpha_j = 1, \; \alpha_j = \frac{6}{5} + \frac{1}{8}\gamma_j\,.$$

Let us consider the asymptotic form of the Green's tensor, as $r \to \infty$. It is easily seen that this is determined by the asymptotic behavior of the function $h_j(r)$. The latter has the form

$$h_j(r) \approx -\frac{r}{8\pi}\left(1 - \frac{2\gamma_j}{\kappa^2 r^2}\right). \tag{2.9.19}$$

Substitution in (2.9.12) and (2.9.13) yields the dominant terms of the asymptotic form of $G_{\alpha\beta}(r)$ to be of the order of r^{-1} and r^{-3}, which, as it should, coincides with the asymptotic form of the Green's tensor for the elastic continuum with action-at-a-distance examined in the preceeding section. As $\kappa \to \infty$, the Green's tensor is transformed into the Green's tensor of the local theory of elasticity and (2.9.17) diverges linearly with κ.

2.10 Boundary-Value Problems and Surface Waves

The results obtained in [Ref. 1.12, Sects. 2.6, 11] can be transferred to the three-dimensional case but here some specific three-dimensional effects arise; in particular, special types of surface waves appear.

Let us restrict ourselves to a consideration of the simplest type of model of a medium with binary interaction. In this case the operator of the elastic bonds $\Psi^{\alpha\beta}$ has the form of (2.3.8). Repeating almost literally the arguments carried out earlier for the one-dimensional case, one can introduce the three-dimensional operators $\Phi_V^{\alpha\beta}$, $\Gamma^{\alpha\beta}(\omega)$, etc., write down the Green's formula and formulate the basic boundary value problems. In particular, in the obvious notation (see the notation [Ref. 1.12, Sect. 2.6]), the Green's formula has the form

$$u_\alpha = G_{\alpha\beta}(\omega)q_D^\beta + G_{\alpha\beta}(\omega)q_S^\beta - G_{\alpha\beta}(\omega)\Gamma^{+\beta}_{\cdot\lambda}(\omega)u_S^\lambda\,. \tag{2.10.1}$$

The equations of the first basic problem take the form

$$D\Phi^{\alpha\beta}(\omega)u_\beta \stackrel{\text{def}}{=} (-\,\omega^2\rho_D\delta^{\alpha\beta} + D\Phi^{\alpha\beta})u_\beta = q_D^\alpha, \qquad (\Phi^{\alpha\beta} = \Phi^{\beta\alpha})\,, \tag{2.10.2}$$

and the boundary conditions are written in the form

$$\Gamma^{\alpha\beta}(\omega)u_\beta \stackrel{\text{def}}{=} -\omega^2\rho_S\delta^\alpha_\beta u_S^\beta + \Gamma^{\alpha\beta}u_\beta = q_S^\alpha, \qquad (\Gamma^{\alpha\beta} = \Gamma^{\beta\alpha})\,, \tag{2.10.3}$$

where, for example, the kernel

$$\Gamma^{\alpha\beta}(\mathbf{r}, \mathbf{r}') = S(\mathbf{r})\delta(\mathbf{r} - \mathbf{r}')\int_V \Psi^{\alpha\beta}(\mathbf{r}, \mathbf{r}'')d\mathbf{r}'' - S(\mathbf{r})\Psi^{\alpha\beta}(\mathbf{r}, \mathbf{r}')V(\mathbf{r}') \tag{2.10.4}$$

corresponds to the operator $\Gamma^{\alpha\beta}$.

Note that in the above formulation of the problem, the conditions of invariance with respect to rigid translations and rotations of the medium, are automatically satisfied.

Since the radius of the action-at-a-distance is usually small in comparison with other characteristic dimensions, the main interest in the theory of the elastic medium with action-at-a-distance is in the boundary-value problems for a half space (half plane).

As an example, let us consider the problem of propagation of surface waves in a homogeneous isotropic medium with binary interaction, for which

$$\Psi^{\alpha\beta}(\mathbf{r}, \mathbf{r}') = \Psi^{\alpha\beta}(\mathbf{r} - \mathbf{r}'),\ \Psi^{\alpha\beta}(\mathbf{r}) = \frac{x^\alpha x^\beta}{r^2}\Psi(r)\,. \tag{2.10.5}$$

The equations of the problem can be also represented in the form

$$-\omega^2 \rho u_{\mathrm{j}} + \Phi_{\mathrm{j}} u = 0\,, \tag{2.10.6}$$

where u_{j} denotes either the longitudinal or the transverse component u_{l} or u_{t}, and Φ_{j} is the corresponding operator Φ_{l} or Φ_{t}.

Problem 2.10.1. Show that for the model under consideration.

$$\Phi_{\mathrm{j}}(k) = 4\pi \int_0^l r^1 \Psi_{\mathrm{j}}(r)\left(1 - \frac{\sin kr}{kr}\right) dr\,, \tag{2.10.7}$$

where the functions Ψ_{l}, Ψ_{t} and Ψ are connected by the relations

$$\Psi(r) = \Psi_{\mathrm{l}}(r) + 2\Psi_{\mathrm{t}}(r), \qquad \Psi_{\mathrm{t}}(r) = \int_r^l r^{-1}\Psi(r)\, dr\,. \tag{2.10.8}$$

Problem 2.10.2. Show that the interaction of binary type implies the equality of the Lamé constants

$$\lambda_0 = \mu_0 = \frac{2\pi}{15}\int_0^l r^4 \Psi(r)\, dr\,. \tag{2.10.9}$$

Let us now assume that the medium occupies the upper half space and choose the coordinate system so that the z-axis is directed normally to the surface. The boundary conditions, corresponding to the absence of forces in the surface layer $S(z)$, have the form

$$-\omega^2 \rho u^\alpha(\mathbf{r}) + \int \Gamma^{\alpha\beta}(\mathbf{r}, \mathbf{r}') u_\beta(\mathbf{r}')\, d\mathbf{r}' = 0,\ \mathbf{r} \in S(z)\,. \tag{2.10.10}$$

Let us consider a plane surface wave propagating along the x-axis, i.e. let us seek a solution of the form

$$u(x, z) = \mathrm{e}^{\mathrm{i}px}\, v(z)\,, \tag{2.10.11}$$

where $v(z) \to 0$ as $z \to \infty$, and p is a real parameter. It is easy to see that in this case two types of surface waves can propagate independently in the medium. For a wave of the first type the displacement vector lies in the plane, which is drawn in the direction of propagation perpendicularly to the surface, and which satisfies the boundary conditions

$$\begin{aligned} -\omega^2 \rho u_x + \Gamma^{11} u_x + \Gamma^{13} u_z &= 0\,, \\ -\omega^2 \rho u_z + \Gamma^{13} u_x + \Gamma^{33} u_z &= 0\,. \end{aligned} \tag{2.10.12}$$

For a wave of the second type, the displacement vector is parallel to the z-axis and satisfies the condition

$$-\omega^2 \rho u_y + \Gamma^{22} u_y = 0\,. \tag{2.10.13}$$

The formal decomposition of the boundary conditions is a consequence of the fact that, for the given model, the following integrals vanish:

$$\int \Psi^{\alpha 2}(r)\, dy - x^{\alpha} \int \frac{y}{r^2}\, \Psi(r)\, dy = 0 \quad (\alpha \neq 2)\,. \tag{2.10.14}$$

We carry out the further analysis, restricting ourselves to an approximate consideration of the problem. In the zeroth approximation the conditions (2.10.12) are transformed into usual boundary conditions for Rayleigh waves. The condition (2.10.13), as $l \to 0$, yields

$$\frac{\partial u_y}{\partial z} = 0\,. \tag{2.10.15}$$

On account of (2.10.11) it follows that in the zeroth approximation

$$u_y = 0\,. \tag{2.10.16}$$

Thus the wave of the second type does not have any classical analogy.

Let us proceed to the approximation with respect to the first root. The equations of motion in this approximation take the form

$$-\omega^2 u + c^2 \Delta \left(1 + \frac{\Delta}{\lambda^2}\right)\left(1 + \frac{\Delta}{\lambda^2}\right) u = 0\,, \tag{2.10.17}$$

where c^2 is equal to $3\mu_0/\rho$ or to μ_0/ρ, $\lambda = k_{l,2}$ or $\lambda = k_{t,2}$ depending upon the type of the waves under consideration (longitudinal or transverse).

Let us seek $v(z)$ in the form of a linear combination of exponentials $\exp(i\kappa z)$, where κ is obtained from the condition that $u(x, z) = \exp[i(p\lambda + \kappa z)]$ is a solution of (2.10.17). For u to be a surface wave, the fulfillment of the additional condition $\operatorname{Im} \kappa > 0$ is necessary.

Substitution in (2.10.17) leads to an equation of sixth order with respect to κ

$$\omega^2 - c^2(p^2 + \kappa^2)\left(1 - \frac{p^2 + \kappa^2}{\lambda^2}\right)\left(1 - \frac{p^2 + \kappa^2}{\bar{\lambda}^2}\right) = 0\,. \tag{2.10.18}$$

At small values of ω, for the roots of the above equation we obtain

$$\begin{aligned} &\kappa_0^2 \approx -\left(p^2 - \frac{\omega^2}{c^2}\right) \quad \left(p^2 - \frac{\omega^2}{c^2} > 0\right), \\ &\kappa_1^2 \approx \lambda^2 - p^2 + \frac{\omega^2}{c^2}\frac{\bar{\lambda}^2}{\lambda^2 - \bar{\lambda}^2}\,, \quad \kappa_2^2 = \bar{\kappa}_1^2 \approx \bar{\lambda}^2 - p^2 - \frac{\omega^2}{c^2}\frac{\lambda^2}{\lambda^2 - \bar{\lambda}^2}\,. \end{aligned} \tag{2.10.19}$$

The root κ_0 corresponds to the usual Rayleigh wave and the roots $\kappa_{1,2}$ correspond to the new surface waves, which are absent in the classical theory. With $p \sim \omega/c$, we have

$$\operatorname{Im} \kappa_0 \sim \frac{\omega}{c}, \ \operatorname{Im} \kappa_1 = \operatorname{Im} \kappa_2 \sim \operatorname{Im} \lambda \sim l^{-1}\,. \tag{2.10.20}$$

For the complete solution of the problem it is necessary to write down the boundary conditions (2.10.12, 13) in the appropriate approximation. This will enable us to find the coefficients of $\exp(i\kappa z)$ and to obtain the dispersion law of the surface waves in the given approximation. We omit these cumbersome computations. However, even from the relations shown above, it follows that in the nonlocal theory, new types of surface waves exist which, in contrast to the Rayleigh waves, decay in the case of large wavelength, at a distance of the order of the radius of interaction.

2.11 Notes

The considered model of quasicontinuum introduced independently by Krumhansl [B3.19], Kunin [B3.22] and Rogula [B6.36] can be used in linear field theories only. A space quantization which can be applied to nonlinear field theories was considered in [B8.28, 29].

Various problems of the nonlocal elasticity were considered in the works of Kröner [B3.13–16], Kröner and Datta [B3.17, 18], Kunin [B3.20–21], Mindlin [B.3.29], Edelen, Green and Laws [B3.1], Eringen and Edelen [B3.2], Eringen [B3.3–6]. Note the work of Soos [B3.30], where the case of slowly decreasing (according to the Coulomb law) forces of action-at-a-distance is considered.

The existence of surface waves, which have no analogy in classical elasticity was pointed out in the work of Kunin and Vaisman [B3.28, 31] (see also Kaliski and Rymarz [B3.8]).

3. Medium of Complex Structure

The medium of complex structure is considered in detail with the main attention being focused on specific three-dimensional effects.

Starting from the harmonic model of the crystal lattice with a basis, we introduce the collective variables: the mass center displacement, microrotations and microdeformations. In terms of these variables, the general nonlocal equations of the medium of complex structure are obtained. The physical requirements of stability and invariance with respect to rigid body motions lead to the existence of acoustical and optical modes of vibrations. It is shown that in the corresponding acoustical region the microrotations and microdeformations can be excluded from the equations of motion. As a result, the medium of complex structure can be described in the acoustical region in terms of the medium of simple structure with space and time dispersion (dynamical elastic moduli).

The detailed comparison of the theory with approximate couple-stress theories is provided. It is shown that the latter have a very restricted range of applicability. As an example of the medium of complex structure the multibar system is considered in detail.

3.1 Equations of Motion

The general formalism of the three-dimensional medium of complex structure is very similar to the one-dimensional case. The most important distinctions are connected with replacing the scalar variables and operators by tensor quantities. To avoid repetition, we shall take into account this circumstance when generalizing the results, obtained in [Ref.1.12, Chap. 3] for the one-dimensional model.

As a starting point for a micro-model of the medium of complex structure, let us consider a model of an infinite complex crystal lattice in the harmonic approximation. The introduction of collective cell variables will enable us later on to include this model, as a particular case, in the general phenomenological scheme.

A geometrically complex lattice is a three-dimensional discrete periodic structure with an elementary cell, constructed on the basis of the vector $e_\alpha (\alpha = 1, 2, 3)$. The positions of N particles with masses $m_j (j = 1, \ldots, N)$ inside the cell are given by a set of vectors $\xi_j = \xi_j^\alpha e_\alpha$. The interaction between particles both inside the cell and with particles belonging to other cells is given by the force constant $\Phi^{\alpha\beta}(n, n', j, j')$, where $n = n^\alpha e_\alpha$ is a vector which characterizes the number of the cell. In the oblique system of coordinates x^α with the basis

e_α and the metric tensor $g_{\alpha\beta} = e_\alpha \cdot e_\beta$, the vector n has the integer component n^α.

In the absence of initial forces, the Lagrangian of the lattice has the form

$$2L = g^{\alpha\beta} \sum_{nj} m_j \dot{w}_\alpha(n, j)\dot{w}_\beta(n, j) - \sum_{nn'jj'} w_\alpha(n, j)\, \Phi^{\alpha\beta}(n, n', j, j')\, w_\beta(n', j') + 2 \sum_{nj} f^\alpha(n, j) w_\alpha(n, j)\,, \quad (3.1.1)$$

where $w_\alpha(n, j)$ is the displacement of the jth particle in the nth cell, and $f^\alpha(n, j)$ is the external force which acts on it.

Let us generalize the model in such a way that it would be possible to take into consideration both discrete and continuous periodic structures. To this end, let us introduce the density $\rho(\xi)$ of the mass distribution within the cell and a force matrix $\Phi^{\alpha\beta}(n, n', \xi, \xi')$ which determines the interaction. Here, $\xi = \xi^\alpha e_\alpha$ is the position vector of a point of a cell with the number n in the local system of coordinates with the origin situated at the center of mass of the cell. Then the Lagrangian is given by

$$\begin{aligned} 2L = {} & g^{\alpha\beta} \sum_n \rho(\xi)\dot{w}_\alpha(n, \xi)\dot{w}_\beta(n, \xi) d\xi \\ & - \sum_{nn'} \int\int w_\alpha(n, \xi)\Phi^{\alpha\beta}(n, n', \xi, \xi') w_\beta(n', \xi')\, d\xi\, d\xi' \\ & + 2 \sum_n \int f^\alpha(n, \xi) w_\alpha(n, \xi)\, d\xi\,. \end{aligned} \quad (3.1.2)$$

In order to pass to the discrete structure it is necessary to set

$$\begin{aligned} \rho(\xi) &= \sum_j m_j \delta(\xi - \xi_j)\,, \\ \Phi^{\alpha\beta}(n, n', \xi, \xi') &= \sum_{jj'} \Phi^{\alpha\beta}(n, n', j, j')\delta(\xi - \xi_j)\delta(\xi' - \xi_{j'})\,. \end{aligned} \quad (3.1.3)$$

Using the formalism of the quasicontinuum let us write down the Lagrangian (3.1.2) in the x representation

$$\begin{aligned} 2L = {} & g^{\alpha\beta} \int\int \rho(\xi)\dot{w}_\alpha(x, \xi)\dot{w}_\beta(x, \xi)\, d\xi\, dx \\ & - \int\int\int\int w_\alpha(x, \xi)\Phi^{\alpha\beta}(x, x', \xi, \xi') w_\beta(x', \xi')\, d\xi\, d\xi'\, dx\, dx' \\ & + 2 \int\int f^\alpha(x, \xi) w_\alpha(x, \xi)\, d\xi\, dx\,. \end{aligned} \quad (3.1.4)$$

As in the case of a one-dimensional structure, it is convenient to introduce collective cell parameters in order to pass to the theory of elasticity. To this end let us define a set of moment-of-inertia tensors of order s ($s = 0, 1, \ldots$) of the mass distribution in the cell

$$\rho^{\lambda_1 \ldots \lambda_s} \stackrel{\text{def}}{=} \frac{1}{v_0} \int \rho(\xi) \xi^{\lambda_1} \cdots \xi^{\lambda_s} d\xi, \tag{3.1.5}$$

where v_0 is the cell volume. In an abreviated form

$$\rho^s = \frac{1}{v_0} (\rho, \xi^s). \tag{3.1.6}$$

Here, s denotes the multiindex $\lambda_1 \cdots \lambda_s$ and the parentheses denote the corresponding scalar product.

Let us denote by ρ_{2s}^{-1} the tensor inverse to ρ_{2s} and introduce the two quasi-diagonal matrices

$$I^{ss'} \stackrel{\text{def}}{=} \rho^{2s} \delta^{ss'}, \; I_{ss'}^{-1} = \rho_{2s}^{-1} \delta_{ss'}, \tag{3.1.7}$$

where $\delta^{ss'}$ and $\delta_{ss'}$ are Kronecker deltas.

By a method analogous to the one considered in [Ref. 1.12, Chap. 3], let us construct an orthonormalized system of basis polynominals $e^s(\xi) \equiv e^{\lambda_1 \ldots \lambda_s}(\xi)$ with weight $\rho(\xi)$ and a reciprocal system of functions $e_{s'}(\xi) \equiv e_{\mu_1 \ldots \mu_{s'}}(\xi)$, defining them by the relations

$$(\rho e^s, e^{s'}) = I^{ss'}, \; (e_s, e^{s'}) = \delta_s^{s'}. \tag{3.1.8}$$

Problem 3.1.1. Verify that

$$e_s(\xi) = \rho(\xi) \, I_{ss'}^{-1} \, e^{s'}(\xi). \tag{3.1.9}$$

Problem 3.1.2. Taking into account the fact that the origin of the coordinate system coincides with the mass center of a cell, show that for the first two elements of the bi-orthogonal basis, the following relations are valid:

$$\begin{aligned} e^0(\xi) &= 1, & e^\lambda(\xi) &= \xi^\lambda, \\ e_0(\xi) &= \rho_0^{-1} \rho(\xi), & e_\mu(\xi) &= \rho_{\mu\lambda}^{-1} \, \xi^\lambda \rho(\xi). \end{aligned} \tag{3.1.10}$$

Let us expand the functions of the variables ξ, ξ' entering (3.1.4) in terms of the bi-orthogonal basis

$$\begin{aligned} w_\beta(x, \xi) &= w_{s\beta}(x) e^s(\xi), \qquad f^\alpha(x, \xi) = f^{s\alpha}(x) e_s(\xi), \\ \Phi^{\alpha\beta}(x, x', \xi, \xi') &= \Phi^{s\alpha s'\beta}(x, x') e_s(\xi) e_{s'}(\xi'). \end{aligned} \tag{3.1.11}$$

For the coefficients of the expansion, taking into account (3.1.8) we find

$$\begin{aligned} w_{s\beta}(x) &= (e_s, w_\beta(x)), \qquad f^{s\alpha}(x) = (e^s, f^\alpha(x)), \\ \Phi^{s\alpha s'\beta}(x, x') &= \frac{1}{v_0^2} \iint \Phi^{\alpha\beta}(x, x', \xi, \xi') e^s(\xi) e^{s'}(\xi') \, d\xi \, d\xi'. \end{aligned} \tag{3.1.12}$$

In the new set of variables, the Lagrangian (3.1.4) can be written in a form which is invariant with respect to the n, x and k representations:

$$2L = \langle \dot{w}_{s\alpha} | I^{s\alpha s'\beta} | \dot{w}_{s'\beta} \rangle - \langle w_{s\alpha} | \Phi^{s\alpha s'\beta} | w_{s'\beta} \rangle + 2\langle f^{s\alpha} | w_{s\alpha} \rangle , \qquad (3.1.13)$$

where $I^{s\alpha s'\beta} \overset{\text{def}}{=} g^{\alpha\beta} I^{ss'}$.

If we forget about the quasicontinuum, then this Lagrangian describes the most general linear model of a three-dimensional medium of complex structure. The corresponding equations of motion have the form [Ref. 1.12, Eq. (3.3.4.]

$$\int I^{s\alpha s'\beta}(x, x')\, w_{s'\beta}(x')\, dx' + \iint \Phi^{s\alpha s'\beta}(x, x')\, w_{s'\beta}(x')\, dx' = f^{s\alpha}(x) , \qquad (3.1.14)$$

where, in the general case, $I^{s\alpha s'\beta}(x, x')$ is the kernel of a nonlocal moment-of-inertia operator.

Let us now consider the physical meaning of the collective cell variables $w_{\alpha s}(x)$ and $f^{s\alpha}(x)$. In view of (3.1.10, 12) we find that $w_{0\alpha}$ is the displacement of the mass center of the cell and $f^{0\alpha}$ is the average body force density. It is clear that these quantities will have a considerable significance when the medium is described macroscopically. Hence, let us introduce the special notation

$$u_\alpha \overset{\text{def}}{=} w_{0\alpha}, \quad q^\alpha \overset{\text{def}}{=} f^{0\alpha} . \qquad (3.1.15)$$

The remaining kinematic and force variables, which correspond to the internal degrees of freedom, will be denoted by

$$\eta_{p\alpha} \overset{\text{def}}{=} w_{p\alpha}, \; \mu^{p\alpha} \overset{\text{def}}{=} f^{p\alpha}, p = 1, 2, \dots \qquad (3.1.16)$$

Let

$$\begin{aligned} \eta_{1\alpha} &\equiv \eta_{\lambda\alpha} = \varepsilon'_{\lambda\alpha} + \Omega'_{\lambda\alpha} , \\ \varepsilon'_{\lambda\alpha} &\overset{\text{def}}{=} \eta_{(\lambda\alpha)}, \quad \Omega'_{\lambda\alpha} \overset{\text{def}}{=} \eta_{[\lambda\alpha]} . \end{aligned} \qquad (3.1.17)$$

It is easy to verify that $\varepsilon'_{\lambda\alpha}$ is the average microdeformation of a cell and $\Omega'_{\lambda\alpha} = -\Omega'_{\alpha\lambda}$ is the average microrotation. Analogously,

$$\mu^{1\alpha} \equiv \mu^{\lambda\alpha} = \mu^{(\lambda\alpha)} + m^{\lambda\alpha} , \qquad (3.1.18)$$

where $\mu^{(\lambda\alpha)}$ is the average density of force dipoles and $m^{\lambda\alpha} = -m^{\alpha\lambda}$ is the average density of micromoments. It is clear that the $\mu^{(\lambda\alpha)}$ and $m^{\lambda\alpha}$ are generalized forces, which correspond to the generalized displacements $\varepsilon'_{\lambda\alpha}$ and $\Omega'_{\lambda\alpha}$. The quantities $\eta_{p\alpha}$ and $\mu^{p\alpha}$, for $p > 1$, describe the microdeformations and micromoments of higher orders.

Note that the number of independent elements of the basis is determined by the number of degrees of freedom of a cell. If the number of degrees of freedom does not exceed 12 (in the case of a discrete structure, this means that the number of particles in a cell does not exceed 4), then the basis is automatically restricted to the first two elements. In this case, the role of kinematic variables will be played by displacements of the center of mass u, microdeformation ε' and microrotation Ω'; and the role of force variables will be assumed by the body force density q, the dipole force density μ and the micromoments density m.

If the cell consists of one particle with a finite moment of inertia, then the internal degrees of freedom are restricted to a microrotation Ω'. This simplest model of the medium of complex structure, known as the Cosserat medium, will be considered in Sect 3.5.

In terms of the new variables, the equations of motion (3.1.14) take the form

$$\begin{aligned} \rho g^{\alpha\beta}\ddot{u}_\beta + \Phi^{\alpha\beta}u_\beta + \Phi^{+'\alpha p'\beta}\eta_{p'\beta} &= q^\alpha, \\ I^{p\alpha p'\beta}\eta_{p'\beta} + \Phi^{p\alpha\beta}u_\beta + \Phi^{p\alpha p'\beta}\eta_{p'\beta} &= \mu^{p\alpha}, \end{aligned} \tag{3.1.19}$$

where the operator coefficients are connected with the operators in (3.1.14) in an obvious way.

3.2 Energy Operator

From (3.1.13, 14) it follows that the properties of the medium are completely determined by the kinetic energy operators $I^{s\alpha s'\beta}$ and the elastic energy operators $\Phi^{s\alpha s'\beta}$. In this section we consider the structure of these operators.

According to a reasoning analogous to that of [Ref. 1.12, Sect, 3.3] let us assume that the operator $I^{s\alpha s'\beta}$ is determined by a set of constant moment-of-inertia tensors, the following ones having the principal significance:

$$I^{0\alpha 0\beta} = \rho g^{\alpha\beta}, \quad I^{1\alpha 1\beta} \equiv I^{\alpha\lambda\mu\beta}. \tag{3.2.1}$$

From this assumption, in view of (3.1.13) we obtain the symmetry property

$$I^{s\alpha s'\beta} = I^{s'\beta s\alpha}. \tag{3.2.2}$$

It is evident that it is also necessary to require the positive-definiteness of the kinetic energy.

Let us now consider the properties of the elastic-energy operator and their consequences.

Hermiticity: From (3.1.13), we have

$$\Phi^{s\alpha s'\beta}(x, x') = \Phi^{s'\beta s\alpha}(x', x) \tag{3.2.3}$$

or in the representation (3.1.19)

$$\Phi^{\alpha\beta}(x, x') = \Phi^{\beta\alpha}(x', x), \qquad \Phi^{p\alpha p'\beta}(x, x') = \Phi^{p'\beta p\alpha}(x', x),$$
$$\Phi^{+\alpha p\beta}(x, x') = \Phi^{p\beta\alpha}(x', x) . \tag{3.2.4}$$

Boundedness of action-at-a-distance is assumed as before. The analytical properties of the kernels in various representations, which follow from this, are already well-known.

Invariance with respect to translation and rotation. Let the corresponding displacements be $w^0_{s\alpha}$. Then for any $w_{s\alpha}$ the condition of invariance of energy must be satisfied:

$$\Phi(w_{s\alpha} + w^0_{s\alpha}) = \Phi(w_{s\alpha}) . \tag{3.2.5}$$

Together with (3.1.13) this yields

$$\langle w^0_{s\alpha} | \Phi^{s\alpha s'\beta} | w_{s'\beta} \rangle = 0 , \tag{3.2.6}$$

or, due to the arbitrariness of $w_{s\beta}$:

$$\int w^0_{s\alpha}(x)\, \Phi^{s\alpha s'\beta}(x, x')\, dx = 0 . \tag{3.2.7}$$

For a translation by the vector a_α

$$w^0_\alpha(x) = a_\alpha,\ w^0_{p\alpha}(x) = 0,\ p = 1, 2, \ldots . \tag{3.2.8}$$

For a rotation determined by the tensor $a_{\lambda\alpha} = -a_{\alpha\lambda}$,

$$w^0_\alpha(x) = -\, a_{\alpha\lambda} x^\lambda, \quad w^0_{[\lambda\alpha]}(x) = \Omega'_{\lambda\alpha}(x) = a_{\lambda\alpha} \tag{3.2.9}$$

with all other $w_{s\alpha}(x)$ remaining unchanged.

Problem 3.2.1. Obtain (3.2.8, 9) from the laws of transformation of $w_\alpha(x, \xi)$, taking into account (3.1.10, 12).

From this it follows that the conditions

$$\int \Phi^{0\alpha s\beta}(x, x')\, dx' = 0 , \tag{3.2.10}$$

$$\int [x^{[\lambda}\Phi^{0\alpha]s\beta}(x, x') + \Phi^{[\lambda\alpha]\, s\beta}(x, x')]\, dx = 0 \tag{3.2.11}$$

are to be safisfied. In the k-representation ($\partial^\lambda = \partial/\partial k_\lambda$), these become

$$\Phi^{0\alpha s\beta}(0, k') = 0 ,$$
$$i\partial^{[\lambda}\Phi^{0\alpha]\, s\beta}(0, k') + \Phi^{[\lambda\alpha] s\beta}\,(0, k') = 0 . \tag{3.2.12}$$

This enables us to represent the matrix $\Phi^{s\alpha s'\beta}(k, k')$ in (3.1.19) in the form [Ref. 1.12, Eq. (3.3.15)]

$$\Phi^{s\alpha s'\beta}(k, k') = \begin{pmatrix} k_\lambda k'_\mu \gamma^{\lambda\alpha\mu\beta}(k, k') & -\mathrm{i}k_\lambda \chi^{+\lambda\alpha p'\beta}(k, k') \\ \mathrm{i}k'_\mu \chi^{p\alpha\mu\beta}(k, k') & \Gamma^{p\alpha p'\beta}(k, k') \end{pmatrix}, \tag{3.2.13}$$

where the functions χ and χ^+ are connected by the relation

$$\chi^{+\lambda\alpha p\beta}(k, k') = \overline{\chi^{p\beta\lambda\alpha}(k', k)} \tag{3.2.14}$$

and satisfy the conditions

$$\begin{aligned} &\gamma^{\lambda\alpha[\mu\beta]}(k, 0) + \chi^{+\lambda\alpha[\mu\beta]}(k, 0) = 0, \\ &\chi^{p\alpha[\mu\beta]}(k, 0) + \Gamma^{p\alpha[\mu\beta]}(k, 0) = 0. \end{aligned} \tag{3.2.15}$$

Problem 3.2.2. Prove (3.2.13). [Hint: Use the Hermitian adjointness and the condition (3.2.12).]

The equations of motion (3.1.12), after taking into account (3.2.13), become

$$\begin{aligned} &\rho g^{\alpha\beta}\ddot{u}_\beta - \partial_\lambda \gamma^{\lambda\alpha\mu\beta}\partial_\mu u_\beta - \partial_\lambda \chi^{+\lambda\alpha p\beta}\eta_{p\beta} = q^\alpha, \\ &I^{p\alpha p'\beta}\ddot{\eta}_{p'\beta} + \chi^{p\alpha\mu\beta}\partial_\mu u_\beta + \Gamma^{p\alpha p'\beta}\eta_{p'\beta} = \mu^{p\alpha}, \end{aligned} \tag{3.2.16}$$

where γ, χ, Γ are the corresponding operators.

Stability is equivalent to the requirement of the positive definiteness of the elastic energy for all generalized displacements except translations and rotations. As a consequence, the positive definiteness of the matrices $\gamma^{\lambda\alpha\mu\beta}$ and $\Gamma^{p\alpha p'\beta}$ follows. In fact, the elastic energy Φ can be represented in the form of a sum of three terms

$$\Phi = \Phi_0 + \Phi_1 + \Phi_{\mathrm{int}}, \tag{3.2.17}$$

$$\begin{aligned} &\Phi_0 \stackrel{\mathrm{def}}{=} \frac{1}{2}\langle \partial_\lambda u_\alpha | \gamma^{\lambda\alpha\mu\beta} | \partial_\mu u_\beta \rangle, \\ &\Phi_1 \stackrel{\mathrm{def}}{=} \frac{1}{2}\langle \eta_{p\lambda} | \Gamma^{p\alpha p'\beta} | \eta_{p'\beta} \rangle, \\ &\Phi_{\mathrm{int}} \stackrel{\mathrm{def}}{=} \langle \eta_{p\alpha} | \chi^{p\alpha\mu\beta} | \partial_\mu u_\beta \rangle. \end{aligned} \tag{3.2.18}$$

Here, Φ_0 and Φ_1 are the elastic energies, corresponding to the displacements of the centers of masses and to the internal degrees of freedom and Φ_{int} is the interaction energy. It is clear that the terms Φ_0 and Φ_1 are to be positive definite forms.

Homogeneous Medium. As we know, in this case all the operators have difference kernels in the x representation and are reduced to a product of functions of k in the k representation [Ref. 1.12, Eq. (3.9.7)]. The matrix (3.2.12) becomes

$$\Phi^{s\alpha s'\beta}(k) = \begin{pmatrix} k_\lambda k_\mu \gamma^{\lambda\alpha\mu\beta}(k) & -\,\mathrm{i}k_\lambda \chi^{+\lambda\alpha p'\beta}(k) \\ \mathrm{i}k_\mu \chi^{p\alpha\mu\beta}(k) & \Gamma^{p\alpha p'\beta} \end{pmatrix}. \tag{3.2.19}$$

When writing down the above matrix, the conditions of invariance of the energy with respect to translation and rotation were used. However, additional conditions, analogous to (3.2.15), are also to be fulfilled. One of them is obtained immediately from the second relation in (3.2.15)

$$\chi_0^{p\alpha[\mu\beta]} + \Gamma_0^{p\alpha[\mu\beta]} = 0\,. \tag{3.2.20}$$

The simple form of the limiting transition is here due to the fact that the tensors $\chi^{p\alpha\mu\beta}$ and $\Gamma^{p\alpha\mu\beta}$ do not possess any symmetry, which would be destroyed by alternating with respect to the indices $\mu\beta$.

The analog of the first relation of (3.2.15) produces more difficulties. When passing to the limiting case of a homogeneous medium, a new symmetry with respect to the indices $\lambda\mu$ arises in the tensor $\gamma^{\lambda\alpha\mu\beta}$, which does not commute with the alternation with respect to $\mu\beta$. Therefore, as in the case of a medium with simple structure, we have to obtain this condition directly for a homogeneous medium. The corresponding discussion was carried out in details in Sect. 2.5. This enables us to restrict ourselves to a few principal points. Hence, it may be worthwhile, with a view to brevity, to stress only the most salient features.

We start with a requirement of invariance of the energy density of the zeroth approximation $\varphi_0(x)$ with respect to rotation. Taking into account (3.2.19), the density of the Lagrangian in the zeroth approximation can be represented in the form

$$\begin{aligned} \varphi_0'(x) &= \frac{1}{2}\,\partial_\lambda u_\alpha(x)\,\gamma_0^{\lambda\alpha\mu\beta}\partial_\mu u_\beta(x) \\ &\quad + \eta_{p\alpha}(x)\,\chi_0^{p\alpha\mu\beta}\partial_\mu u_\beta(x) + \frac{1}{2}\,\eta_{p\alpha}(x)\,\Gamma_0^{p\alpha p'\beta}\,\eta_{p'\beta}(x)\,, \end{aligned} \tag{3.2.21}$$

which is already invariant with respect to translation. The energy density $\varphi_0(x)$ differs from $\varphi_0'(x)$ by divergence terms, which are to be chosen from the conditions of the invariance of $\varphi_0(x)$ with respect to rotation.

It is not difficult to verify that the most general expression which can be represented as a divergence and, which is invariant with respect to translation, has the form

$$\varphi_0(x) - \varphi_0'(x) = \partial_\lambda[u_\alpha(x) b^{\lambda\alpha\mu\beta}\partial_\mu u_\beta(x)]\,, \tag{3.2.22}$$

where $b^{\lambda\alpha\mu\beta} = -b^{\mu\alpha\lambda\beta}$ is a constant tensor.

For $\varphi_0(x)$ being invariant with respect to rotation, it is necessary and sufficient to require that

$$(b^{\lambda\alpha\mu\beta} + b^{\mu\beta\lambda\alpha} + \gamma_0^{\lambda\alpha\mu\beta} + \chi_0^{\mu\beta\lambda\alpha})_{[\mu\beta]} = 0 . \tag{3.2.23}$$

At last, analogously to the treatment given in [Ref. 1.12, Sect. 1.5], it can be shown that the condition of solvability of (3.2.23) with respect to $b^{\lambda\alpha\mu\beta}$ has the form

$$[\lambda\alpha, \mu\beta] = [\alpha\lambda, \beta\mu] , \tag{3.2.24}$$

where

$$[\lambda\alpha, \mu\beta] = \gamma_0^{\lambda\alpha\mu\beta} + (\chi_0^{\lambda\alpha\mu\beta} + \chi_0^{\mu\beta\lambda\alpha} + \Gamma_0^{\lambda\alpha\mu\beta})_{(\lambda\mu)} . \tag{3.2.25}$$

Under the condition (3.2.24) the solution of (3.2.23) is unique

$$2b^{\lambda\alpha\mu\beta} = (2\gamma_0^{\alpha\lambda\mu\beta} - \chi_0^{\lambda\mu\alpha\beta} - \chi_0^{\lambda\alpha\mu\beta} + \chi_0^{\alpha\lambda\mu\beta})_{[\lambda\mu]} . \tag{3.2.26}$$

Thus, for the homogeneous medium, the additional conditions concerning the invariance of the energy with respect to rotation are reduced to (3.2.20, 24). These conditions can be clearly interpreted, if together with the micro-rotation one introduces the macroscopic rotation $\Omega_{\mu\beta} = \partial_{[\mu}u_{\beta]}$. Then it can be easily shown that the conditions obtained above are equivalent to the requirement that the energy density $\varphi_0(x)$ depends on the difference $\Omega_{\mu\beta} - \Omega'_{\mu\beta}$ only.

The equations of motion (3.2.16) in the (k, ω) representation can be written in the form of an algebraic system

$$\begin{aligned} &[-\omega^2\rho g^{\alpha\beta} + k_\lambda k_\mu \gamma^{\lambda\alpha\mu\beta}(k)]\, u_\beta(k, \omega) \\ &\qquad - \mathrm{i}k_\lambda \chi^{+\lambda\alpha p\beta}(k)\, \eta_{p\beta}(k, \omega) = q^\alpha(k, \omega) , \\ &[-\omega^2 I^{p\alpha p'\mu} + \Gamma^{p\alpha p'\beta}(k)]\eta_{p'\beta}(k, \omega) \\ &\qquad + \mathrm{i}\chi^{p\alpha\mu\beta}(k) k_\mu u_\beta(k, \omega) = \mu^{p\alpha}(k, \omega) . \end{aligned} \tag{3.2.27}$$

Quasicontinuum. Restricting to the class of admissible functions of the types described in Sect. 2.1, the operators and the equations of motion can be written in the discrete form, too. Particular models are obtained either with a special choice of kinematic variables (for example, the Cosserat model, which is considered in Sect. 3.5), or in the presence of or by imposing additional groups of internal symmetry. Discrete symmetry groups for crystals or the rotation group for an isotropic medium can serve as examples.

Problem 3.2.3. Write the three-dimensional analog of [Ref. 1.12, Eq. (3.2.35)] for the case, in which the elementary cell has a centre of inversion.

Problem 3.2.4. Formulate the three-dimensional analog of the conditions which were considered in I.3.7 concerning the existence of an equivalent model of medium of simple structure.

In the general case the matrix $\Phi^{s\alpha s'\beta}(k)$ does not possess any symmetry with respect to α, β. However, if this symmetry is present, then it can be given a simple interpretation. Let us call the interaction in a discrete model binary (trinary), if the elastic energy can be represented in the form of a sum, each term of which depends on the difference of displacements of two (three) particles[1]. The generalization to the case of a continuous medium is obvious.

In [3.1] it was shown that, in the harmonic approximation, the trinary interaction is the most general form of interaction in a crystal lattice. It can be proved [3.2] that the symmetry of $\Phi^{\alpha\beta}(n - n', \xi, \xi')$ and hence also of $\Phi^{s\alpha s'\beta}$ with respect to α, β, is a necessary and sufficient condition of the interaction being binary (in above-described sense). From this, in particular, it follows that in any one-dimensional model the interaction is always binary, and we saw this also directly. In the three-dimensional models of simple structure, invariance with respect to inversion

$$\Phi^{\alpha\beta}(x) = \Phi^{\alpha\beta}(-x) = \Phi^{\beta\alpha}(x) , \tag{3.2.28}$$

insures that the conditions of the interaction being binary, are fulfilled.

Let us point out an interesting case, when the conditions of invariance of the equations of motion with respect to a homogeneous deformation

$$u_\beta = a_{\beta\mu}x^\mu, \; \varepsilon_{\mu\beta} = a_{\mu\beta} \quad (a_{\mu\beta} = a_{\beta\mu}) . \tag{3.2.29}$$

are satisfied. To this end, as follows from (3.2.27), it is necessary and sufficient to satisfy the following conditions:

$$\chi_0^{p\alpha(\mu\beta)} + \Gamma_0^{p\alpha(\mu\beta)} = 0 . \tag{3.2.30}$$

Together with (3.2.20) they give [Ref. 1.12, Eq.(3.8.7)]

$$\chi_0^{p\alpha\mu\beta} + \Gamma_0^{p\alpha\mu\beta} = 0 . \tag{3.2.31}$$

In Sect. 3.4, we shall see that this condition considerably simplifies the computation of the macroscopic tensor of elastic constants.

3.3 Approximate Models and Comparison with Couple-Stress Theories

As in the case of a medium of simple structure, various methods of constructing appoximate models are possible. Let us restrict ourselves to the simplest model of the long-wavelength approximation.

[1]This definition is more general than the one given in Sect. 2.3, since the invariance of the terms of the sum with respect to rotation is not required. In connection with this, the definition has a formal nature.

Let us expand the functions $\gamma(k)$, $\chi(k)$ and $\Gamma(k)$ entering (3.2.22) in powers of k and consider only a finite number of terms. With $k = 0$, we have the zeroth approximation, which is considered in the next section. When proceeding to successive approximations, it is expedient to distinguish two cases (Sect. 2.6). For a gyrotropic medium, the next approximation is the first one and the operators (3.2.17) have the form

$$\begin{aligned} &\gamma^{\lambda\alpha\mu\beta} = \gamma_0^{\lambda\alpha\mu\beta} + \gamma_1^{\lambda\alpha\mu\beta\tau}\partial_\tau, \ \chi^{p\alpha\mu\beta} = \chi_0^{p\alpha\mu\beta} + \chi_1^{p\alpha\mu\beta\tau}\partial_\tau\,, \\ &\Gamma^{p\alpha p'\beta} = \Gamma_0^{p\alpha p'\mu} + \Gamma_1^{p\alpha p'\beta\tau}\partial_\tau\,. \end{aligned} \tag{3.3.1}$$

As was pointed out above, this corresponds to the rotation of the polarization plane of a wave, which is typical of a gyrotropic medium. If the medium is nongyrotropic then the approximation following the zeroth, is the second order approximation where

$$\begin{aligned} &\gamma^{\lambda\alpha\mu\beta} = \gamma_0^{\lambda\alpha\mu\beta} + \gamma_2^{\lambda\alpha\mu\beta\sigma\tau}\partial_\sigma\partial_\tau, \chi^{p\alpha\mu\beta} = \chi_0^{p\alpha\mu\beta} + \chi_2^{p\alpha\mu\beta\sigma\tau}\partial_\sigma\partial_\tau\,, \\ &\Gamma^{p\alpha p'\beta} = \Gamma_0^{p\alpha p'\beta} + \Gamma_2^{p\alpha p'\beta\sigma\tau}\partial_\sigma\partial_\tau\,. \end{aligned} \tag{3.3.2}$$

By the indices 1, 2 in (3.3.1, 2) we denote the corresponding coefficients in the expansions of the functions $\gamma(k)$, $\chi(k)$ and $\Gamma(k)$ in powers of $\mathrm{i}k$, the coefficients obviously being real tensors.

Replacing the integral operators (3.2.17) by the differential operators (3.3.1, 2) [i.e. transforming (3.2.17) into (3.3.1)] or by analogous operators of higher order, corresponds to the transition to a model of a medium in the approximation of weak dispersion. In other words, this means taking into account nonlocal effects. From considerations of a dimensional analysis it follows that the coefficients of the higher-order derivatives contain indirectly a scale parameter l, which, due to the approximate character of the model, should be considered small in comparison with a characteristic wave length or the size of the body. Hence we conclude that the "resolving power" of the model is limited by the parameter l, i.e. solutions of the approximate equations, in the best case, can pretend to give a qualitatively correct description of nonlocal effects in regions of the order of l, but, generally speaking, should not be extrapolated to smaller regions.

The equations of a medium with weak dispersion contain, as a particular case, the equations of the Cosserat continuum, of an oriented medium, and of couple-stress, multipoles, etc. theories of elasticity. The above considerations point out the possible regions of applicability of these theories.

Usually, such theories are derived on phenomenological grounds, roughly according to the following scheme. It is postulated, that the Lagrangian depends on the strain and its derivatives and on a set of additional kinematic variables such as microrotations and microstrain and their derivatives. By analogy with the classical theory of elasticity the new kinematic variables and their derivatives are put into correspondence with moment stresses of various orders, for which equations of equilibrium (conservation laws) and equations of state of the Hooke's law type, are postulated. This enables us to obtain a closed system of equations and a set of boundary conditions. Further formulation of boundary-value problems usually does not differ from the formulation of analogous problems in the classical theory of elasticity. For example, many papers are devoted to investigations, in the scope of the couple-stress theory of elasticity, concerning stress concentration around holes. In this situation, one would expect that significant differences from the classical theory of elasticity arise for holes with diameters of the order of or less than l.

Let us consider some of the difficulties which arise in such an approach.

1) All the couple-stress theories approximately account for the nonlocal effects, and the parameter l which is contained in them has to be considered to be small. Therefore, when extrapolating the results to the regions of the order of or much smaller than l, one should be very cautious.

2) The presence of a small parameter in terms with higher-order derivatives, as is well-known, leads to effects of the boundary-layer type. In connection with this, it is hardly reasonable to consider boundary-value problems for complicated regions in the scope of couple-stress theories. Moreover, it is also unknown how to interpret the boundary conditions physically and from where to take the values of the moments on the boundary.

3) The theories contain new constants (sometimes in large numbers) but usually it is not clear, from what sort of experiments these could be determined. It is necessary to know at least the qualitative relationship between these constants and the parameters of the micromodel.

4) In an axiomatic construction of the theories without taking into account their approximate nature, the equations of motion are often written down with different degrees of accuracy with respect to different variables. We shall consider this aspect with the example of the Cosserat model in Sect 3.5.

5) The connection between the couple-stress theories and the classical theory of elasticity is usually discussed in a simplified manner: in order to proceed to the theory of elasticity, it is necessary to set the constants contained in the equation equal to zero. As we have observed in the example of the one-dimensional model of a medium of complex structure, the actual connection is more involved: for sufficiently long wavelengths and for sufficiently low frequencies the internal degrees of freedom are excluded from the equations of motion, but they do make some contribution to the effective macroparameters. In the next section this question will be considered for the three-dimensional medium.

6) We observed that, in the nonlocal theory of a medium of simple structure, the stress tensor is not a physically uniquely determined quantity. This holds even more strongly for the couple stress in the medium of complex structure. For its formally unique determination, it is necessary to postulate conservation laws,[2] which do not have any group basis. As is well-known in field theory

[2]This is also pointed out in [B3.1].

[3.3], the conservation laws are a result of the invariance of the Lagrangian with respect to the corresponding groups and in our case they are the groups of translations and rotations. However, only displacements and microrotations are essentially with this fundamental group; this leads to a special divergence form of the equations of motion, which ensures the equilibrium of forces and moments. The remaining internal degrees of freedom (microdeformations) are invariant with respect to the fundamental group, and additional laws for couple stress have to be postulated rather arbitrarily.

Note also that couple stresses do not satisfy the natural principle of correspondence, i.e. they do not become the usual stresses in the limiting case of long waves, when a transition to the conventional theory of elasticity should take place.

Taking into account these arguments, one should not be surprised that the rare attempts of experimental determination of couple stress in an nonhomogeneous medium have not led to any results.

Let us sum up. The couple-stress theories of elasticity and, generally, models of a medium with weak dispersion approximately take into account comparatively delicate effects of nonlocality and of internal degrees of freedom. Their application is expedient, first of all, when they yield not merely small quantitative corrections, but describe qualitatively new phenomena. Therefore the physically justified field of applicability of these theories is narrower than that of the classical theory of elasticity, though the latter is the limiting case.

3.4 Exclusion of Internal Degrees of Freedom in the Acoustic Region

As in the one-dimensional case, we have the dispersion equation for free vibrations

$$\det\{-\omega^2 I^{s\alpha s'\beta} + \Phi^{s\alpha s'\beta}(k)\} = 0 \,, \tag{3.4.1}$$

which connects the frequency ω and the wave vector k. However this equation now yields three acoustic and 3N-3 optical modes $\omega(k)$, where 3N is the number of degrees of freedom of a cell.

As pointed out in [Ref. 1.12, Chap. 3], in the acoustic region of frequencies the main kinematic variable is the displacement of the center of the masses of a cell $u_\alpha = w_{0\alpha}$. The distinctive role of this variable is connected with the fact that in the acoustic region the internal degrees of freedom $\eta_{p\alpha}$ can be excluded from the equations of motion by transforming the latter into an equation, which contains only one variable u_α, and into equations which express explicitly the remaining variables in terms of displacements. Recall that the possibility of such an operation is connected with the special law of transformation of the kinematic variables under translation and with the stability of the medium.

Let us consider, for the beginning, the equations of the zeroth approximation [Ref. 1.12, Eq. (3.6.1)]

$$\begin{aligned}(-\omega^2\rho g^{\alpha\beta} + k_\lambda k_\mu \gamma_0^{\lambda\alpha\mu\beta})\, u_\beta(k,\omega) - \mathrm{i}k_\lambda \chi_0^{\lambda\alpha p\beta}\eta_{p\beta}(k,\omega) &= q^\alpha(k,\omega)\,,\\ (-\omega^2\, I^{p\alpha p'\beta} + \Gamma_0^{p\alpha p'\beta})\,\eta_{\mu'\beta}(k,\omega) + \mathrm{i}\chi_0^{p\alpha\mu\beta}k_\mu u_\beta(k,\omega) &= \mu^{p\alpha}(k,\omega)\,.\end{aligned} \tag{3.4.2}$$

Problem 3.4.1. Show that the optical frequencies at $k = 0$ are determined by the equation

$$\det(-\omega^2 I^{p\alpha p'\beta} + \Gamma_0^{p\alpha p'\beta}) = 0\,, \tag{3.4.3}$$

and, moreover, from the conditions of stability, the optical frequencies cannot be zero.

We see that the matrix $(-\omega^2 I + \Gamma_0)$ has an inverse in the region of acoustic frequencies. This means that in the acoustic region the second equation of the system (3.4.2) can be solved with respect to $\eta_{p'\beta}$, and $\eta_{p'\beta}$ can be eliminated from the first equation. From here it also follows that there exists a finite neighborhood of the origin of the k, ω-space, in which the internal degrees of freedom can be eliminated in such a manner.

In the direct matrix notations (3.2.22) coincide with [Ref. 1.12, Eq. 3.3.24)] for the one-dimensional medium. This enables us to describe the final results without going into details.

Let us introduce the matrix

$$A(k,\omega) \stackrel{\text{def}}{=} [-\omega^2 I + \Gamma(k)]^{-1} = A^+(k,\omega)\,. \tag{3.4.4}$$

The neighborhood of the origin of the k, ω-space, in which $A(k, \omega)$ exists, will be called an admissible acoustic region.

With the help of the matrix $A(k, \omega)$ the system (3.2.22) can be transformed into the form

$$-\omega^2\rho g^{\alpha\beta}u_\beta + k_\lambda \Phi^{\lambda\alpha\mu\beta}(k,\omega)k_\mu u_\beta = q_*^\alpha\,, \tag{3.4.5}$$

$$\eta_{p\alpha} = -\,\mathrm{i}a_{p\alpha}^{\cdot\cdot\mu\beta}(k,\omega)\,k_\mu u_\beta + A_{p\alpha p'\beta}(k,\omega)\,\mu^{p'\beta}\,. \tag{3.4.6}$$

Here, the following notations

$$\begin{aligned}\Phi^{\lambda\alpha\mu\beta} &= \gamma^{\lambda\alpha\mu\beta} - (\chi^{+\lambda\alpha p\sigma} A_{p\sigma p'\tau}\chi^{p'\tau\mu\beta})_{(\lambda\mu)}\,,\\ a_{p\alpha}^{\cdot\cdot\mu\beta} &= A_{p\alpha p'\tau}\,\chi^{p'\sigma\mu\beta},\\ q_*^\alpha &= q^\alpha - \mathrm{i}k_\lambda\mu^{\lambda\alpha},\quad \mu^{\lambda\alpha} = -\,a^{+\lambda\alpha}{}_{\cdot\cdot p'\beta}\mu^{p'\beta}\,,\end{aligned} \tag{3.4.7}$$

have been introduced.

The form of (3.4.5) coincides with that of the equation of motion (2.5.25) of the medium of simple structure, but now $\Phi^{\lambda\alpha\mu\beta}$ depends not only on k but also on ω, or, in other words, there exists a time dispersion besides with the spatial dispersion, the former being unconnected with the dissipation of energy.

Problem 3.4.2. Show that, when k and ω are small, the time dispersion appears only beginning with the second approximation.

The right-hand side of (3.4.5) contains the equivalent density of external forces q_*^α, which is equal to the difference between the density of external forces q^α and the divergence of the density of micromoments $\mu^{\lambda\alpha}$, as is the case in the classical theory of elasticity in the presence of distributed moments.

Repeating reasonings analogous to those described in Sect. 2.5, we can rewrite (3.4.5) in the form

$$-\omega^2 \rho g^{\alpha\beta} u_\beta + k_\lambda c^{\lambda\alpha\mu\beta}(k, \omega)\, k_\mu u_\beta = q_*^\alpha , \qquad (3.4.8)$$

where $c^{\lambda\alpha\mu\beta}(k, \omega)$ possesses the symmetry of the tensor of elastic moduli (2.3.19) and is connected with $\Phi^{\lambda\alpha\mu\beta}(k, \omega)$ by the operator S. This operator is defined by the set of operations described in Sect. 2.5

$$c^{\lambda\alpha\mu\beta}(k, \omega) = S\Phi^{\lambda\alpha\mu\beta}(k, \omega) . \qquad (3.4.9)$$

We can now introduce a symmetric stress tensor

$$\sigma^{\lambda\alpha} \overset{\text{def}}{=} c^{\lambda\alpha\mu\beta}(k, \omega)\, \varepsilon_{\mu\beta} \qquad (3.4.10)$$

and write the equations of motion (3.4.8) in the (x, t)-representation in the usual form

$$\rho \ddot{u} - \operatorname{div} \sigma = q_* . \qquad (3.4.11)$$

As distinct from the case of a medium with simple structure, the operator Hooke's law has the form

$$\sigma^{\lambda\alpha}(x, t) = \iint c^{\lambda\alpha\mu\beta}(x - x', t - t')\varepsilon_{\mu\beta}(x', t')\, dx'\, dt', \qquad (3.4.12)$$

i.e. the strain tensor $\varepsilon_{\mu\beta}$ and the stress tensor $\sigma^{\lambda\alpha}$ have nonlocal connections not only with respect to the spatial variables but also with respect to time.

It is important to emphasize that in the zeroth long-wavelength approximation, we obtain the equations of the classical theory of elasticity, and the stress tensor $\sigma^{\lambda\alpha}$ transforms into the usual stress tensor, which is connected with strain by the tensor of the elastic constants $c_0^{\lambda\alpha\mu\beta}$. The latter, according to (2.5.4), is given by the expression[3] [Ref. 1.12, Eq. (3.6.12)]

[3]For the crystal lattice, this expression coincides with that obtained by Born and Huang [3.5, 6].

$$\begin{aligned} c_0^{\lambda\alpha\mu\beta} &= S\Phi_0^{\lambda\alpha\mu\beta} = \Phi_0^{\lambda\mu\alpha\beta} + \Phi_0^{\alpha\mu\lambda\beta} - \Phi_0^{\lambda\alpha\mu\beta}, \\ \Phi_0^{\lambda\alpha\mu\beta} &\stackrel{\text{def}}{=} \Phi^{\lambda\alpha\mu\beta}(0, 0) = \gamma_0^{\lambda\alpha\mu\beta} - [\chi_0^{+\lambda\alpha p\tau} A_{p\sigma p'\tau}(0, 0)\, \chi_0^{p'\tau\mu\beta}]_{(\lambda\mu)}\,. \end{aligned} \tag{3.4.13}$$

Problem 3.4.3. Show that for the tensor $c_0^{\lambda\alpha\mu\beta}$ to possess the required symmetry, it is necessary and sufficient to have symmetry of $\Phi^{\lambda\alpha\mu\beta}$ with respect to the indices $\lambda\mu$ and $\alpha\beta$, as well as with respect to the permutation of these pairs of indices, and this symmetry is ensured by (3.2.20).

Problem 3.4.4. Verify that if the conditions of homogeneous deformation (3.2.26) are satisfied, then

$$\Phi_0^{\lambda\alpha\mu\beta} = \gamma_0^{\lambda\alpha\mu\beta} - \frac{1}{2}(\Gamma_0^{\mu\beta\lambda\alpha} + \Gamma_0^{\lambda\beta\mu\alpha}), \tag{3.4.14}$$

i.e. for the determination of $c_0^{\lambda\alpha\mu\beta}$, it is not necessary to invert the matrix Γ_0.

Problem 3.4.5. On the basis of (3.2.17), show that in the long-wave length approximation the elastic energy Φ can be written in the usual form (with $\mu = 0$)

$$\Phi = \frac{1}{2}\langle \varepsilon_{\lambda\alpha} | c_0^{\lambda\alpha\mu\beta} | \varepsilon_{\mu\beta} \rangle, \tag{3.4.15}$$

where $\varepsilon_{\lambda\alpha}$ is the strain tensor.

Problem 3.4.6. Let us write $c_0 \geq 0$ in (3.4.15) if the corresponding form is positive. Show that the following inequality is valid:

$$0 \leq c_0 \leq S\gamma_0\,. \tag{3.4.16}$$

Eq. (3.4.13) and the analogous formulae for the elastic constants of higher approximations enable us to express these constants in terms of the parameters of the micromodel. In the following section we shall illustrate this point with the example of the Cosserat model.

When proceeding to the higher approximations in (3.4.8) it is expedient to distinguish the cases of gyrotropic and nongyrotropic media. However, it is necessary to note that the property of gyrotropy in the above sense is not invariant with respect to the transformation of the equations carried out. A medium can be gyrotropic with respect to the initial system of equations (3.2.22) but then the first approximation in (3.4.8) may coincide with the zeroth one.

Problem 3.4.7. Show that, if the conditions of homogeneous deformation (3.2.26) are fulfilled and the interaction is binary, then generally speaking, the medium is gyrotropic, but the first approximation in (3.4.8) coincides with the zeroth one.

The time dispersion appears in the second approximation only (Problem 8.4.2) and, for a nongyrotropic medium, the operator of elastic moduli in this approximation has the form

$$c^{\lambda\alpha\mu\beta} = c_0^{\lambda\alpha\mu\beta} + c_2^{\lambda\alpha\mu\beta\rho\sigma}\partial_\rho\partial_\sigma + c_2'^{\lambda\alpha\mu\beta}\partial_t^2 , \tag{3.4.17}$$

where the constant tensors c_2 and c_2' are determined as the corresponding coefficients of the expansion of the function $c(k, \omega)$ with respect to k and ω.

For an isotropic medium, the structure of $c(k, \omega)$ in the general case of strong dispersion coincides with (2.8.4), but $\lambda(k, \omega)$ and $\mu(k, \omega)$ are even analytic functions of $|k|$ and ω. The evenness with respect to ω is obvious, and evenness with respect to $|k|$ follows from the fact that isotropic tensors with odd parity do not exist.

The equation of motion for an isotropic medium is written in the (x, t)-representation in the second approximation, as [Ref. 1.12, Eq. (3.6.18)]

$$J\ddot{u} - (\mu_0 + l^2\mu_2\Delta)\Delta u - [\lambda_0 + \mu_0 + l^2(\lambda_2 + \mu_2)\Delta] \operatorname{grad}\operatorname{div} u = q_* , \tag{3.4.18}$$

$$J \stackrel{\text{def}}{=} \rho - \tau^2[\mu_2'\Delta + (\lambda_2' + \mu_2')\operatorname{grad}\operatorname{div}] . \tag{3.4.19}$$

Here l and τ are the characteristic length and time, λ_0 and μ_0 are the corresponding coefficients of the expansion of the functions $\lambda(k, \omega)$ and $\mu(k, \omega)$, and J can be considered to be an inertial operator characteristic of the medium.

In the static case the equations for an isotropic medium coincide with the equations for a medium of simple structure and, hence, for solving them, one can use the static Green's tensor constructed in Sect. 2.8.

Thus, in the admissible acoustic region, the equations of a medium with complex structure are reduced to the equations for an equivalent medium with simple structure with spatial and time dispersions. In the long-wavelength approximation this corresponds to the transition to a macroscopic description of the medium with microstructure, when the internal degrees of freedom are excluded from explicit consideration, but contribute to the effective elastic constants.

Note that, conversely, in the region of high frequencies it is possible to eliminate the displacements of the mass centers from the equations, analogously to what was done in the one-dimensional case [Ref. 1.12, Sect. 3.6].

3.5 Cosserat Model

Let us consider a medium for which the internal degrees of freedom are reduced to microrotation. The latter can be characterized by an antisymmetric tensor $\Omega'_{\mu\beta}$ or by the pseudovector

$$\Omega'^\tau = \varepsilon^{\tau\mu\beta}\Omega'_{\mu\beta} , \tag{3.5.1}$$

where in the cartesian coordinate system $\varepsilon^{\tau\mu\beta}$ is the unit antisymmetric pseudotensor.[4] Analogously, the force micromoments are given by the antisymmetric tensor $m^{\lambda\alpha}$ or by the pseudovector m_τ. Eq. (3.2.17), after obvious simplifications, takes the form

$$\begin{aligned} &\rho g^{\alpha\beta}\ddot{u}_\beta - \partial_\lambda \gamma^{\lambda\alpha\mu\beta}\partial_\mu u_\beta - \partial_\lambda \chi^{+\lambda\alpha\beta}\Omega'_\beta = q^\alpha , \\ &I^{\alpha\beta}\ddot{\Omega}'_\beta + \chi^{\alpha\mu\beta}\partial_\mu u_\beta + \Gamma^{\alpha\beta}\Omega'_\beta = m^\alpha . \end{aligned} \tag{3.5.2}$$

A more particular model is obtained if it is assumed that the operator $\chi^{\alpha\mu\beta}$ is antisymmetric with respect to $\mu\beta$. In this case the second equation contains only the microrotation Ω' and the macrorotation $\Omega = 1/2 \text{ rot } u$. Finally, if we restrict the obtained equations to the second approximation and an isotropic medium, then we arrive at the well-known model of a Cosserat continuum

$$\begin{aligned} \rho\ddot{u} - (\gamma_0 + l^2\gamma_2\Delta)\,\Delta u - [\beta_0 + \gamma_0 + l^2\,(\beta_2 + \gamma_2)\Delta]\text{ grad div } u& \\ - (\chi_0 + l^2\chi_2\Delta)\text{ rot }\Omega' = q\,,& \\ I\ddot{\Omega}' + (\chi_0 + l^2\chi_2\Delta)\Omega + (\Gamma_0 + \Gamma_2\Delta)\,\Omega' = m\,.& \end{aligned} \tag{3.5.3}$$

More precisely, in order to obtain the usual equations of the Cosserat continuum [3.8], it is necessary to equate the constants β_2, γ_2, χ_2 to zero in (3.5.3), but to preserve the nonzero[5] constant Γ_2. The equations thus obtained appear to be slightly inconsistent with respect to the order of approximation. The first equation is of the zeroth order with respect to the scale parameter l, while the second equation is of second order. This inconsistency is a consequence of the "symmetrical" assumption, that the Lagrangian depends on the displacements u, microrotations Ω' and their first order derivatives.

In reality, the variables u and Ω' enter the equations nonsymmetrically, in view of the differences in the laws of their transformations with respect to the translation group. Therefore, in order to have consistent accuracy, it is necessary to assume that, in the Lagrangian, the displacement derivatives have an order higher by one than the order of the derivatives of microrotation.

We shall not continue the general investigation of the Cosserat medium, but shall instead proceed to a concrete model which is interesting for its own sake. Let us consider a system of rods which forms a cubic lattice and assume that, at the knots of the lattice, masses connected by the weightless rods are located, and that the masses possess finite moments of inertia. Nearest-neighbor interaction is assumed for simplicity.

Let us introduce the following characteristics of the rods being parallel to the axis x^α of the lattice coordinate system: length l, longitudinal stiffness

[4]In an arbitrary affine or lattice coordinate system this is the corresponding tensor density [B8.25].

[5]Other possible variants are connected with different definitions of an isotropic medium.

EF_α, transverse stiffness EJ_α (which is assumed to be independent of the axes), torsional rigidity GJ'_α. The masses have a cubical symmetry with respect to the axes, which coincide with those of the lattice, and hence, the moment of inertia is diagonal $I_0^{\alpha\beta} = I_0 g^{\alpha\beta}$. Let us also point out that, although the lattice possesses a cubical symmetry in the geometrical sense, in view of different characteristics of the rods, the system has lower symmetry.

The role of the kinematic variables is played by the displacement of the centers of mass u_α and the pseudovector of rotation of the masses Ω'_α. The role of the force variables is played by the force f^α and the moment μ^α, which act on the masses. The equations of motion have the form

$$\begin{aligned} m g^{\alpha\beta}\ddot{u}_\beta(n, t) &+ \sum_{n'} \Phi_{00}^{\alpha\beta}(n - n')u_\beta(n', t) \\ &+ \sum_{n'} \Phi_{01}^{\alpha\beta}(n - n')\Omega'_\beta(n', t) = f^\alpha(n, t)\,, \\ I_0 g^{\alpha\beta}\Omega'_\beta(n, t) &+ \sum_{n'} \Phi_{10}^{\alpha\beta}(n - n')u'_\beta(n\,, t) \\ &+ \sum_{n'} \Phi_{11}^{\alpha\beta}(n - n')\,\Omega'_\beta(n', t) = \mu^\alpha(n, t)\,. \end{aligned} \tag{3.5.4}$$

The above equations enable us to give an interpretation of the force constants $\Phi(n)$. Thus, if we fix all the masses except the mass at a point n' and give the latter a unit displacement in the direction of the axis β, then $\Phi_{00}^{\alpha\beta}(n - n')$ and $\Phi_{10}^{\alpha\beta}(n - n')$ are equal, respectively, to the force and the moment, which act in the direction α on the mass situated at the point n and compensate the reactions of the elastic bonds. One can draw analogous conclusions about $\Phi_{01}^{\alpha\beta}(n)$ and $\Phi^{\alpha\beta}(n)$.

We now make use of the well-known formulae of the strength of materials for bending. It is convenient to solve these expressions with respect to the displacement u and the angle of rotation Ω' of the end of the rod in one plane, representing them in the form

$$\begin{pmatrix} f \\ \mu \end{pmatrix} = \frac{12EJ}{l^3}\begin{pmatrix} 1 & -\frac{1}{2}l \\ \frac{1}{2}l & -\frac{1}{6}l^2 \end{pmatrix}\begin{pmatrix} u \\ \Omega' \end{pmatrix}, \tag{3.5.5}$$

where f and μ are the corresponding force and moment at the clamped end.

Taking into account analogous relations for tension and twisting, we obtain explicit expressions for the force constants in terms of the parameters of the rods. The nonzero components of the force matrices have the form

$\Phi(n)$ \ n	$\begin{matrix} 1 & 0 & 0 \\ -1 & 0 & 0 \end{matrix}$	$\begin{matrix} 0 & 1 & 0 \\ 0 & -1 & 0 \end{matrix}$	$\begin{matrix} 0 & 0 & 1 \\ 0 & 0 & -1 \end{matrix}$
$-l^{-1}g_{\beta\lambda}\Phi_{00}^{\alpha\lambda}$	$\begin{matrix} A_1 & & \\ & B_1 & \\ & & B_1 \end{matrix}$	$\begin{matrix} B_2 & & \\ & A_2 & \\ & & B_2 \end{matrix}$	$\begin{matrix} B_3 & & \\ & B_3 & \\ & & A_3 \end{matrix}$
$2l^{-2}g_{\beta\lambda}\Phi_{10}^{\alpha\lambda}$	$\begin{matrix} & \mp B_1 \\ \pm B_1 & \end{matrix}$	$\begin{matrix} & \pm B_2 \\ \mp B_2 & \end{matrix}$	$\begin{matrix} & \mp B_3 \\ \pm B_3 & \end{matrix}$
$6l^{-3}g_{\beta\lambda}\Phi_{11}^{\alpha\lambda}$	$\begin{matrix} -6C_1 & & \\ & B_1 & \\ & & B_1 \end{matrix}$	$\begin{matrix} B_2 & & \\ & -6C_2 & \\ & & B_2 \end{matrix}$	$\begin{matrix} B_3 & & \\ & B_3 & \\ & & -6C_3 \end{matrix}$

(3.5.6)

Here, the following notations have been adopted:

$$A_\alpha = l^{-2}EF_\alpha,\ B_\alpha = 12l^{-4}\,EJ_\alpha,\ C_\alpha = l^{-4}GJ'_\alpha\,. \tag{3.5.7}$$

Problem 3.5.1. Verify that the force constants of self-action, found from the conditions of invariance of (3.5.4) with respect to translation and rotation, see (3.2.10) and (3.2.11), have the form (do not use the summation convention!)

$$\Phi_{00}^{\alpha\alpha}(0) = 2l^{-1}(A_\alpha + B_\beta + B_\gamma),\ \Phi_{10}^{\alpha\beta}(0) = 0,$$
$$\Phi_{11}^{\alpha\alpha}(0) = 2l\left(C_\alpha + \frac{1}{3}B_\beta + \frac{1}{3}B_\gamma\right),$$
$$(\alpha \neq \beta \neq \gamma)\,. \tag{3.5.8}$$

In the usual manner, passing to the k representation, we obtain the nonzero components of the tensors $\gamma(k)$, $\chi(k)$ and $\Gamma(k)$ in the form (no summation)

$$\gamma^{\alpha\alpha}_{\cdot\cdot\alpha\alpha} = A_\alpha s^2(k_\alpha),\ \gamma^{\alpha\beta}_{\cdot\cdot\alpha\beta} = B_\alpha s^2(k_\alpha),\ \chi^{\alpha\mu\beta} = \varepsilon^{\alpha\beta\mu}B_\mu s(2k_\mu),$$
$$\Gamma^{\alpha}_{\cdot\alpha} = C_\alpha k_\alpha^2 s^2(k_\alpha) + B_\beta\left[1 - \frac{1}{6}k_\beta^2 s^2(k_\beta)\right] + B_\gamma\left[1 - \frac{1}{6}k_\gamma^2 s^2(k_\gamma)\right],$$
$$(\alpha \neq \beta \neq \gamma),\ s(k_\alpha) \overset{\text{def}}{=} \frac{2}{k_\alpha}\sin\frac{k_\alpha}{2}\,. \tag{3.5.9}$$

Problem 3.5.2. Write down the conditions of invariance with respect to rotation (3.2.20, 24) taking into account (3.5.1), and verify that these conditions are fulfilled for the system under consideration.

Problem 3.5.3. Write down the equations of motion (3.5.2) in the zeroth and second approximations expressing the tensor coefficients in terms of the parameters of the system.

Problem 3.5.4. Verify that the conditions of homogeneous deformation (3.2.26) are satisfied only for rods having the same stiffness.

In the admissible acoustic region, it is possible to eliminate the microrotation Ω'_α from the equations of motion (3.5.2) by transforming them to the form (3.4.5, 6). The tensor $\Phi^{\lambda\alpha\mu\beta}(k, \omega)$ entering this equation possesses for $k = 0$ and $\omega = 0$, the required symmetry with respect to the indices and in particular, satisfies the conditions of invariance with respect to the rotation (2.5.8). Recall that this condition was not fulfilled in the consideration of the rod system (Sect. 2.7), in the framework of a medium of simple structure (without taking into account microrotations).

Let us now present the final expressions for the operator of the elastic moduli $c^{\lambda\alpha\mu\beta}(k, \omega)$ confining ourselves (for simplicity) to the case of identical rods and setting $A_\alpha = A$, $B_\alpha = B$, $C_\alpha = C$. For the essential components of $c^{\lambda\alpha\mu\beta}(k, \omega)$ we have ($\alpha \neq \beta \neq \gamma$)

$$\begin{aligned} c^{\alpha\alpha}_{\cdot\cdot\alpha\alpha} &= As^2(k_\alpha)\,, \\ c^{\alpha\alpha}_{\cdot\cdot\beta\beta} &= \frac{1}{2} B\{-s^2(k_\alpha) - s^2(k_\beta) + [s(2k_\alpha) + s(2k_\beta)]\,p_\gamma(k, \omega)\}\,, \\ c^{\alpha\beta}_{\cdot\cdot\alpha\beta} &= B\Big\{-1 + s^2(k_\alpha) + s^2(k_\beta) \\ &\qquad + \frac{1}{2}[1 - s^2(2k_\alpha) - s^2(2k_\beta)]\,p_\gamma(k, \omega)\Big\}\,, \\ c^{\alpha\beta}_{\cdot\cdot\beta\beta} &= \frac{1}{2} B\left[k_\alpha \frac{1 - s^2(k_\beta)}{k_\beta} + \frac{1}{2} k_\alpha \frac{1 - s^2(2k_\beta)}{k_\beta}\, p_\gamma(k, \omega)\right], \end{aligned} \tag{3.5.10}$$

where $p_\gamma(k, \omega)$ is determined by the expression

$$\frac{1}{p_\gamma(k, \omega)} = 1 + \frac{C}{2B} k_\gamma^2 s^2(k_\gamma) - \frac{1}{12} k_\alpha^2 s^2(k_\alpha) - \frac{1}{12} k_\beta^2 s^2(k_\beta) - \frac{I\omega^2}{2B}\,, \tag{3.5.11}$$

where $I = v_0^{-1} I_0$ is the density of moments of inertia.

From here, by expanding in k and ω, the elastic constants of the zeroth and successive approximations are found. In particular, for the tensor of the elastic constants of the zeroth approximation, we have [see (2.7.12)]

$$c^{\lambda\alpha}_{0\cdot\cdot\mu\beta} = \frac{1}{2} B(\delta^\lambda_\mu \delta^\alpha_\beta + \delta^\lambda_\beta \delta^\alpha_\mu) + (A - B) \sum_i \delta^{i\lambda} \delta^{i\alpha} \delta_{i\mu} \delta_{i\beta}\,. \tag{3.5.12}$$

It is obvious that, under tension along a coordinate axis, there is no side thrust, as one should expect (Sect. 2.7). Note that this expression does not contain the constant C, i.e. the stiffness of the rod with respect to twisting does not contribute to the elastic constants in zeroth approximation. This can be easily explained, if we take into account that twisting is connected with an inhomogeneous deformation.

Let us give the expressions for the essential components of the elastic-constants tensor in the case of nonidentical rods:

$$c_{0\cdot\cdot\alpha\alpha}^{\alpha\alpha} = A_\alpha,\ c_{0\cdot\cdot\alpha\beta}^{\alpha\beta} = \frac{B_\alpha B_\beta}{B_\alpha + B_\beta},\ c_{0\cdot\cdot\beta\beta}^{\alpha\alpha} = 0 \quad (\alpha \neq \beta)\,. \tag{3.5.13}$$

For the rod system, let us consider the dispersion equation (3.4.1), which connects the frequency ω and the wavevector k. For some characteristic directions of the cubic lattice, in particular along the diagonals (1, 1, 1) and (1, 1, 0) and along the axis (1, 0, 0), it is possible to obtain the solution $\omega(k)$ of the dispersion equation explicitly. In all these cases degeneracy takes place: there exists one acoustical and one optical mode, these being the analogs of longitudinal waves, as well as two coinciding acoustical and two coinciding optical modes, these being the analogs of transverse waves.

Let us write down the dispersive modes for the direction (1, 0, 0,), i.e. for the wave vector $(k_1, 0, 0)$. The acoustic modes are given by $(\rho = v_0^{-1} m)$

$$\begin{aligned} \omega^2 &= \omega_1^2(k_1) = \frac{4A}{\rho l^2} \sin^2\frac{k_1}{2}\,, \\ \omega^2 &= \omega_{2,3}^2(k_1) = \frac{B}{4I}\left[f(k_1) - \sqrt{f^2(k_1) - g(k_1)}\right], \end{aligned} \tag{3.5.14}$$

and the optical modes by

$$\begin{aligned} \omega^2 &= \omega_1^2(k_1) = \frac{2B}{I}\left(1 + \frac{8C}{B}\sin^2\frac{k_1}{2}\right), \\ \omega^2 &= \omega_{2,3}^2(k_1) = \frac{B}{4I}\left[f(k_1) + \sqrt{f^2(k_1) - g(k_1)}\right]. \end{aligned} \tag{3.5.15}$$

Here

$$\begin{aligned} f(k_1) &= 1 + 4\left(\frac{2I}{\rho l^2} - \frac{1}{3}\right)\sin^2\frac{k_1}{2}\,, \\ g(k_1) &= \frac{4I}{\rho l^2}\left(8\sin^2\frac{k_1}{2} - \frac{32}{3}\sin^4\frac{k_1}{2} - k_1^2\right). \end{aligned} \tag{3.5.16}$$

Problem 3.5.5. Write down the solutions of the dispersion equation for the directions along the diagonals.

3.6 Notes

Extensive literature is devoted to different types of couple-stress theories, see e.g. [B2.1–38]. All these theories can be obtained as long-wavelength approximations for nonlocal elasticity developed in [B3.1–34].

The theory of media of complex structure considered in this chapter is based on [B3.23, 34] and is closely connected with the theory of crystals with a basis [B5.2, 10]. The method of calculating the elastic constants for crystals with a basis is due to Huang [B5.4], see also [B5.2, 5, 9–12]. The nonlocal Cosserat model was considered in [B3.10].

4. Local Defects

In this chapter, we consider local inhomogeneities as defects in a homogeneous medium. We use systematically the Green's function technique which seems to be most adequate to the problem.

We start from a general scheme for arbitrary inhomogeneities. Then we consider impurity atoms in a lattice and show that the problem is reduced to solving a finite system of linear algebraic equations. A smilar method is used for point defects in a quasicontinuum.

The second part of the chapter is devoted to inhomogeneities in elasticity. Our approach is essentially different from that adopted in classical elasticity. We formulate boundary-value problems for inhomogeneities in terms of a special class of global pseudo-differential operators (projection operators and Green's operators for stress and strain). It is shown that the developed technique can be efficiently applied to ellipsoidal inhomogeneities and cracks. Finally a similar technique is used in the theory of internal stress, dislocations and stochastic medla.

4.1 General Scheme

Let Φ_0 be a linear operator for which the Green's function (inverse operator) G_0 is known and satisfies certain boundary conditions, i.e., $G_0\Phi_0 = I$, where I is the identity operator. If Φ_1 is a perturbation of the operator Φ_0 such that a Green's function G of the operator $\Phi = \Phi_0 + \Phi_1$ exists, then one can show that the Green's function G has the representation

$$G = G_0 - G_0 P G_0 \,, \tag{4.1.1}$$

where the operator P is defined by

$$P = \Phi_1(\Phi_1 + \Phi_1 G_0 \Phi_1)^{-1}\, \Phi_1 \,. \tag{4.1.2}$$

In fact, substituting P into (4.1.1) and multiplying by Φ from the left, we have

$$\begin{aligned}\Phi G &= (\Phi_0 + \Phi_1)\,[G_0 - G_0\Phi_1(\Phi_1 + \Phi_1 G_0 \Phi_1)^{-1}\,\Phi_1 G_0] \\ &= I + \Phi_1 G_0 - \Phi_1(\Phi_1 + \Phi_1 G_0\Phi_1)^{-1}\,\Phi_1 G_0 - \Phi_1 G_0 \Phi_1(\Phi_1 + \Phi_1 G_0 \Phi_1)^{-1}\,\Phi_1 G_0 \\ &= I + [I - (\Phi_1 + \Phi_1 G_0 \Phi_1)\,(\Phi_1 + \Phi_1 G_0 \Phi_1)^{-1}]\,\Phi_1 G_0 = I \,.\end{aligned}$$

Thus, the construction of the Green's function G is reduced to finding P from (4.1.2) or, equivalently from

$$P + \Phi_1 G_0 P = \Phi_1 . \tag{4.1.3}$$

Later, as a rule, the operator Φ_0 will correspond to an infinite homogeneous medium and the operator Φ_1 will describe defects of the type of inhomogeneities which are localized in small regions. It is easy to see that the kernel of the operator P, as distinct from that of the Green's function G, is also localized in these regions and due to this its construction both analytically and by numerical methods appears to be more convenient.

From the Hermiticity of the operators G_0 and G, the Hermiticity of P follows immediately, this being evident from (4.1.1 and 4.1.2).

In those cases for which the operator Φ_1^{-1} exists, (4.1.2) can also be written in the form

$$P = (\Phi_1^{-1} + \Gamma_0)^{-1} , \tag{4.1.4}$$

where Γ_0 is the restriction of the operator G_0 to the region of defects.

Let a distribution of external forces q be given and let $u_0 = G_0 q$ be the corresponding external field in the homogeneous medium without defects. Then, for the field $u = Gq$ in the medium with defects, taking into account (4.1.1) we find

$$u = u_0 - \Gamma_0 P u_0 , \tag{4.1.5}$$

where the second term is the perturbation caused by the defects. It is obvious that this equation is valid also in the case when the external field u_0 is given directly.

For the elastic energy Φ, according to (2.3.22), we have

$$2\Phi = \langle q|G|q\rangle = \langle u_0|\Phi_0|u_0\rangle - \langle u_0|P|u_0\rangle . \tag{4.1.6}$$

The first term corresponds to the energy Φ_0 of the external field in the medium without defects and the second term is the energy Φ_{int} of interaction of the defects with the external field and among themselves. This enables us to interpret P as an operator for interaction energy.

Problem 4.1.1. Show that from the invariance of the energy with respect to translation, it follows that the operator P can be represented in the form

$$P = \nabla \tilde{P} \nabla . \tag{4.1.7}$$

Substituting P into (4.1.6) we obtain the expression for the energy of interaction in terms of the deformation ε_0 of the external field and the operator $\tilde{P}$:

$$\Phi_{\text{int}} = \frac{1}{2}\langle \varepsilon_0 | \tilde{P} | \varepsilon_0 \rangle \,. \tag{4.1.8}$$

4.2 Impurity Atom in a Lattice

Let $\Phi_0^{\alpha\beta}(n - n')$ be a matrix of the force constants for a homogeneous lattice, and let $G^0{}_{\alpha\beta}(n - n')$ be the corresponding static Green's tensor. Let the impurity atom be located at the point $n = 0$, this atom causing a local perturbation of the force constants. Then the perturbed matrix has the form

$$\Phi^{\alpha\beta}(n, n') = \Phi_0^{\alpha\beta}(n - n') + \Phi_1^{\alpha\beta}(n, n') \,, \tag{4.2.1}$$

where $\Phi_1^{\alpha\beta}(n, n')$ differs from zero only for atoms of the first coordinate spheres.

According to the above, the Green's tensor of the lattice with a defect can be represented in the form

$$G_{\alpha\beta}(n, n') = G^0_{\alpha\beta}(n - n') - \sum_{m, m'} G^0_{\alpha\lambda}(n - m) P^{\lambda\mu}(m, m') G^0_{\mu\beta}(m' - n') \,, \tag{4.2.2}$$

where the summation is carried out over the region of perturbation and the matrix $P^{\lambda\mu}(mm')$ is defined by

$$P^{\lambda\mu}(m, m') = \{[\Phi_1^{\lambda\mu}(m, m')]^{-1} + G^0_{\lambda\alpha}(m - m')\}^{-1} \,. \tag{4.2.3}$$

Since the points m, m' here belong to the finite region of perturbed bonds, the problem is reduced to the solution of a finite system of linear algebraic equations.

When inverting the matrices, it is necessary to take into account that, due to the conditions of invariance with respect to translation and rotation, the inverse matrices exist only in some special vector subspaces, which can be distinguished by constructing the corresponding projection operators. Also, in the general case, expressions of the type (4.1.2) should be used instead of (4.2.3).

Thus, an effective analytical solution of the problem can be constructed only for the simplest models, which we shall not dwell upon. In more complicated situations, numerical methods are needed.

In analogous manner (by numerical methods) the problem of the dynamic Green's tensor $G_{\alpha\beta}(n, n', \omega)$ can be solved for the case of a single defect as well as for the case of finite set of defects.

In order to obtain tractable analytical expressions, it is expedient to accept another model for the defect, which will be considered in the next section.

4.3 Point Defects in a Quasicontinuum

We assume the existence of an elementary unit of length a in the medium; this is equivalent, as was pointed out above, to a corresponding truncation of the Fourier spectra of the admissible functions, i.e. to a transition to the elastic quasicontinuum. This enables us to use the δ-function model for the description of local inhomogeneities, introducing δ-functions and their derivatives directly into the coefficients of the equations. Such a procedure would be incorrect for an elastic continuum.

Suppose that there is a single defect at a point $x = x_0$. Let us first consider the static case. The simplest model of a defect of the type of local inhomogeneity of the operator for the elastic moduli in the neighborhood of the point x_0 can be described by an expression of the form

$$c^{\lambda\alpha\mu\beta}(x, x') = c_0^{\lambda\alpha\mu\beta}(x - x') + v_0 c_1^{\lambda\alpha\mu\beta}\delta(x - x_0)\delta(x' - x_0) . \tag{4.3.1}$$

Here, the constant tensor $c_1^{\lambda\alpha\mu\beta}$ has the dimension and symmetry of the elastic-constants tensor and characterizes the variation of the elastic properties in the neighborhood of the point x_0. A more complicated model of a defect can be obtained if we add additional terms with derivatives of δ-functions.

Straightforward computations according to the scheme of Sect. 4.1 give for the Green's tensor $G_{\alpha\beta}(x, x')$ of the medium with the defect, the expression

$$G_{\alpha\beta}(x, x') = G^0_{\alpha\beta}(x - x') - \partial_\nu G^0_{\alpha\mu}(x - x_0) P^{\nu\mu\lambda\tau} \partial_\lambda G^0_{\tau\beta}(x_0 - x') . \tag{4.3.2}$$

The constant tensor $P^{\nu\mu\lambda\tau}$ appearing here has the symmetry of the tensor of elastic constants and is determined by

$$P^{\nu\mu\lambda\tau} = -(g^0_{\nu\mu\lambda\tau} + v_0^{-1} b_{\nu\mu\lambda\tau})^{-1} , \tag{4.3.3}$$

where $b_{\nu\mu\lambda\tau}$ is the tensor which is inverse to $c_1^{\nu\mu\lambda\tau}$, and

$$g^0_{\nu\mu\lambda\tau} \overset{\text{def}}{=} -[\partial_\nu \partial_\lambda G^0_{\mu\tau}(0)]_{(\nu\mu)(\lambda\tau)} . \tag{4.3.4}$$

The existence of $g^0_{\nu\mu\lambda\tau}$ follows from the analyticity of $G^0_{\alpha\beta}(x)$ for the quasicontinuum. From here it also follows that $G_{\alpha\beta}(x, x')$ is an analytic function of x and x'. However $G_{\alpha\beta}(x, x')$ depends on the parameter a nonanalytically.

Problem 4.3.1. Show that $g^0 \sim v_0^{-1}$, $P \sim v_0$ and, as $a \to 0$, the second term in (4.3.2) is of the order of v_0 for the points $(x \neq x_0,\ x' \neq x_0)$, of the order of a for points $(x = x_0,\ x' \neq x_0)$ or $(x \neq x_0,\ x' = x_0)$, and tends to infinity as a^{-1} for the point $(x = x' = x_0)$.

The construction of the dynamic Green's tensor for a single defect can be carried out in an analogous way. Here it is reasonable to consider two cases.

If the density $\rho(x) = \rho_0 = \text{const}$ and if the defect is caused only by a change of the elastic properties of the medium, then the expression for the dynamic Green's tensor $G_{\alpha\beta}(x, x', \omega)$ coincides with (4.3.2), after replacing $G^0_{\alpha\beta}(x)$ by the dynamic Green's tensor for the homogeneous medium $G^0_{\alpha\beta}(x, \omega)$. Similarly, $P = P(\omega)$ is determined by (4.3.3) with $g^0 = g^0(\omega)$.

In the other case, when $c(x, x') = c_0(x - x')$, and the density has the form:

$$\rho(x) = \rho_0 + v_0\rho_1\delta(x - x_0)\,, \tag{4.3.5}$$

we find for the Green's tensor

$$G_{\alpha\beta}(x, x', \omega) = G^0_{\alpha\beta}(x - x', \omega) - G^0_{\alpha\lambda}(x - x_0, \omega)P^{\lambda\mu}(\omega)G^0_{\mu\beta}(x_0 - x', \omega)\,, \tag{4.3.6}$$

$$P^{\lambda\mu}(\omega) = -v_0\rho_1\omega^2[\delta_{\lambda\mu} - v_0\rho_1\omega^2 G^0_{\lambda\mu}(0, \omega)]^{-1}\,. \tag{4.3.7}$$

In the general case of a defect of mass and elastic moduli, the expression for the Green's tensor is constructed in an analogous way, but has a more complicated structure.

Let us now consider a perturbation of the external field $u^0_\alpha(x)$, caused by the defect, and let us confine ourselves to the static case. Taking into account (4.3.2), we find

$$u_\alpha(x) = u^0_\alpha(x) + u^1_\alpha(x)\,, \tag{4.3.8}$$

where $u^1_\alpha(x)$ is the perturbation which coincides, as can be easily proved, with the displacement in the homogeneous medium, caused by a force dipole at the point x_0 with density

$$q^\nu(x) = -Q^{\nu\mu}\partial_\mu\delta(x - x_0) \tag{4.3.9}$$

and with the moment

$$Q^{\nu\mu} = P^{\nu\mu\lambda\tau}\varepsilon^0_{\lambda\tau}(x_0)\,, \tag{4.3.10}$$

where $\varepsilon^0_{\lambda\tau}(x_0)$ is the deformation at the point x_0 without the defect.

Thus, the perturbation caused by a defect can be obtained, if we replace the defect by an equivalent force dipole having the moment (4.3.10). Roughly speaking, a local inhomogeneity generates a force dipole which is proportional to the field. By analogy with the electric field in a dielectric, one can also speak about the polarization effect caused by a local inhomogeneity.

Let us represent the elastic energy Φ as a sum of the energy of the field Φ_0 in the absence of the defect and the energy of interaction Φ_{int} of the defect with the external field. For Φ_{int}, analogously to (4.1.8), we find

$$2\Phi_{\text{int}} = \varepsilon^0_{\nu\mu}(x_0)P^{\nu\mu\lambda\tau}\varepsilon^0_{\lambda\tau}(x_0)\,. \tag{4.3.11}$$

This energy depends on the position of the defect. In turn, it enables us to introduce naturally the concept of a force f, with which the external field acts on the defect. By definition

$$f_\lambda = -\frac{\partial}{\partial x_0^\lambda}\Phi_{\text{int}} = Q^{\nu\mu}\partial_\lambda \varepsilon^0_{\nu\mu}(x_0)\,. \tag{4.3.12}$$

As an example, let us consider a spherically symmetric defect located at the origin in an isotropic quasicontinuum. The tensor c_1 in this case has the form (see (2.8.4))

$$c_1 = 2\mu_1 E_1 + \lambda_1 E_2\,,$$
$$E_1^{\lambda\alpha\mu\beta} \overset{\text{def}}{=} \frac{1}{2}(\delta^{\alpha\beta}\delta^{\lambda\mu} + \delta^{\alpha\lambda}\delta^{\beta\mu}),\; E_2^{\lambda\alpha\beta\mu} \overset{\text{def}}{=} \delta^{\lambda\alpha}\delta^{\mu\beta}\,. \tag{4.3.13}$$

Using the general expression for the Green's tensor of the Debye quasicontinuum with spatial dispersion (2.9.11) and taking into account (4.3.3, 4 and 13) we obtain

$$g^0 = \frac{1}{15v_0}\left[\left(\frac{\beta_l}{\lambda_0 + 2\mu_0} - \frac{\beta_t}{\mu_0}\right)E_2 + \left(\frac{2\beta_l}{\lambda_0 + 2\mu_0} + \frac{3\beta_t}{\mu_0}\right)E_1\right], \tag{4.3.14}$$

$$P = -\frac{1}{3}(p_1 - p_2)E_2 - p_2 E_1\,, \tag{4.3.15}$$

where

$$p_1 = v_0\left[\frac{\beta_l}{3(\lambda_0 + 2\mu_0)} + \frac{1}{3\lambda_1 + 2\mu_1}\right]^{-1},$$
$$p_2 = v_0\left[\frac{1}{15}\left(\frac{2\beta_l}{\lambda_0 + 2\mu_0} + \frac{3\beta_t}{\mu_0}\right) + \frac{1}{2\mu_1}\right]^{-1}. \tag{4.3.16}$$

The dimensionless constants β_j depend on the dispersion law

$$\beta_j = \frac{3v_j^2}{\mathcal{H}^3}\int_0^{\mathcal{H}} \frac{k^4}{\omega_j^2(k)}\,dk \quad (j = l, t)\,. \tag{4.3.17}$$

For the Debye model and for the model (2.9.8) we have, respectively,

$$\beta_j = 1, \quad \beta_j = \frac{10}{7} + \frac{3}{5}\gamma_j\,. \tag{4.3.18}$$

Let us assume in addition that with the defect there is associated a force dipole with the defect there is associated a force dipole with the density [1]

[1]Recall that for an isotropic medium we use the notations **r** and r instead of x and $|x|$.

$$q_\lambda(\mathbf{r}) = -Q\partial_\lambda\delta(\mathbf{r}) . \tag{4.3.19}$$

The internal stresses due to a vacancy, an interstitial atom and so on are often modeled in such a manner. In the given case, the model corresponds to a center of dilatation.

It was pointed out above that the displacement field $u(\mathbf{r})$ can be represented in the form of a sum of the unperturbed and perturbed fields $u^0(\mathbf{r})$ and $u^1(\mathbf{r})$. In the model under consideration, $u^0(\mathbf{r})$ is a field of the force dipole (4.3.19) in the homogeneous medium. The deformation of the unperturbed field at the defect point, as can easily be verified, is equal to

$$\varepsilon^0_{\alpha\beta}(0) = Q\delta^{\lambda\mu}g^0_{\alpha\beta\lambda\mu} = \frac{Q\beta_l}{3v_0(\lambda_0 + 2\mu_0)}\delta_{\alpha\beta} . \tag{4.3.20}$$

For the moment of the effective dipole, according to (4.3.10) and taking into account (4.3.15, 16 and 20), we have

$$Q'^{\alpha\beta} = -Q'\delta^{\alpha\beta},\ Q' = \frac{Q}{1 + \dfrac{\lambda_0 + 2\mu_0}{\beta_l(3\lambda_1 + 2\mu_1)}} . \tag{4.3.21}$$

The expression for the total deformation $\varepsilon = \varepsilon^0 + \varepsilon^1$ now has the form

$$\varepsilon_{\alpha\beta}(\mathbf{r}) = Q''\partial_{(\alpha}\partial^\lambda G^0_{\beta)\lambda}(r),\ Q'' = \frac{Q}{1 + \dfrac{\beta_l(3\lambda_1 + 2\mu_1)}{\lambda_0 + 2\mu_0}} \tag{4.3.22}$$

or, if (2.9.12) is taken into account,

$$\varepsilon_{\alpha\beta}(\mathbf{r}) = \frac{Q''}{\lambda_0 + 2\mu_0}\partial_\alpha\partial_\beta h_l(r) . \tag{4.3.23}$$

At the point of the defect

$$\varepsilon_{\alpha\beta}(0) = \frac{Q''\beta_l}{3v_0(\lambda_0 + 2\mu_0)}\delta_{\alpha\beta} , \tag{4.3.24}$$

and at large distances from the defect

$$\varepsilon_{\alpha\beta}(\mathbf{r}) \simeq \frac{Q''}{4\pi(\lambda_0 + 2\mu_0)}\partial_\alpha\partial_\beta\frac{1}{r} . \tag{4.3.25}$$

It is also easy to calculate the corresponding stresses $\sigma^{\alpha\beta}(0)$ and $\sigma^{\alpha\beta}(r)$.

It is interesting to compare the model under consideration with the classical problem of a spherical inhomogeneity in an isotropic elastic medium with the Lamé constants λ_0, μ_0. Let the elastic constants of a small sphere of the volume

v_0 be equal to $\lambda_0 + \lambda_1$ and $\mu_0 + \mu_1$ and let, on its boundary, radial forces with the surface density Q/v_0 be applied. Then it can be shown that for the Debye model (i.e., when spatial dispersion is absent) the values of $\varepsilon(0)$ and $\sigma(0)$ coincide with the strain and stress within the sphere and the asymptotic values of $\varepsilon(\mathbf{r})$ and $\sigma(\mathbf{r})$ coincide with the strain and stress outside the sphere. The total elastic energies for these cases also coincide exactly.[2]

For the interaction energy Φ_{int} of the defect with the external filed $\varepsilon^0_{\alpha\beta}(\mathbf{r})$, we find, according to (4.3.11, 15)

$$\Phi_{\text{int}} = \frac{p_1}{6}\left[\mathrm{Tr}\left\{\varepsilon^0(0)\right\}\right]^2 + \frac{p_2}{2}\left[\mathrm{Dev}\left\{\varepsilon^0(0)\right\}\right]^2, \tag{4.3.26}$$

where $\mathrm{Tr}\{\varepsilon^0\}$ and $\mathrm{Dev}\{\varepsilon^0\}$ are the spherical and deviatoric parts of ε^0.

From this it follows that the force which acts upon the defect is equal to

$$f_\lambda = -\frac{p_1}{3}\mathrm{Tr}\,\{\varepsilon^0(0)\}\,\partial_\lambda \mathrm{Tr}\{\varepsilon^0(0)\} - p_2\,\mathrm{Dev}\{\varepsilon^0(0)\}\,\partial_\lambda\,\mathrm{Dev}\{\varepsilon^0(0)\}\,. \tag{4.3.27}$$

4.4 System of Point Defects

Let us now consider the case of a system of point defects; for definiteness, we take it to be a system of defects of the elastic moduli. Let us also restrict ourselves to the static case (for the simplicity of notation). We have

$$c^{\lambda\alpha\mu\beta}(x, x') = c_0^{\lambda\alpha\mu\beta}(x - x') + v_0 \sum_i c_i^{\lambda\alpha\mu\beta}\delta(x - x_i)\delta(x' - x_i)\,. \tag{4.4.1}$$

Omitting simple but lengthy computations, we present the final result for the static Green's tensor

$$\begin{aligned} G_{\alpha\beta}(x, x') &= G^0_{\alpha\beta}(x - x') \\ &\quad - \sum_{ij}{}' \partial_\nu G^0_{\alpha\mu}(x - x_i)\, P^{\nu\mu\lambda\tau}_{ij} \partial_\lambda G^0_{\tau\beta}(x_j - x')\,. \end{aligned} \tag{4.4.2}$$

The matrix P_{ij} with tensor components entering the above expression, is the inverse of the matrix

$$R^{ij}_{\nu\mu\lambda\tau} = -\, g^{ij}_{\nu\mu\lambda\tau} - v^{-1} b^j_{\nu\mu\lambda\tau}\delta^{ij}\,, \tag{4.4.3}$$

where

$$g^{ij}_{\nu\mu\lambda\tau} = -\,[\partial_\nu\partial_\lambda G^0_{\mu\tau}(x_i - x_j)]_{(\nu\mu)\,(\lambda\tau)} \tag{4.4.4}$$

and the $b^j_{\nu\mu\lambda\alpha}$ are tensors inverse to $c^j_{\nu\mu\lambda\alpha}$.

---

[2]Such a coincidence is, of course, caused by the high symmetry of the model. In the general case, the correspondence between point defects and local inhomogeneities is more complicated. We return to this question later.

The dynamic Green's tensor for the system of point mass defects has an analogous structure.

In many cases, the dominant term of the asymptotics of the perturbed elastic and or interaction energy is of principal interest. It may be shown that, in such a situation, the problem is reduced to the binary interaction of defects. In this connection, let us consider in more detail the case of two defects, which is by itself of interest.

When i and $j = 1, 2$, the computation of the components P_{ij} is conveniently carried out according to (no summation)

$$\begin{aligned} P_{ii} &= (R_{ii} - R_{ij}R_{jj}^{-1}R_{ji})^{-1}, \quad (i \neq j), \\ P_{ij} &= (R_{ji} - R_{jj}R_{jj}^{-1}R_{ii})^{-1} \end{aligned} \tag{4.4.5}$$

whose validity is verified by direct computation.

These formulae also enable us to estimate easily the asymptotics of the matrix P_{ij} for a large distance $\rho = |x_1 - x_2|$ between the defects. In fact, $G^0(\rho) \sim \rho^{-1}$, and hence $g^{ij}(\rho) \sim \rho^{-3}$. Therefore, taking into account (4.4.3, 5), we obtain

$$\begin{aligned} P_{ii}(\rho) &= P_i^0 + O(\rho^{-6}), \\ P_{ij}(\rho) &\sim \rho^{-3} + O(\rho^{-9}), \end{aligned} \tag{4.4.6}$$

where the tensor P_i^0 corresponds to an isolated defect at the point x_i and is defined by (4.3.3).

For the isotropic Debye quasicontinuum, the matrix $P_{ij}(\rho)$ can be computed explicitly. Let us also present the generalization of (4.3.8) to the case of a system of defects in the external field $u_\alpha^0(x)$:

$$u_\alpha(x) = u_\alpha^0(x) + \sum_i u_\alpha^i(x). \tag{4.4.7}$$

Here, $u_\alpha^i(x)$ is the displacement due to a defect at the point x_i. However, the moments of the equivalent force dipoles

$$Q_i^{\nu\mu} = \sum_j P_{ij}^{\nu\mu\lambda\tau} \varepsilon_{\lambda\tau}^0(x_j) \tag{4.4.8}$$

now depend on the values of the field at all points x_j as well as on the elastic characteristics of all defects. Therefore, in the general case, the perturbation is not the superposition of perturbations caused by individual defects, i.e. the matrix P_{ij} depends on the perturbations of the elastic constants of individual defects in a complicated manner. In the limiting case of large distances between the defects it follows from formulae of the type of (4.4.6) that in the zeroth approximation, the perturbations of individual defects can be linearly superposed; in the next approximation we have the superposition of binary interactions.

The calculation of the energy of interaction of defects with the external field and between themselves can be carried out according to (4.1.6), which in the present case takes the form

$$2\Phi_{\text{int}} = \sum_{ij} \varepsilon^0_{\nu\mu}(x_i) P^{\nu\mu\lambda\tau}_{ij} \varepsilon_{\lambda\tau}(x_j) . \tag{4.4.9}$$

The force, which acts on the point defect at a point x_k is given by the expression

$$\begin{aligned} f^k_\lambda &\stackrel{\text{def}}{=} -\frac{\partial}{\partial x^\lambda_k} \Phi_{\text{int}} \\ &= Q^{\nu\mu}_k \partial_\lambda \varepsilon^0_{\nu\mu}(x_k) - \frac{1}{2} \sum_{ij} \varepsilon^0_{\nu\mu}(x_i) \left[\frac{\partial}{\partial x^\lambda_k} P^{\nu\mu\kappa\tau}_{ij} \right] \varepsilon^0_{\kappa\tau}(x_j) . \end{aligned} \tag{4.4.10}$$

For defects situated far enough from each other, the second term becomes vanishingly small and the first term coincides with the force which acts on the isolated defect at the point x_k. Conversely in a homogeneous external field $\varepsilon^0 = \text{const}$, only the second term contributes to the force.

4.5 Local Inhomogeneity in an Elastic Medium

Strictly speaking, the problem of inhomogeneity in the elastic medium is related not to the theory of a medium with microstructure but to the conventional theory of elasticity. However, we shall be interested in the asymptotic behavier of perturbed fields and of the interaction energy. We shall see that in such a formulation of the problem, a far-reaching analogy with the model of point defects considered above exists.

Let the tensor of the elastic moduli of the medium be of the form

$$c^{\alpha\beta\lambda\mu}(x) = c^{\alpha\beta\lambda\mu}_0 + \sum_i c^{\alpha\beta\lambda\mu}_i(x) , \tag{4.5.1}$$

where $c^{\alpha\beta\lambda\mu}_0$ is the tensor of the elastic constants of the homogeneous medium, $c^{\alpha\beta\lambda u}_i$ is the perturbation caused by the defect localized in a (small) region V_i. The case $c_i \to \infty$, corresponds to a rigid inclusion and the case $c_i \to -c_0$ corresponds to a cavity. Let us consider, for simplicity, an infinite elastic medium.

Using the general scheme described in Sect. 4.1, one can show without difficulty that the Green's tensor can be represented in the form

$$\begin{aligned} G_{\alpha\beta}(x, x') &= G^0_{\alpha\beta}(x - x') \\ &\quad - \iint G_{\alpha(\nu}(x, y) \partial_{\lambda)} P^{\lambda\nu\sigma\tau}(y, y') \partial_{(\sigma} G_{\tau)\beta}(y', x) dy\, dy' \end{aligned} \tag{4.5.2}$$

or, in operator form

$$G = G^0 - G^0 \nabla P \nabla G^0 , \tag{4.5.3}$$

where the operator P for the interaction energy is a sum of operators

$$P = \sum_{ij} P^{ij} . \tag{4.5.4}$$

The matrix P^{ij} and its inverse matrix R_{ij} have the structure analogous to (4.4.3):

$$\begin{aligned} P^{ij} &= c_i(c_j \nabla G^0 \nabla c_i - c_j \delta_{ij})^{-1} c_j = (\nabla G^0 \nabla - c_i^{-1} \delta_{ij})^{-1} = R_{ij}^{-1} , \\ R_{ij}(y, y') &= -\nabla \nabla' G^0(y - y') - c_i^{-1}(y) \delta(y - y') \delta_{ij} , \\ y &\in V_i, \; y' \in V_j' . \end{aligned} \tag{4.5.5}$$

It can easily be seen that the operator P is self-adjoint; its kernel satisfies the symmetry conditions

$$P^{\lambda\nu\sigma\tau}(y, y') = P^{\nu\lambda\sigma\tau}(y, y') = P^{\sigma\tau\lambda\nu}(y', y) \tag{4.5.6}$$

and is localized in the region $\bigcup_{ij} V_i \otimes V_j$.

For the displacement $u(x)$ in the external field and for the energy of the interaction Φ_{int} we obtain

$$u_\alpha(x) = u_\alpha^0(x) - \int G_{\alpha\nu}^0(x - y) \partial_\lambda P^{\lambda\nu\sigma\tau}(y, y') \varepsilon_{\sigma\tau}^0(y') \, dy \, dy' , \tag{4.5.7}$$

$$\Phi_{\text{int}} = \frac{1}{2} \langle \varepsilon_{\alpha\beta}^0 | P^{\alpha\beta\lambda\mu} | \varepsilon_{\lambda\mu}^0 \rangle . \tag{4.5.8}$$

Taking into account that $P(y, y')$ implicitly depends on the coordinates of the centers of mass of the defects, we have for the force f_λ^k, which acts on the kth defect, due to the external field and other defects,

$$f_\lambda^k \overset{\text{def}}{=} -\frac{\partial \Phi_{\text{int}}}{\partial x_k^\lambda} = -\frac{1}{2} \langle \varepsilon^0 | \frac{\partial}{\partial x_k^\lambda} P | \varepsilon^0 \rangle . \tag{4.5.9}$$

In an analogous manner one can find the moment which acts on a defect.

In the general case, the kernel $P(y, y')$ may be found by numerical methods, and use of the latter is facilitated by the fact that, unlike $G(x, x')$, $P(y, y')$ is localized in a bounded region.

The problem is considerably simplified and, in a number of cases, can be solved analytically if the defects are located at distances, which are large in comparison with their size and if only the asymptotic form of the perturbed field is of interest. Eventually, this is equivalent to replacing local inhomogeneities by equivalent point defects.

First, let us consider approximations with respect to the distance $r_{ij} = |x_i - x_j|$ between defects. In the zeroth approximation the defects do not interact and $P^{ij} = P^i\delta^{uj}$, where P^i is the operator of the i-th defect. One can show that in the first approximation the problem is reduced to binary interaction. Therefore let us restrict ourselves to the case of two defects. For two defects, according to (4.5.5), we have

$$R_{ij} = \begin{pmatrix} (P^1)^{-1}\nabla G^0\nabla \\ \nabla G^0\nabla(P^2)^{-1} \end{pmatrix}. \tag{4.5.10}$$

The calculation of the components P^{ij} in terms of R_{ij} is conveniently carried out according to (4.4.5). Keeping only dominant terms with respect to $r_{12} = |x_1 - x_2|$, we find the explicit expressions for the components P^{ij} in terms of the operators P^i of individual defects

$$P^{ij} = \begin{pmatrix} P^1 & -P^1\nabla G^0\nabla P^2 \\ -P^2\nabla G^0\nabla P^1 & P^2 \end{pmatrix} + O(r_{12}^{-6}). \tag{4.5.11}$$

Let us now proceed to approximations of another type, which enable us to find the asymptotics of the perturbed field. With this in mind, let us approximate kernel $P(y, y')$ by first terms of its multipole expansion in the neighborhood of each defect

$$P(y, y') = \sum_{ijmn} (-1)^{m+n} P^{ij}_{mn}(r_{ij})\delta^{(m)}(y - x_i)\delta^{(n)}(y' - x_j) \tag{4.5.12}$$
$$(x_i \in V_i,\ x_j \in V_j').$$

Substituting this expansion in the equation for P of the type of (4.1.3), it is possible to obtain finally a linear system of equations for the moments P^{ij}_{mn}.

It is important that the dominant terms with respect to r_{ij} of the first moment of the multipoles can be determined explicitly, if the first moment P^i_m for each defect, in the absence of all the other defects, is known.

Note that, in the case of a homogeneous external field $\varepsilon^0_{\alpha\beta} = \text{const}$, we have for the interaction energy Φ_{int} (for any r_{ij}) the exact formula, see (4.4.9),

$$\Phi_{\text{int}} = \frac{1}{2}\,\varepsilon^0_{\alpha\beta} P^{\alpha\beta\lambda\mu}_{00}(r_{ij})\varepsilon^0_{\lambda\mu},\quad P_{00} = \sum_{ij} P^{ij}_{00}. \tag{4.5.13}$$

From here it follows that only the asymptotics of the matrix $P^{ij}_{00}(r_{ij})$ contributes to the dominant term of the asymptotics of Φ_{int} in a homogeneous external field. In the case of an arbitrary external field, it is necessary to take into account also the contribution to the asymptotics of Φ_{int} from the asymptotics of the diagonal components of the matrices $P^{ij}_{01}(r_{ij})$ and $P^{ij}_{10}(r_{ij})$. This contribution is equal to zero if the defects possess central symmetry or if the external field changes slowly over a distance of the order of the size of a defect.

Let us sum up. For the description of the interaction of defects in an elastic medium, it is necessary to know, in some approximation, the Green's function for the medium with defects, or, equivalently, one has to be able to solve the corresponding problems of the classical theory of elasticity, exactly or approximately. The following sections of the present chapter will be devoted to these problems, which are of interest in themselves. The results obtained will be used in an essential way, particularly in Chap. 7. This chapter is devoted to elastic media with random distribution of defects.

Note that the approach to problems of the classical theory of elasticity, which is developed further, differs from that presented in standard textbooks.

4.6 Homogeneous Elastic Medium

The equations of classical elasticity were written above in the form of (2.6.4). In this section we confine ourselves to statics. Then the equations become

$$-\partial_\lambda c^{\lambda\alpha\mu\beta}(x)\partial_\mu u_\beta(x) = q^\alpha(x) . \tag{4.6.1}$$

We assume additionally that the medium is unbounded and homogeneous; hence, the tensor $c^{\lambda\alpha\mu\beta}$ is constant. The density of external forces $q(x)$ is assumed to be a generalized function, decreasing sufficiently rapidly at infinity; for example, it can have a bounded support.

Under these conditions, the solution of (4.6.1) which tends to zero at infinity, is unique and is representable in the form

$$u = Gq , \tag{4.6.2}$$

where $G = G^+$ is the Green's operator for displacements. In a more detailed expression

$$u_\alpha(x) = \int G_{\alpha\beta}(x - x')q^\beta(x')\,dx' , \tag{4.6.3}$$

where $G_{\alpha\beta}(x)$ is the tensor Green's function for displacements. Obviously, in the k representation,

$$G(k) = [\Phi(k)]^{-1} , \tag{4.6.4}$$

where

$$\Phi^{\alpha\beta}(k) = c^{\alpha\lambda\mu\beta}k_\lambda k_\mu . \tag{4.6.5}$$

For the isotropic medium, the tensor of elastic moduli becomes, see (2.8.4),

$$C = 2\mu E^1 + \lambda E^2 , \tag{4.6.6}$$

where E_1 and E_2 are constant isotropic tensors defined in (A.1), and λ, μ are the Lamé constants. The expression for $G(k)$ coincides with (2.8.6), where λ and μ are taken to be constants.

For an arbitrary anisotropic medium

$$G_{\alpha\beta}(k) = \frac{g_{\alpha\beta}(n)}{k^2}, \tag{4.6.7}$$

where $n_\alpha = k_\alpha/|k|$ and $g_{\alpha\beta}(n)$ is a function defined on the unit sphere in k-space, which is a determined explicitly by (4.6.4, 5).

The explicit expression for $G_{\alpha\beta}(x)$ is known only for an isotropic medium and media with transversly isotropic and hexagonal symmetries [4.1]. However, we shall see that in a number of important cases it is sufficient to know general properties of $G_{\alpha\beta}(x)$ and its expression in the k-representation.

In particular, note that it follows from (4.6.4, 5) that

$$g_{\alpha\beta}(n) = g_{\alpha\beta}(-n), \; G_{\alpha\beta}(x) = G_{\alpha\beta}(-x). \tag{4.6.8}$$

Let us consider briefly the behavior of the solution (4.6.3) for a few important types of the density of external forces $q^\alpha(x)$. Let $q^\alpha(x) = \tilde{q}^\alpha(x)V(x)$, where $V(x)$ is the characteristic function of a bounded region V with a sufficiently smooth boundary Ω, and $\tilde{q}^\alpha(x)$ is a continuous function. Then $u_\alpha(x)$ is a continuous function, which is twice differentiable inside V, analytic outside V and tends to zero as r^{-1}, when $r = |x| \to \infty$. Its first derivatives and consequently, the strain and stress tensors, have a finite discontinuity on Ω.

If $q^\alpha(x) = q_0^\alpha\delta(x)$, then the solution is proportional to $G_{\alpha\beta}(x)$ and has, as follows from (4.6.6), a singularity of the order of r^{-1} as $x \to 0$. From here, it is easy to find also the behavior of $u_\alpha(x)$ for external forces concentrated on surfaces.

Proofs of all these statements can be obtained in a manner analogous to that used in the potential theory, we shall not therefore dwell on this.

We shall see later, that in a number of applications, in particular for a medium with defects, it is convenient to have an equivalent form of the equations of elasticity and to have their solutions not in terms of the displacement u_α, but in terms of the strain $\varepsilon_{\alpha\beta}$ or stress $\sigma^{\alpha\beta}$. For this purpose, let us rewrite (4.6.1) in the form of the usual system

$$\partial_\beta\sigma^{\alpha\beta} = -q^\alpha, \; \sigma^{\alpha\beta} = c^{\alpha\lambda\beta\mu}\varepsilon_{\lambda\mu}, \; \varepsilon_{\lambda\mu} = \partial_{(\lambda}u_{\mu)} \tag{4.6.9}$$

or, in the operator form,

$$\operatorname{div}\sigma = -q, \; \sigma = C\varepsilon, \; \varepsilon = \operatorname{def} u, \tag{4.6.10}$$

where def is the operator for the symmetrized gradient.

Our first task is to write down these equations and their solutions in a form which contains the strain tensor only.

Let us represent the density of external forces in the form

$$q^{\alpha}(x) = -\partial_{\mu} p^{\alpha\mu}(x) \tag{4.6.11}$$

where $p^{\alpha\mu} = p^{\mu\alpha}$ is the density of force moments, the latter being defined non-uniquely. After some additional conditions are imposed, the correspondence $q^{\alpha} \leftrightarrow p^{\alpha\mu}$ becomes one-to-one, and it is not difficult to find $p^{\alpha\mu}$ explicitly. It is to be emphasized that, in problems concerning inhomogeneities in an external field, $p^{\alpha\mu}$, and not q^{α} is the naturally given quantity (see below).

Substituting into (4.6.3) the expression (4.6.11) for q^{α} and applying the operator def to both sides of the equality, we obtain

$$\varepsilon_{\alpha\beta}(x) = \int K_{\alpha\beta\lambda\mu}(x - x') p^{\lambda\mu}(x') \, dx' , \tag{4.6.12}$$

where

$$K_{\alpha\beta\lambda\mu}(x) = -\left(\partial_{\alpha}\partial_{\mu} G_{\beta\lambda}\right)_{(\alpha\beta)\,(\lambda\mu)} . \tag{4.6.13}$$

In operator notations,

$$\varepsilon = Kp, \; K = -\operatorname{def} G \operatorname{def} . \tag{4.6.14}$$

From the preceding, it follows that K is a correctly defined self-adjoint operator, $K = K^{+}$. Its kernel $K(x)$ is a generalized function, whose Fourier transform $K(k)$ is, as follows from (4.6.7, 13), a homogeneous function of the zeroth degree, i.e. $K(k)$ is, in fact, a function of the unit vector $n_{\alpha} = k_{\alpha}/|k|$

$$K_{\alpha\beta\lambda\mu}(k) = K_{\alpha\beta\lambda\mu}(n) = [n_{\alpha} n_{\mu} g_{\beta\lambda}(n)]_{(\alpha\beta)\,(\lambda\mu)} . \tag{4.6.15}$$

Problem 4.6.1. Show that, for an isotropic medium,

$$K(n) = \frac{1}{\mu} \left[E^{5}(n) - \frac{\lambda+\mu}{\lambda+2\mu} E^{6}(n)\right] , \tag{4.6.16}$$

where $E^{5}(n)$ and $E^{6}(n)$ are defined in (A.1).

From the theory of generalized functions, it follows [4.2] that the inverse Fourier transform $K(x)$ of the generalized homogeneous function of the zeroth degree $K(k)$, is defined with the help of the corresponding regularization. The latter is reduced to a decomposition of $K(x)$ into a regular part and a singular one, which is proportional to $\delta(x)$. Properties of the generalized function $K(x)$ are considered in detail in Appendix A.3.

From (4.6.12), it follows that the desired strain field is obtained as a convolution of the function $K(x)$ with the density of force moments, $p(x)$, i.e. K is the Green's operator for strain. Let us consider a few features of this operator.

From (4.6.14), it is obvious that the operator K is annihilated by all operators which annul the operator def. In particular, this is true for the operator Rot, which is considered in detail in the next chapter. Thus, we have

$$\mathrm{Rot}\, K = 0\,. \tag{4.6.17}$$

Further, taking into account (4.6.4, 14) and the symmetry of the tensor of elastic constants, we have

$$\begin{aligned} KCK &= \mathrm{def}\, G\nabla C\nabla G\, \mathrm{def} \\ &= -\mathrm{def}\, G\Phi G\, \mathrm{def} = -\mathrm{def}\, G\, \mathrm{def} \end{aligned} \tag{4.6.18}$$

or

$$KCK = K\,. \tag{4.6.19}$$

In particular, it follows from here that

$$\overrightarrow{\Pi} = KC,\ \overleftarrow{\Pi} = CK \tag{4.6.20}$$

are projection operators (projections), i.e.

$$\overrightarrow{\Pi}^2 = \overrightarrow{\Pi},\ \overleftarrow{\Pi}^2 = \overleftarrow{\Pi} \tag{4.6.21}$$

while

$$\overleftarrow{\Pi}^+ = \overrightarrow{\Pi}\,. \tag{4.6.22}$$

Let us introduce a dimensionless density of moments

$$m = Bp\,, \tag{4.6.23}$$

where, as usual, $B = C^{-1}$. Then, from (4.6.14), it follows that

$$\varepsilon = \overrightarrow{\Pi} m,\ \sigma = \overleftarrow{\Pi} p\,, \tag{4.6.24}$$

i.e. the projections $\overrightarrow{\Pi}$ and $\overleftarrow{\Pi}$ give one of the representations of the Green's operators for strain and stress, respectively.

Problem 4.6.2. Show that

$$\partial_\lambda \overleftarrow{\Pi}^{\lambda\alpha}{}_{\cdot\cdot\mu\beta} = \partial_{(\mu}\delta^{\alpha}_{\beta)}\,, \tag{4.6.25}$$

from which follows the equation of equilibrium in (4.6.9).

In direct notation, in view of (4.6.17, 25), we have

$$\mathrm{Rot}\, \overrightarrow{\Pi} = 0,\ \mathrm{div}\, \overleftarrow{\Pi} = \mathrm{div}\,. \tag{4.6.26}$$

Another representation of the Green's operators in the form of self-adjoint operators follows from the relations

$$\varepsilon = Kp, \sigma = CKCm\,. \tag{4.6.27}$$

This representation is closely connected with the elastic energy Φ. In fact, taking into account (4.6.19), we have

$$2\Phi = \langle \sigma | \varepsilon \rangle = \langle p | K | p \rangle = \langle m | CKC | m \rangle \tag{4.6.28}$$

This gives an energy interpretation of the Green's operators in terms of the corresponding field sources.

When considering inhomogeneities in an external field, we shall also need an operator S, which is connected with the operator K by the relations

$$S = C - CKC, K = B - BSB\,. \tag{4.6.29}$$

The connection of this operator with internal stress will be found in the next chapter. Here, we consider only the relations which are analogous to those for the operator K presented above.

Problem 4.6.3. Taking into account (4.6.19), show that

$$SBS = S\,. \tag{4.6.30}$$

From here, it follows that the projections

$$\overleftarrow{\Theta} = SB, \overrightarrow{\Theta} = BS \tag{4.6.31}$$

are connected with the operator S, while

$$\overleftarrow{\Theta}^{+} = \overrightarrow{\Theta}\,. \tag{4.6.32}$$

Problem 4.6.4. Taking into account (4.6.17, 26), show that

$$\operatorname{div} S = 0, \quad \operatorname{Rot} BS = \operatorname{Rot}\,, \tag{4.6.33}$$

$$\operatorname{div} \overleftarrow{\Theta} = 0, \quad \operatorname{Rot} \overrightarrow{\Theta} = \operatorname{Rot}\,. \tag{4.6.34}$$

Problem 4.6.5. Verify that

$$I = \overrightarrow{\Pi} + \overrightarrow{\Theta} = \overleftarrow{\Pi} + \overleftarrow{\Theta}\,. \tag{4.6.35}$$

Thus, the strain and the stress can be obtained by projecting the field sources onto the corresponding subspaces of the Hilbert space. To the projections $\vec{\Pi}$ and $\overleftarrow{\Pi}$, there correspond the usual stress σ (caused by external forces q, i.e. $\operatorname{div}\sigma = -q$) and the usual strain ε (which satisfies the compatibility conditions $\operatorname{Rot}\varepsilon = 0$ and is represented in terms of the displacement, as $\varepsilon = \operatorname{def} u$). Conversely, the internal stress σ and the internal strain ε, for which $\operatorname{div}\sigma \equiv 0$, but $\operatorname{Rot}\varepsilon \neq 0$ correspond to the projections $\vec{\Theta}$ and $\overleftarrow{\Theta}$. In this case, the deformation ε is not represented in terms of a displacement. A detailed consideration of the internal stress and the corresponding field sources will be given in the next two chapters.

Unfortunately, the projection operators introduced above are not self-adjoint, i.e. the corresponding projections are not orthogonal. This leads to some mathematical complications, since arbitrary projections do not possess a number of nice properties inherent to orthogonal ones. This difficulty can be easily overcome by a renormalization of the field variables, after which the projections will become orthogonal. The loss of the usual physical meaning of the field variables, is an inevitable price.

Positive definite tensors such as C and $B = C^{-1}$, have uniquely determined self-adjoint square roots C' and B'. Let

$$\tau = B'\sigma = C'\varepsilon\,, \ \mu = B'p = C'm\,. \tag{4.6.36}$$

Then, (4.6.24) is equivalent to one single equation

$$\tau = \Pi\mu\,, \tag{4.6.37}$$

where

$$\Pi = C'KC' = C'\vec{\Pi}B' = B'\overleftarrow{\Pi}C'\,. \tag{4.6.38}$$

It is easy to see that

$$\Pi^2 = \Pi,\ \Pi^+ = \Pi\,, \tag{4.6.39}$$

i.e. we have obtained the desired result.

In terms of the new variables, the expression for the elastic energy takes the form

$$2\Phi = \langle\tau|\tau\rangle = \langle\mu|\Pi|\mu\rangle \tag{4.6.40}$$

and for the internal stress we have

$$\Theta = B'SB' = C'\vec{\Theta}B' = B'\overleftarrow{\Theta}C'\,, \tag{4.6.41}$$

$$\Theta^2 = \Theta,\ \Theta^+ = \Theta\,, \tag{4.6.42}$$

while

$$I = \Pi + \Theta . \tag{4.6.43}$$

Thus, the solution of the problems of elasticity for both external and internal stress is reduced to the construction of the orthogonal projections Π and Θ in the Hilbert space with the scalar product, defined by the elastic energy (4.6.40). The projections were explicitly expressed above by means of the Green's operator G for displacement. It can be shown that the converse is also valid, namely the operator G can be expressed in terms of Π and Θ.

It is essential that if one admits the definition of the operator K in the form of (4.6.14), then all the relations, starting with (4.6.17), remain valid for an arbitrary inhomogeneous elastic medium. In connection with this, a closed system of equations for the projections is of interest. This system will be deduced in Sect. 4.8.

4.7 The Interface of Two Media

Before we proceed to boundary value problems for elastic media with inhomogeneities, let us consider a fundamental local problem about a discontinuity of stress and strain at a point on the interface between two media.

Let two elastic media with tensors of elastic moduli $c_0^{\alpha\beta\lambda\mu}$ and $c^{\alpha\beta\lambda\mu}$ be joined. Assume that, on the (sufficiently smooth) interface Ω between them, the usual conditions of continuity of displacement and of the normal stress vector

$$u_\alpha^0 = u_\alpha, \quad n_\alpha \sigma_0^{\alpha\beta} = n_\alpha \sigma^{\alpha\beta} , \tag{4.7.1}$$

are satisfied. Let us find an expression for the discontinuity of the stress and strain at a point on the interface Ω.

From the continuity of u on Ω, follows the continuity of the tangential component of the tensor ∇u on Ω. Introducing the projection operators onto the normal and onto the tangent plane

$$\pi_{\alpha\beta} = n_\alpha n_\beta, \quad \vartheta_{\alpha\beta} = \delta_{\alpha\beta} - n_\alpha n_\beta , \tag{4.7.2}$$

let us write down this condition in the form

$$\vartheta_\lambda^\varrho \partial_\varrho u_\mu^0 = \vartheta_\lambda^\varrho \partial_\varrho u_\mu . \tag{4.7.3}$$

To transform the second condition (4.7.1) we have

$$\sigma^{\alpha\beta} = c^{\alpha\beta\lambda\mu} \varepsilon_{\lambda\mu} = c^{\alpha\beta\lambda\mu} \partial_\lambda u_\mu , \tag{4.7.4}$$

where replacing $\varepsilon_{\lambda\mu}$ by $\partial_\lambda u_\mu$ is possible due to the symmetry of the tensor of elastic moduli. Decomposing ∇u^0 into the sum of normal and tangent components, we obtain

$$n_\alpha c_0^{\alpha\beta\nu\sigma} \pi_\nu^\varrho \partial_\rho u_\sigma^0 + n_\alpha c_0^{\alpha\beta\nu\sigma} \vartheta_\nu^\varrho \partial_\rho u_\sigma^0 = n_\alpha c^{\alpha\beta\rho\tau} \partial_\rho u_\tau . \tag{4.7.5}$$

Using (4.7.3), we obtain the equation for $n \cdot \nabla u^0$

$$L_0^{\beta\sigma}(n) n^\rho \partial_\rho u_\sigma^0 = L_0^{\beta\sigma}(n) n^\rho \partial_\rho u_\sigma + n_\nu [c^{\nu\beta\rho\tau}] \partial_\rho u_\tau , \tag{4.7.6}$$

where

$$L_0^{\beta\sigma}(n) = c_0^{\alpha\beta\nu\sigma} n_\alpha n_\sigma , \; [c^{\alpha\beta\rho\tau}] = c^{\nu\beta\rho\tau} - c_0^{\nu\beta\rho\tau} . \tag{4.7.7}$$

Applying to both sides of (4.7.6) a matrix $G^0(n)$, which is inverse to $L_0(n)$ [the matrix $G^0(n)$ exists due to the positive definiteness of the elastic energy], we solve (4.7.6) with respect to $n \cdot \nabla u^0$. Then, multiplying (tensorially) the result obtained by n, we find

$$\pi_\lambda^\varrho \partial_\rho u_\mu^0 = \pi_\lambda^\varrho \partial_\rho u_\mu + n_\lambda G_{\mu\beta}^0(n) n_\nu [c^{\nu\beta\rho\tau}] \partial_\rho u_\tau . \tag{4.7.8}$$

Adding this to (4.7.3) and symmetrizing with respect to the indices λ, μ, we have

$$[\varepsilon_{\lambda\mu}] = \varepsilon_{\lambda\mu} - \varepsilon_{\lambda\mu}^0 = - K_{\lambda\mu\nu\alpha}^0 [c^{\nu\sigma\rho\tau}] \varepsilon_{\rho\tau} = - K_{\lambda\mu\nu\sigma} [c^{\nu\sigma\rho\tau}] \varepsilon_{\rho\tau}^0 , \tag{4.7.9}$$

where

$$\begin{aligned} K_{\lambda\mu\nu\sigma}^0(n) &= [n_\lambda G_{\mu\sigma}^0(n) n_\nu]_{(\lambda\mu)\,(\nu\sigma)} , \\ K_{\lambda\mu\nu\sigma}(n) &= [n_\lambda G_{\mu\sigma}(n) n_\nu]_{(\lambda\mu)\,(\nu\sigma)} . \end{aligned} \tag{4.7.10}$$

Using direct notations, one can rewrite (4.7.9), in the form

$$[\varepsilon] = - K_0[C]\varepsilon = - K[C]\varepsilon_0 . \tag{4.7.11}$$

or, equivalently,

$$\varepsilon = (I - K[C])\varepsilon_0, \quad \varepsilon_0 = (I + K_0[C])\varepsilon . \tag{4.7.12}$$

From here it follows in particular that

$$K - K_0 + K[C]K_0 = 0 . \tag{4.7.13}$$

Let us present explicit formulae for the discontinuity $[\varepsilon]$ of the strain on the interface between two isotropic media for the case, when the x^3-axis coincides with the normal:

$$\begin{aligned} &[\varepsilon_{11}] = [\varepsilon_{12}] = [\varepsilon_{22}] = 0 , \\ &[\varepsilon_{13}] = - ([\mu]/\mu_0)\varepsilon_{13}, \; [\varepsilon_{23}] = - ([\mu]/\mu_0)\varepsilon_{23} , \\ &[\varepsilon_{33}] = - \{[\lambda](\varepsilon_{11} + \varepsilon_{22} + \varepsilon_{33}) + 2[\mu]\varepsilon_{33}\}/(\lambda_0 + 2\mu_0) . \end{aligned} \tag{4.7.14}$$

In this simplest case the formulae for $[\varepsilon]$ can be obtained directly, by obvious reasoning. The continuity of the tangent component of the tensor ∇u follows from the first condition in (4.7.1). In the coordinate system chosen, this means the continuity of six components of $\partial_\alpha u_\beta$ with indices 1α and 2β and, hence, of the three components of strain ε_{11}, ε_{12} and ε_{22}. The continuity of stress components σ^{13}, σ^{23} and σ^{33} follows from the second condition in (4.7.1). Using the derived conditions on σ and ε and also Hooke's law for an isotropic medium, we arrive at (4.7.14).

Calculations of such a kind, in a fixed basis, have an essential defect, namely they do not enable one to find an invariant meaning of the coefficients in the formulae for the discontinuities. We shall see how useful it is to have a physical interpretation of these coefficients.

Notice. The formulae obtained for the discontinuity of the strain are a consequence of the interface conditions (4.7.1), but are not completely equivalent to the latter, since they are valid under weaker assumptions. In particular, for the case of the isotropic media considered, the component u^3 can have a discontinuity on the interface that will lead to a δ-function term in ε_{33}. However, this δ-function will not affect the discontinuity in the limiting values of ε_{33}.

It is not difficult to obtain formulae for the discontinuity of the stress from the formulae for the discontinuity in the strain. We present these formulae in a form, which is completely symmetric to (4.7.11, 13),

$$[\sigma] = -S_0[B]\sigma = -S[B]\sigma_0 . \tag{4.7.15}$$

$$\sigma = (I - S[B])\sigma_0, \; \sigma_0 = (I + S_0[B])\sigma \tag{4.7.16}$$

$$S - S_0 + S[B]S_0 = 0 . \tag{4.7.17}$$

Here

$$S(n) = C - CK(n)C , \tag{4.7.18}$$

$$[B] = B - B^0, \; B = C^{-1} , \tag{4.7.19}$$

and similarly for S_0.

Later, we shall need an explicit relation between σ_0 and ε. From (4.7.16) after simple transformations, we obtain

$$\sigma_0^{\alpha\beta} = S_*^{\alpha\beta\lambda\mu}(n)\varepsilon_{\lambda\mu} , \tag{4.7.20}$$

where

$$S_*(n) = C_0 + C_0K_0(n)[C] . \tag{4.7.21}$$

Let us agree that, when considering a defect in an inhomogeneous medium, we shall denote by C and C_0 the moduli of the medium with and without the inhomogeneity, respectively. At the same time, we introduce the notations

$$C = C_0 + C_1, \quad B = B_0 + B_1, \tag{4.7.22}$$

while an evident relation

$$B_1 = (C_0 + C_1)^{-1} - C_0^{-1}. \tag{4.7.23}$$

holds.

Note, that $C_1 = -C_0$ or $B_1 \to \infty$ corresponds to a cavity whereas $C_1 \to \infty$ or $B_1 = -B_0$ corresponds to a rigid inclusion.

In the notations adopted above,

$$S_*(n) = C_0 + C_0 K_0(n) C_1, \tag{4.7.24}$$

while for the cavity, as follows from (4.7.18), $S_*(n) = S_0(n)$.

A critical reader has probably noticed a remarkable correspondence between local coefficients in the formulae for the discontinuities of stress and strain and the Green's functions introduced in the previous section. This correspondence was reflected above in the adopted notations. In fact, the matrix $G_0(n)$ is, by definition, the inverse or the matrix $L_0(n)$, which is given by (4.7.7). Comparison with (4.6.4 and 5) shows that $G_0(n)$ coincides with the value of the Green's function $G_0(k)$ of the homogeneous medium on the unit sphere in k-space. As regards $K_0(n)$ and $S_0(n)$, they coincide exactly with the Green's functions of the homogeneous medium for strain and stress in the k representation. This correspondence will be used below, in the problem about an ellipsoidal inclusion.

Problem 4.7.1. Verify that the analogs of (4.6.19, 30) are particular cases of (4.7.13, 17). (Hint: Set formally $C = -C_0$.)

Problem 4.7.2. Show that (4.7.13, 17) are equivalent to the relations

$$\vec{\Theta}_0(n)\vec{\Pi}(n) = 0, \quad \overleftarrow{\Pi}_0(n)\,\overleftarrow{\Theta}(n) = 0 \tag{4.7.25}$$

or

$$\Pi_0(n) B_0' C' \Theta(n) = 0, \quad \Theta_0(n) C_0' B' \Pi(n) = 0 \tag{4.7.26}$$

where the projection operators are defined by (4.6.20, 31, 38 and 41).

4.8 Integral Equations for an Inhomogeneous Medium

In this section we shall obtain integral equations for inhomogeneous elastic media in terms of stress and strain. As we shall see this form of the equations is the most convenient one for investigating problems about inclusions in a homogeneous medium.

Let us consider an unbounded elastic medium described by the elastic moduli tensor $C(x)$, which is representable in the form

$$C(x) = C_0 + C_1(x)\,, \tag{4.8.1}$$

where C_0 is a constant tensor and $C_1(x)$ is a perturbation. It is assumed that $C(x)$ and C_0 are positive definite and $C_1(x) \to 0$ as $|x| \to \infty$.

It is also convenient to introduce the tensor of elastic compliance $B(x)$, connected with $C(x)$ by the relations

$$B(x) = B_0 + B_1(x)\,, \tag{4.8.2}$$

$$B(x) = C^{-1}(x), \quad B_0 = C_0^{-1}\,. \tag{4.8.3}$$

It is obvious that $B(x)$ and B_0 are positive definite, too.

The cases of a cavity and a rigid inclusion are to be described separately, since, for them, $C(x)$ and $B(x)$ are not positive definite. In these cases, generally speaking, additional assumptions are needed, which ensure the uniqueness of a solution of the corresponding equations. To avoid this we consider these cases only as limiting ones: for a cavity V

$$C_1(x) \to -C_0, \quad B_1(x) \to \infty \qquad (x \in V) \tag{4.8.4}$$

and for a rigid inclusion,

$$C_1(x) \to \infty, \quad B_1(x) \to -B_0 \qquad (x \in V). \tag{4.8.5}$$

The equations of elasticity in displacements are written in the form

$$Lu = (L_0 + L_1)u = q\,, \tag{4.8.6}$$

where

$$L_0 = -\nabla C_0 \nabla, \quad L_1 = -\nabla C_1 \nabla\,. \tag{4.8.7}$$

The quantity q denotes an external force, decreasing sufficiently rapidly as $|x| \to \infty$. For these equations to have a unique solution, the conditions at infinity must be added.

We shall understand (4.8.6) in the sense of generalized functions, so that we can consider also piecewise continuous functions $C_1(x)$. The displacement $u(x)$ and the normal stress vector $n \cdot \sigma(x)$ will be automatically continuous on the surface Ω, where $C_1(x)$ has a discontinuity, if $q(x)$ does not contain singularities of the type of the single or double layer on Ω, the latter being hereafter assumed. As is shown in potential theory, under the conditions pointed out above, the (generalized) solution of (4.8.6) exists and is unique.

In order to proceed to the integral equation, let us first introduce additional restrictions, which we shall drop later. Suppose that $C_1(x)$ be a continuous function with bounded support, $q(x)$ be a piecewise continuous function and the required solution be in the class of functions which tend to zero at infinity. Let us apply the Green's operator G_0 of the homogeneous medium to both sides of (4.8.6). We have

$$u - G_0 \nabla C_1 \nabla u = u_0 , \tag{4.8.8}$$

where $u_0 = G_0 q$ is an external field which vanishes at infinity, i.e. the field which would be present in the homogeneous medium under the action of the force q. By virtue of our assumptions, $u(x)$ and $u_0(x)$ are continuously differentiable functions.

Let us apply the operator def to both sides of (4.8.8). Taking into account the symmetry of C_1, we obtain the equation for the strain $\varepsilon = \operatorname{def} u$

$$\varepsilon + K_0 C_1 \varepsilon = \varepsilon_0 , \tag{4.8.9}$$

where $\varepsilon_0 = \operatorname{def} u_0$ is an external strain field and

$$K_0 = -\operatorname{def} G_0 \operatorname{def} \tag{4.8.10}$$

is the Green's operator for strain in the homogeneous medium, which was introduced in Sect. 4.6.

Problem 4.8.1. Show that the equation for stress

$$\sigma + S_0 B_1 \sigma = \sigma_0 , \tag{4.8.11}$$

follows from (4.8.9), where $\sigma = C\varepsilon$, $\sigma_0 = C_0 \varepsilon_0$ and

$$S_0 = C_0 - C_0 K_0 C_0 \tag{4.8.12}$$

is the Green's operator for internal stress introduced in Sect. 4.6, see (4.6.29).

Let us consider now, how far one can extend the field of applicability of (4.8.9, 11). We set temporarily

$$\varepsilon = \varepsilon_0 + \varepsilon_1 \tag{4.8.13}$$

where ε_1 is a perturbed field, caused by the inhomogeneity $C_1(x)$, satisfying

$$\varepsilon_1 + K_0 C_1 \varepsilon_1 = -K_0 C_1 \varepsilon_0 . \tag{4.8.14}$$

The right hand side contains only the restriction ε_0^+ of the external field ε_0 to the support V of the function $C_1(x)$ with bounded support. Thus ε_1 will not change, if one extends ε_0^+ outside V in any (smooth) way. Therefore, the external field ε_0 appearing in (4.8.9) can satisfy any conditions at infinity, $\varepsilon \to \varepsilon_0$ as $|x| \to \infty$ being automatically satistied. The same conclusion is also valid for equation (4.8.11).

Let us now assume that $C_1(x)$ has a discontinuity on a certain surface Ω, and that ε_0 is continuous on Ω. Then, by virtue of the properties of the operator K_0 (Appendix A.3) the solution also has a discontinuity, which satisfies the conditions (4.7.11), i.e. the necessary conditions at the interface are fulfilled. This enables us to extend (4.8.9) to piecewise continuous $C_1(x)$'s, in particular, to functions of the type of a product of a constant tensor and the characteristic function of a region. Consequently, (4.8.11) can also be extended to piecewise continuous functions $B_1(x)$.

Let us proceed to the extension of (4.8.9, 11) to the case of functions $C_1(x)$ and $B_1(x)$ of the oscillatory type. We confine ourselves to the most interesting case, when $C_1(x)$ and $B_1(x)$ are almost periodic functions, which are representable as series of exponentials with arbitrary wave vectors. Let the external fields be functions of the same type. Then, the solutions of the equations and, consequently, the products $C_1\varepsilon$ and $B_1\varepsilon$ belong to the same class of almost-periodic functions. Thus, we arrive at a problem of extending the operators K_0 and S_0 to functions of the above-mentioned type.

Since the series are understood in the sense of generalized functions, the action of the operators on exponentials can be taken to be termwise. The action of an operator on an exponential with a nonzero wave vector is written in the k representation in the form of a product of the corresponding homogeneous function and a δ-function with support at a nonzero point. This is a well-defined operation. Hence it remains only to complete the definition of the action of the operators on constants, which corresponds to multiplying the δ-function concentrated at the zero point by homogeneous functions, the latter being a quite nonunique operation.

In order to understand how to overcome this difficulty, let us consider a model problem. Let an "inhomogeneity" be given by the tensors C_1 or B_1 which are constant throughout the whole space. Assume first that what is really given is $\sigma_0 = $ const. Then it is obvious that

$$\sigma = \sigma_0, \; \varepsilon = B\sigma_0 . \tag{4.8.15}$$

It is natural to require that (4.8.9, 11) give the same result. This is possible only if K_0 and S_0 act on constants as follows:

$$\int K_0(x - x')\, dx' = B_0, \; \int S_0(x - x')\, dx' = 0 , \tag{4.8.16}$$

while, according to (4.8.12), each of the relations is a consequence of another.

Analogously, if one considers an external field of strain to be really given, then K_0 and S_0 are to act on constants as follows:

$$\int K_0(x - x')\, dx' = 0, \int S_0(x - x')\, dx' = C_0 . \qquad (4.8.17)$$

We emphasize once more that such an extension of the definition is equivalent to certain additional physical conditions.[3] There is no "canonical regularization" of the operators K_0 and S_0.

Note that (4.8.9, 11) obviously remain valid, if one assumes the external medium to be also inhomogeneous, i.e. $C_0 = C_0(x)$. Here, of course, the operators K_0 and S_0 will not be translationaly invariant.

Let us, finally, remove the last restriction. The external field is not to be caused by force sources only. Sources of internal stress may also contribute to it. In other words, no restrictions such as Rot $\varepsilon_0 = 0$ or div $\sigma_0 = 0$, are imposed on the external field. At the same time, from (4.6.17) and (4.6.33), it follows that the conditions

$$\text{Rot}\, \varepsilon = \text{Rot}\, \varepsilon_0, \ \text{div}\, \sigma = \text{div}\, \sigma_0 \qquad (4.8.18)$$

are automatically fulfilled.

Thus, we have shown that the integral equations in the form of (4.8.9, 11) have meaning for a rather wide class of inhomogeneities and external fields, i.e. under the indicated conditions solutions of the equations exist, are unique and coincide with the solutions of the differential equations of elasticity with the corresponding conditions at infinity.

The solutions of (4.8.9, 11) can be represented in a form, which is symmetric to that of the original equations

$$\varepsilon = \varepsilon_0 - KC_1\varepsilon_0 , \qquad (4.8.19)$$

$$\sigma = \sigma_0 - SB_1\sigma_0 , \qquad (4.8.20)$$

where K and S are naturally interpreted as the corresponding Green's operators for a medium with characteristics C or B while $-C_1$ and $-B_1$ can be considered as "perturbations."

Let us emphasize the correspondence between the pairs of integral equations (4.8.9, 19) and (4.8.11, 20) on the one hand and the pairs of local relations at the discontinuities (4.7.12, 16) on the other hand. This correspondence is not accidental. The conditions for the jumps can be obtained from the integral equations, making use of properties of the Green's operators. Conversely, the conditions for the discontinuities imply a natural form of the integral equations.[4]

---

[3]There is an analogy between these asymptotic conditions and the first and second boundary value problems.

[4]One can say that there is a remarkable finite-dimensional representation of the global Green's and projection operators by local ones.

In particular, the global analogs of (4.7.13, 17) are

$$K - K_0 + KC_1K_0 = 0\,, \tag{4.8.21}$$

$$S - S_0 + SB_1S_0 = 0\,. \tag{4.8.22}$$

Problem 4.8.2. Show that (4.6.19, 30) are consequences of these relations.

The fundamental equations for the projection operators

$$\vec{\Theta}_0\vec{\Pi} = 0,\ \vec{\Pi}_0\vec{\Theta} = 0 \tag{4.8.23}$$

are the global analog of (4.7.25). One can prove that these equations together with the relations

$$C\vec{\Pi} = \vec{\Pi}C, \quad C\vec{\Theta} = \vec{\Theta}C \tag{4.8.24}$$

$$\vec{\Pi}^{+} = \vec{\Pi},\ \vec{\Theta}^{+} = \vec{\Theta} \tag{4.8.25}$$

and the analogous ones for $\vec{\Pi}_0$, $\vec{\Theta}_0$, determine the projections $\vec{\Pi}$ and $\vec{\Theta}$ uniquely, if the projections $\vec{\Pi}_0$ and $\vec{\Theta}_0$ are given. For the latter, one can take the standard projections Π_0, Θ_0, which correspond to the case $C_0 = I$. Their explicit expression is presented in (B.7). Then (4.8.23–25) uniquely determine the Green's operators and, hence the solutions of the corresponding problems for an inhomogeneous medium.

Problem 4.8.3. Verify that the system (4.8.23–25) is equivalent to the system

$$\Pi_0B_0'C'\Theta = 0,\ \Theta_0C_0'B'\Pi = 0\,, \tag{4.8.26}$$

see (4.7.26).

In conclusion, let us consider briefly the case, when $C_1(x)$, and consequently, $B_1(x)$ have a bounded support (a local inhomogeneity). Let $V(x)$ be the characteristic function of a bounded region V occupied by the inhomogeneity, and $\bar{V}$ be the complement of V. Let us denote by $+(-)$ the restriction of a field to the region $V\,(\bar{V})$ and introduce the operators

$$K_0^{(+)} = VK_0V,\ K_0^{(-)} = \bar{V}K_0V\,, \tag{4.8.27}$$
$$S_0^{(+)} = VS_0V,\ S_0^{(-)} = \bar{V}S_0V\,,$$

where V and $\bar{V}$ denote the operators of multiplication by the characteristic functions $V(x)$ and $1 - V(x)$.

It is obvious that (4.8.9) is equivalent to the pair of equations

$$\varepsilon^+ + K_0^{(+)}C_1\varepsilon^+ = \varepsilon_0^+ , \tag{4.8.28}$$

$$\varepsilon^- = \varepsilon_0^- - K_0^{(-)}C_1\varepsilon^+ \tag{4.8.29}$$

where the first equation uniquely determines the solution inside V and the second uniquely determines its continuation to $\bar{V}$. Analogously, for (4.8.11), we have

$$\sigma^+ + S_0^{(+)}B_1\sigma^+ = \sigma_0^+ , \tag{4.8.30}$$

$$\sigma^- = \sigma_0^- - S_0^{(-)}B_1\sigma^+ . \tag{4.8.31}$$

The correctness of such a representation of the solutions is justified by the analysis carried out above. In fact, under appropriate conditions on external fields, the solutions do not have singularities on the boundary Ω of the region V and admit the restriction operation. Recall that the functions $C_1(x)$ and $B_1(x)$ can have discontinuities on Ω.

Thus, for local inhomogeneities, the problem is in principle reduced to solving the integral equations (4.8.28) or (4.8.30) inside the region V. We shall see below that these equations can be effectively solved for ellipsoidal inhomogeneities and their limiting cases.

Problem 4.8.4. Show that, for a cavity, (4.8.9, 11) take the form, using (4.8.4),

$$\varepsilon - K_0C_0\varepsilon^+ = \varepsilon_0 , \tag{4.8.32}$$

$$\sigma + S_0\varepsilon^+ = \sigma_0 \tag{4.8.33}$$

and for a rigid inclusion, (4.8.9, 11) take the form, using (4.8.5),

$$\varepsilon + K_0\sigma^+ = \varepsilon_0 , \tag{4.8.34}$$

$$\sigma - S_0B_0\sigma^+ = \sigma_0 . \tag{4.8.35}$$

Note that, in the case of a cavity, ε^+ is finite and $\sigma^+ = 0$. Conversely, for a rigid inclusion, σ^+ is finite and $\varepsilon^+ = 0$.

Let us now consider a field outside the inhomogeneity, e.g. for definiteness the stress field. According to (4.8.31), for the perturbed field $\sigma_1(x)$, when $x \in \bar{V}$, we have

$$\sigma_1(x) = -\int S_0(x - x')m(x')\,dx' \tag{4.8.36}$$

where the function

$$m(x) = B_1(x)\sigma^+(x) \tag{4.8.37}$$

is determined by the solution of (4.8.30). It is convenient to consider the continuation of $m(x)$ outside the region V as being equal to zero.

From (4.6.33) it follows that the field $\sigma_1(x)$ satisfies the condition div $\sigma_1 = 0$ and can be interpreted as a field of internal stress. In the next chapter, we shall see that $m(x)$ then plays the role of the density of dislocation moments.

The asymptotics of the perturbed field, as $|x| \to \infty$, is obtained, if one replaces $m(x)$ by a δ-function source with the dislocation moment

$$M = \int B_1(x)\sigma^+(x)\,dx\,. \tag{4.8.38}$$

In this sense, one can speak about a point inhomogeneity in an elastic medium. However, as distinct from the case of the elastic quasicontinuum, it would here be incorrect to introduce a δ-function directly into the elastic moduli or compliances, since the corresponding equations, as one can easily verify, have no solutions. In connection with this, a model of a point defect in an elastic medium requires a delicate treatment. In particular, the corresponding dislocation moment (4.8.38) cannot be given a priori, but is to be obtained from the solution of the problem for a finite inhomogeneity. In other words, a point defect "knows" about its origin, as distinct from a point defect in a quasicontinuum (Sect. 4.3).

4.9 Ellipsoidal Inhomogeneity

In this section we consider a case which is important for applications, namely

$$C_1(x) = C_1 V(x)\,, \tag{4.9.1}$$

where $C_1 = \text{const}$ and $V(x)$ is a characteristic function of an ellipsoidal domain V. We show that the solutions of the most interesting problems for such ellipsoidal inhomogeneities can be obtained explicitly. This fact is based on the remarkable property of polynomial conservation (the p-property) for the ellipsoidal domain.

Theorem. If the external field $\varepsilon_0(x)$ in the neighborhood of V is a polynomial of degree m, then the field $\varepsilon^+(x)$ induced within V is a polynomial of the same degree.

Proof. Since the statement of the theorem is invariant with respect to nondegenerate linear transformations, it is possible without loss of generality to consider V to be the unit sphere. As usual, we suppose that $C_0 + C_1\,V(x)$ is the positive-definite tensor, i.e. the limiting cases of a cavity, and a rigid inclusion is excluded for the moment. Then the operator $I + K_0^+ C_1$ in (4.8.28) has an inverse, and to prove the theorem it is sufficient to show that a polynomial within V is transformed by the operator $K_0^{(+)}$ into a polynomial of the same degree.

It follows from the definition (4.8.27) of the operator $K_0^{(+)}$ and from the property (C.1) of the generalized function $K_0(x)$ that $K_0^{(+)}$ is decomposed into two parts: the first one is proportional to the identity operator and the second one is determined by the regularized part of $K_0(x)$, an arbitrary region contained in the regularization being naturally identified with V. As a result, it is sufficient to consider the regularized part only. The latter acts on a function $\varphi(x)$ which is smooth within V according to the rule

$$\varphi(x) \to \psi(x) = \int_V K_0(x - x')[\varphi(x') - \varphi(x)]\, dx', \quad x \in V, \tag{4.9.2}$$

where $K_0(x)$ is the formal second derivative of $G_0(x)$ defined by (4.6.13). Taking into account that $G_0(x) \sim |x|^{-1}$ and $G_0(-x) = G_0(x)$ we can rewrite (4.9.2) in the form

$$\psi(x) = \int_V \frac{g_0(n)}{r^3}[\varphi(x') - \varphi(x)]dx', \quad x \in V \tag{4.9.3}$$

where

$$r = |x - x'|, \quad n = \frac{x' - x}{r} \tag{4.9.4}$$

and $g_0(n)$ is a tensor function of fourth order, even in the argument n.

Let us consider the action of this operator on a homogeneous polynomial or, equivalently, on the monomial $x^{(m)} = x^{\lambda_1} \dots x^{\lambda_m}$ where (m) is the tensorial multi-index. Set

$$F^{(m)}(x) = \int_V \frac{g(n)}{r^3}(x'^{(m)} - x^{(m)})\, dx', \quad x \in V. \tag{4.9.5}$$

Substituting from (4.9.4) $x' = x + rn$, we have

$$F^{(m)}(x) = \sum_{k=1}^{m} x^{(m-k)} J^{(k)}(x), \tag{4.9.6}$$

where

$$J^{(k)}(x) = \int_V g_0(n) n^{(k)} r^{k-3}\, dx', \quad x \in V. \tag{4.9.7}$$

Let us set $dx' = r^2 dr d\omega$ and integrate first over an elementary cone $d\omega(n)$ with vertex at the point $x \in V$. Then

$$J^{(k)}(x) = \frac{1}{k}\int_\omega g_0(n) n^{(k)} \rho^k(x, n) d\omega, \quad x \in V, \tag{4.9.8}$$

where

$$\rho(x, n) = -x \cdot n + \sqrt{1 - x \cdot x + (x \cdot n)^2} \tag{4.9.9}$$

is the distance from the point x in the direction n to the boundary of the unit sphere V. For ρ^k we have

$$\rho^k(x, n) = \sum_{l=0}^{k} (-1)^l C_k^l (x \cdot n)^l [\sqrt{1 - x \cdot x + (x \cdot n)^2}]^{k-l}. \tag{4.9.10}$$

Since $g_0(n) = g_0(-n)$ the contribution to $J^{(k)}(x)$ is due only to terms whose product with $n^{(k)}$ is even. It is readily seen that they contain the radical in an even degree and have the form

$$\sum_l A_{(k-2l)} x^{(k-2l)} .$$

As a consequence, $J^{(k)}(x)$ is a polynomial in x of degree k

$$J^{(k)}(x) = \sum_l B^{(k)}_{(k-2l)} x^{(k-2l)}, \quad x \in V, \tag{4.9.11}$$

where

$$B^{(k)}_{(k-2l)} = \frac{1}{k} \int_\omega g_0(n) n^{(k)} A_{(k-2l)}(n) \, d\omega \tag{4.9.12}$$

is a constant tensor.

Substituting (4.9.11) into (4.9.6) we find that $F^{(m)}(x)$ is a polynomial of degree m. Finally, it is easy to show that the term $x^{(m)}$ has a nonzero coefficient. Q.E.D.

Note 1. To extend the theorem to the cases of a cavity and a rigid inclusion we have to make a further assumption. It follows from the proof that for the cavity ($C_1 = -C_0$) a weaker statement is valid: the operator $I - K_0^{(+)} C_0$ transforms a polynomial $\varepsilon^+(x)$ into a polynomial $\varepsilon_0^+(x)$ of the same degree but equation (4.8.28) is not solvable for all $\varepsilon_0^+(x)$. This means that the theorem is valid if and only if conditions of solvability are fulfilled. It is essential that for the most important case, $\varepsilon_0^+ = \text{const}$, (4.8.28) always admits a solution as will be shown below. The extension of the theorem to the case of a rigid inclusion is performed in the same way but the corresponding field variables will be stresses satisfying (4.8.30).

Note 2. It follows from the proof that even (odd) $\varepsilon_0(x)$ corresponds to even (odd) $\varepsilon^+(x)$. In particular, if $\varepsilon_0(x)$ is linear then $\varepsilon^+(x)$ is also linear.

Note 3. Using the Green's functions for homogeneous elliptic operators of the order $2p$ [4.2] it is easy to generalize the theorem to the case $2p < n$ where n is the spatial dimension, n being arbitrary for $p = 1$.

Note 4. It is most likely that the property of polynomial conservation is valid only for the ellipsoid, i.e. that this is a characteristic property of the ellipsoid. However, a proof of this statement is not known to the author.

Before considering the consequences of the theorem let us introduce relations which are useful for what follows. Let us write the equation for the ellipsoid Ω in the form

$$xa^{-2}x = x^{\alpha}(a^{-2})_{\alpha\beta}\, x^{\beta} = 1 \,. \tag{4.9.13}$$

If the Cartesian coordinate system is connected to the ellipsoid semi-axes a_1, a_2, a_3, then the tensor $a_{\alpha\beta}$ is diagonal, the semi-axes being the diagonal components.

Problem 4.9.1. Verify the relations

$$x = \frac{a^2 n}{\sqrt{na^2 n}}, \qquad n = \frac{a^{-2}x}{\sqrt{xa^{-4}x}} \,. \tag{4.9.14}$$

These formulae establish a one-to-one correspondence between a function $f(x)$ on the ellipsoid Ω and a function $f(n)$ on the unit sphere $n^2 = 1$. In particular,

$$\rho(x) \equiv (xa^{-4}x)^{-1/2} \leftrightarrow \rho(n) = (na^2 n)^{-1/2} \,. \tag{4.9.15}$$

We define the mean value of $f(n)$ over the unit sphere $n^2 = 1$ by the expression

$$\langle f(n) \rangle = \frac{\det a}{4\pi} \int f(n)\, \rho^3(n)\, dn \,. \tag{4.9.16}$$

Problem 4.9.2. Show that $\langle f(n) \rangle$ coincides with the mean value of $f(x)$ over the ellipsoid Ω defined by the expression

$$\langle f(x) \rangle = \frac{1}{4\pi \det a} \int_{\Omega} \frac{f(x)}{\rho(x)}\, dx \,, \tag{4.9.17}$$

which permits one to interpret $\langle f(n) \rangle$ as the mean value over the ellipsoid Ω.

Now, let us consider consequences of the theorem. Let the external field ε_0(or σ_0) be homogeneous, i.e. constant. It follows from the theorem that in this case $\varepsilon^+ = \text{const}$ and $\sigma^+ = \text{const}$. To calculate them explicitly we use (4.8.28) and (4.8.30), rewriting them in the form

$$(I + K_0^{(+)}\, C_1)\varepsilon^+ = \varepsilon_0^+ \,, \tag{4.9.18}$$

$$(I + S_0^{(+)}\, B_1)\sigma^+ = \sigma_0^+ \,. \tag{4.9.19}$$

It is seen from here that we need to know the action of the operators $K_0^{(+)}$ and $S_0^{(+)}$ on constants. Let us recall that these operators are represented in the form

$$K_0^{(+)} = AI + \tilde{K}_0^{(+)}, \ S_0'^{(+)} = DI + \tilde{S}_0'^{(+)}, \tag{4.9.20}$$

where A and D are constant tensors and the tilde denotes the regularized parts; these and the constants A and D depend on the choice of the regularization region. In the case under consideration it is natural to identify the regularization region with the ellipsoid. Then we have for the constants A and D (Appendix C)

$$A = \frac{1}{4\pi}\int K_0(a^{-1}\omega)\, d\omega, \ D = \frac{1}{4\pi}\int S_0(a^{-1}\omega)\, d\omega, \tag{4.9.21}$$

where the integration is carried out over the unit sphere $\omega^2 = 1$. Introducing the new variable

$$n = \frac{a^{-1}\omega}{|a^{-1}\omega|} \tag{4.9.22}$$

and taking into account that $K_0(k)$ and $S_0(k)$ are homogeneous functions of zero degree, we find

$$A = \langle K_0(n)\rangle, \ D = \langle S_0(n)\rangle . \tag{4.9.23}$$

It is readily seen from (4.9.2) that in the case $\varepsilon^+(x) = \text{const}$ the contribution of the operator $\tilde{K}_0^{(+)}$ is equal to zero. Hence, the action of $K_0^{(+)}$ on a constant results in multiplication of this constant by A. In the same way the action of $S_0^{(+)}$ results in multiplication by D. Finally, we have

$$(I + AC_1)\varepsilon^+ = \varepsilon_0^+, \tag{4.9.24}$$

$$(I + DB_1)\sigma^+ = \sigma_0^+ . \tag{4.9.25}$$

Thus, to find ε^+ and σ^+ we must invert the corresponding constant matrices. It can be shown that the inverse matrices always exist including the limiting cases of the cavity and the rigid inclusion.

Note that (4.9.23) permits one to rewrite the equations for ε^+ and σ^+ in the form

$$\langle I + K_0(n)C_1\rangle\varepsilon^+ = \varepsilon_0^+, \tag{4.9.26}$$

$$\langle I + S_0(n)B_1\rangle\sigma^+ = \sigma_0^+ . \tag{4.9.27}$$

A comparison with the discontinuity relations (4.7.12, 16) shows that the coefficients in the equations admit an interesting interpretation: they are equal to the mean values over the ellipsoid of the coefficients in the corresponding relations for jumps.

Now let us consider the case for which the external field is a linear function of x

$$\varepsilon^0_{\alpha\beta}(x) = b_{\alpha\beta\lambda}x^\lambda, \quad x \in V. \tag{4.9.28}$$

It follows from the theorem that

$$\varepsilon^+_{\alpha\beta}(x) = d_{\alpha\beta\lambda}x^\lambda, \tag{4.9.29}$$

the tensors b and d being linearly dependent. It is convenient to write this dependence in the form

$$b_{\lambda_1\lambda_2\lambda_3} = 3\Lambda_{\lambda_1\lambda_2\lambda_3}^{\cdot\;\cdot\;\cdot\;\mu_1\mu_2\mu_3}\, d_{\mu_1\mu_2\mu}\,(a^2)^{\mu}_{\cdot\mu_3}. \tag{4.9.30}$$

To find the tensor Λ let us multiply (4.9.18) by x and integrate over V. In a manner analogous to the case of the homogeneous field we obtain

$$\Lambda = \langle \rho(n)n(I + K_0(n)C_1)n\rho(n)\rangle. \tag{4.9.31}$$

The solution of the problem for linear stresses can be deduced in the same way.

If the external field is a polynomial of the second or higher degree, then the p-property permits one to find from (4.9.18, 19) recurrent algebraic equations which connect the coefficients of the polynomials $\varepsilon_0(x)$ and $\varepsilon^+(x)$ or $\sigma_0(x)$ and $\sigma^+(x)$. However, in contrast to the case of the homogeneous field these equations are solvable only under additional conditions. The explanation is as follows: formally there exist external fields (they do not satisfy the equilibrium condition $\operatorname{div} \sigma_0 = 0$) which are not perturbed by the ellipsoidal inhomogeneity. For example, in the isotropic case this property has an external field of the type $\varepsilon_{\alpha\beta} = (c\cdot x)\delta_{\alpha\beta}$ where c is an arbitrary constant vector. As a consequence the matrix Λ can be inverted only on the corresponding subspace.

Let us consider one more problem which is of interest for applications. Let $\sigma(x)$ be the limiting value, from outside, of the stress at the point $x \in \Omega$. As above, instead of $\sigma(x)$ we consider the function $f(n)$ of the normal n. The dependence $\sigma(n)$ upon the external field $\sigma_0(x)$ can be represented in the form

$$\sigma^{\alpha\beta}(n) = F^{\alpha\beta}_{\cdot\cdot\lambda\mu}(n)\,\sigma_0^{\lambda\mu}, \tag{4.9.32}$$

where $F(n)$ is an operator concentration coefficient. For polynomial external fields the operator $F(n)$ can be constructed in an explicit form. For the sake of simplicity we confine ourselves to the case of a homogeneous external field.

It is convenient to construct the operator $F(n)$ in two steps. According to (4.7.20) we have on the boundary Ω of the ellipsoidal inhomogeneity

$$\sigma(n) = S_*(n)\varepsilon^+, \tag{4.9.33}$$

where in this case $\varepsilon^+ = \text{const}$ and

$$S_*(n) = C_0 + C_0 K_0(n) C_1 . \tag{4.9.34}$$

Using (4.9.27), we have

$$\langle S_*(n) \rangle \varepsilon^+ = \sigma_0^+ \tag{4.9.35}$$

and consequently

$$F(n) = S_*(n) \langle S_*(n) \rangle^{-1} . \tag{4.9.36}$$

In particular, for the cavity $C_1 = -C_0$ and $S_*(n) = S_0(n)$. Taking into account (4.9.23) we have

$$F(n) = S_0(n) D^{-1} . \tag{4.9.37}$$

It is interesting to note that this problem as well as the previous ones, were solved for an arbitrary anisotropy without use of the explicit form for the Green's tensor $G_0(n)$. It is sufficient to know its Fourier transform $G_0(k)$. In other words, we find stress concentrations even though we have not been able to solve the corresponding boundary value problem. This is a direct consequence of the property of polynomial conservation.

The representation for $F(n)$ obtained above is especially convenient for the investigation of the dependence of the concentration coefficient on the shape of the ellipsoid. Indeed, the local dependence of $F(n)$ on the normal to the ellipsoid (or equivalently, on the point) is given by the first factor $S_*(n)$, which is independent of the ellipsoid parameters and always remains finite. The second factor $\langle S_*(n) \rangle^{-1}$ does not depend on the point and becomes infinite in the case of the passage to the limit of the elliptical crack.

It follows from (4.9.16) that the entire dependence of $\langle S_*(n) \rangle$ on the ellipsoid parameters is concentrated in the scalar weight factor $\rho(n)$. This is essential since it permits a complete investigation of the stress concentration in the limiting cases of an ellipsoidal crack and a needle for an arbitrary anisotropic medium by a series expansion of $\rho(n)$ for small parameter.

Note the interesting relationships

$$\langle F(n) \rangle = I, \ \langle \sigma(n) \rangle = \sigma_0 \tag{4.9.38}$$

which immediately follow from (4.9.36) and (4.9.32), respectively.

In conclusion, we present some results for the ellipsoidal cavity in an isotropic medium.

The tensors A and D in this case have the symmetry of the ellipsoid and are defined by nine essential components. We have in the coordinate system connected to the ellipsoid axes

$$A_{1111} = \kappa_0\,[3I_{11} + (1 - 4\nu_0)I_1],\; A_{1122} = \kappa_0(I_{21} - I_1)\,,$$
$$A_{1212} = \frac{\kappa_0}{2}\,[I_{21} + I_{12} + (1 - 2\nu_0)\,(I_1 + I_2)]\,, \tag{4.9.39}$$

$$D^{1111} = \gamma_0\,[1 - \frac{1}{8\pi}\,(3I_{11} + I_1)]\,,$$
$$D^{1122} = \gamma_0\left\{\nu_0 - \frac{1}{16\pi}[I_{21} + I_{12} - (1 - 4\nu_0)(I_1 + I_2)]\right\},$$
$$D^{1212} = \gamma_0\left\{\frac{1 - \nu_0}{2} - \frac{1}{16\pi}[I_{21} + I_{12} + (1 - 2\nu_0)(I_1 + I_2)]\right\}, \tag{4.9.40}$$

where

$$\kappa_0 = \frac{1}{16\pi\mu_0(1 - \nu_0)},\; \gamma_0 = \frac{2\mu_0}{1 - \nu_0}\,, \tag{4.9.41}$$

the quantitics

$$I_p = \frac{3}{2}\,v\int_0^\infty \frac{d\xi}{(a_p^2 + \xi)\Delta(\xi)}\,,\quad I_{pq} = \frac{3}{2}\,va_p^2\int_0^\infty \frac{d\xi}{(a_p^2 + \xi)(a_q^2 + \xi)\Delta(\xi)}$$

$$(\Delta(\xi) = \sqrt{(a_1^2 + \xi)(a_2^2 + \xi)(a_3^2 + \xi)}\,;\; p,\, q = 1,\, 2,\, 3) \tag{4.9.42}$$

are expressed in terms of elliptic integrals (v is the ellipsoidal volume). The remaining six tensor components are obtained by a cyclic replacement of the indices 1, 2, 3.

Note that in the particular case of an ellipsoid of revolution only six essential components remain and the formulas do not contain elliptic integrals.

In the case of a cavity according to (4.9.37) we need expressions for $S_0(n)$ and D^{-1}. In the basis E^i (Appendix A)

$$S_0(n) = \gamma_0[\nu_0(E^2 - E^3 - E^4) + (1 - \nu_0)(E^1 - 2E^5) + E^6]\,. \tag{4.9.43}$$

As is customary, in order to evaluate D^{-1} we represent the tensor D of fourth order as a sixth-order matrix. The latter is decomposed into two blocks of third-order matrices. The components $D^{\lambda\lambda\mu\mu}$ form a symmetric martix, and the doubled components $D^{\lambda\mu\lambda\mu}$ $(\lambda \neq \mu)$ a diagonal matrix; consequently

$$(D^{-1})_{\lambda\mu\lambda\mu} = \frac{1}{4D^{\lambda\mu\lambda\mu}}\;(\lambda \neq \mu)\,. \tag{4.9.44}$$

Let us consider two cases. First, let the external field $\sigma_0^{\alpha\beta}$ be the tension along the coordinate axes, i.e. only the diagonal components $\sigma_0^{\lambda\lambda}$ are nonzero. Then the stress concentration is defined by the components

$$F^{\alpha\beta}_{\cdot\cdot\lambda\lambda}(n) = \sum_{\mu=1}^{3} S_0^{\alpha\beta\mu\mu}(n)(D^{-1})_{\mu\mu\lambda\lambda} \,. \tag{4.9.45}$$

Secondly, let the external field $\sigma_0^{\alpha\beta}$ be pure shear. Thd stress concentration is determined by the components

$$F^{\alpha\beta}_{\cdot\cdot\lambda\mu}(n) = \frac{S_0^{\alpha\beta\lambda\mu}(n)}{2D^{\lambda\mu\lambda\mu}} \quad (\lambda \neq \mu) \,. \tag{4.9.46}$$

The formulas obtained permit one to investigate the stress concentration over the whole surface of the inhomogeneity. This is essential because the knowledge of the concentration coefficient just at the vertices of the ellipsoid is not sufficient and can result in incorrect qualitative deductions. For example, for pure shear with the components σ_0^{13}, the stress tensor $\sigma^{\alpha\beta}$ is zero at the vertex (1, 0, 0), but there is a stress concentration in direct proximity to this vertex. Hence a complete investigation of the stress on the whole ellipsoid surface is necessary, and not only at its characteristic points. Such an investigation will be given in the next section for limiting cases of the ellipsoidal inhomogeneity.

4.10 Ellipsoidal Crack and Needle

It is convenient to introduce the dimensionless parameters of an ellipsoid

$$\alpha = \frac{a_2}{a_1}, \quad \varepsilon = \frac{a_3}{a_1}, \quad \xi = \frac{a_3}{a_2} \tag{4.10.1}$$

where $a_1 \geq a_2 \geq a_3$ are the ellipsoid's semiaxes. The case $\alpha \ll 1$, $\xi \sim 1$ corresponds to the needle, while $\xi \ll 1$, $\alpha \sim 1$ to the crack, and $\varepsilon \ll \alpha \ll 1$ to a narrow crack. We note that in all cases we have $\varepsilon = \xi\alpha \ll 1$.

It follows from (4.9.36), that the solution of the stress concentration problems for a needle and a crack reduces to the computation of the principal terms in the expansion of the tensor D^{-1} with respect to the corresponding small parameters. From (4.9.23) and (4.9.16) we have

$$D = \frac{\det a}{4\pi} \int S_0(n)\rho^3(n)\, dn \,. \tag{4.10.2}$$

First we carry out the computation for a needle. Eq. (4.10.2) is transformed to the spherical coordinates φ, θ with the polar axis along the axis of the needle, i.e. along x^1. We perform the change of variable $\cos\theta = t$ and we set

$$\alpha_1 = \alpha\sqrt{\cos^2\varphi + \xi^2 \sin^2\varphi}, \quad D(\varphi, t) = S_0(n(\varphi, t)) \,. \tag{4.10.3}$$

Without loss of generality we can assume that $D(\varphi, t)$ is an even function of

t (only this component contributes to the integral (4.10.2)). We have

$$D = \frac{\xi}{2\pi} \int_0^{2\pi} \frac{d\varphi}{\cos^2 \varphi + \xi^2 \sin^2 \varphi} \int_{-1}^{1} D(\varphi, t) f_1(t, \alpha_1)\, dt\,, \tag{4.10.4}$$

where

$$f_1(t, \alpha_1) = \frac{\alpha_1^2}{2[\alpha_1^2 + (1 - \alpha_1^2)t^2]^{3/2}}\,. \tag{4.10.5}$$

For a needle $\alpha \to 0$, $\xi \sim 1$ and $f_1(t, \alpha_1) \to \delta(t)$. Consequently, the principal term in the expansion of D with respect to α has the form

$$D_0 = \frac{\xi}{2\pi} \int_0^{2\pi} \frac{D(\varphi, 0)}{\cos^2 \varphi + \xi^2 \sin^2 \varphi}\, d\varphi\,. \tag{4.10.6}$$

Thus, for an arbitrary anisotropic medium the problem is reduced to the computation of a simple integral. If the tensor D_0 has an inverse ($\det D_0 \neq 0$), then the coefficient of the concentration tends to a constant value when $\alpha \to 0$, i.e. it has no singularity. The computations show that $\det D_0 \neq 0$ if the symmetry of the medium is not below the rhombic symmetry (orthotropic, hexagonal, cubic and others[5]). Obviously, this holds also in the case of an arbitrary anisotropy. The tensor of the elastic constants of the indicated medium has nine nonzero components, denoted according to the usual rule by

$$\begin{aligned} &c_0^{\alpha\alpha\beta\beta} = c_{\alpha\beta} \quad (\alpha, \beta = 1, 2, 3) \\ &c_0^{2323} = c_{44}, \quad c_0^{1313} = c_{55}, \quad c_0^{1212} = c_{66}\,. \end{aligned} \tag{4.10.7}$$

For an orthotropic medium all the nine components are essential, while for a transversely isotropic medium only five components play a role:

$$c_{11} = c_{22},\ c_{12},\ c_{13} = c_{23},\ c_{33},\ c_{44} = c_{55},\ c_{66} = 1/2(c_{11} - c_{12})\,. \tag{4.10.8}$$

For the case of cubic symmetry we have three constants

$$c_{11} = c_{22} = c_{33},\ c_{12} = c_{13} = c_{23},\ c_{44} = c_{55} = c_{66}\,, \tag{4.10.9}$$

and, finally, for an isotropic medium

$$c_{12} = \lambda_0,\ c_{44} = 1/2(c_{11} - c_{12}) = \mu_0,\ c_{11} = \lambda_0 + 2\mu_0\,. \tag{4.10.10}$$

In all these cases the tensor D_0^{-1} can be found in explicit form. For its

[5]Here and in the following we assume that the symmetry axes of the medium coincide with the axes of the ellipsoid.

computation it is necessary to take $n_1 = 0$ in the tensor $S_0(n)$, insert the expression obtained into (4.10.6), integrate and invert the tensor D_0. However for the component of D_0^{-1} with the indices $(\alpha\alpha\beta\beta)$ corresponding to extension along the axes, the expressions become cumbersome because it is necessary to invert a third-order matrix. Therefore here we give only the nonzero shear components

$$(D_0^{-1})_{1212} = \frac{\sqrt{c_{55}} + \xi\sqrt{c_{66}}}{c_{66}\sqrt{c_{55}}}, \quad (D_0^{-1})_{1313} = \frac{\sqrt{c_{55}} + \xi\sqrt{c_{66}}}{\xi c_{55}\sqrt{c_{66}}} \tag{4.10.11}$$

$$(D_0^{-1})_{2323} = \frac{c_{33}}{c_{22}c_{33} - c_{23}{}^2} \frac{1}{\xi(\xi A_1\sqrt{u_1 u_2} + B_1\sqrt{|u_2|} + C_1\sqrt{|u_1|})}, \tag{4.10.12}$$

$$A_1 = -\frac{1}{(1 + u_1\xi^2)(1 + u_2\xi^2)}, \quad B_1 = -\frac{u_1}{(1 + u_1\xi^2)(u_2 - u_1)},$$

$$C_1 = -\frac{u_2}{(1 + u_2\xi^2)(u_1 - u_2)}.$$

Here u_1 and u_2 are the roots of the quadratic equation

$$c_{33}c_{44}u^2 + (c_{22}c_{33} - 2c_{23}c_{44} - c_{23}^2)u + c_{22}c_{44} = 0. \tag{4.10.13}$$

The computations are considerably simplified in the isotropic case. Taking into account (9.43), we obtain the zero components of $D_0^{-1}(\eta_0 = [2\mu_0(1 + \nu_0)]^{-1})$

$$\begin{aligned}
&(D_0^{-1})_{1111} = \eta_0, \ (D_0^{-1})_{1122} = (D_0^{-1})_{1133} = -\nu_0\eta_0 \\
&(D_0^{-1})_{2233} = -\eta_0(1 - 2\nu_0^2), \ (D_0^{-1})_{2222} = \eta_0[1 + 2(1 - \nu_0^2)\xi] \\
&(D_0^{-1})_{3333} = \eta_0\left(1 + \frac{2(1 - \nu^2)}{\xi}\right), \quad (D_0^{-1})_{1212} = \frac{1 + \xi}{4\mu_0} \\
&(D_0^{-1})_{1313} = \frac{1}{4\mu_0}\left(1 + \frac{1}{\xi}\right), \ (D_0^{-1})_{2323} = \frac{1}{2\kappa_0}\frac{(1 + \xi)^2}{\xi}.
\end{aligned} \tag{4.10.14}$$

Inserting D_0^{-1} into (4.9.37) we can obtain, according to (4.9.32) the stress $\sigma(n)$ at the surface of the needle for an arbitrary external field σ_0.

The expressions (4.10.12, 14) give an obvious mechanism for the appearance of the singularity in the conversion from a needle to a narrow crack, i.e. for $\xi \to 0$: a singularity of order ξ^{-1} appears only for external stresses σ_0 containing components with the index 3.

Now let us consider an arbitrary crack $\xi \ll 1$, $\alpha \sim 1$. In the same way as in the case of a needle, we transform (4.10.2) to spherical coordinates φ, θ, but with the polar axis oriented along x^3. Setting

$$\xi_1 = \frac{\xi\alpha}{\sqrt{\cos^2\varphi + \alpha^2\sin^2\varphi}}, \quad t = \cos\theta \tag{4.10.15}$$

and making use of the definition (4.10.3) for $D(\varphi, t)$, we have

$$D = \frac{\alpha}{2\pi} \int_0^{2\pi} \frac{d\varphi}{\cos^2 \varphi + \alpha^2 \sin^2 \varphi} \int_{-1}^{1} D(\varphi, t) f_2(t, \xi_1)\, dt \qquad (4.10.16)$$

with

$$f_2(t, \xi_1) = \frac{\xi_1}{2[1 - (1 - \alpha^2 \xi_1^2) t^2]^{3/2}} \,.$$

We write the expansion of D as a function of ξ in the form

$$D = D_0 + \xi D_1 + O(\xi^2) \,. \qquad (4.10.17)$$

In contrast to the case of the needle, for the computation of the principal term in the expansion of D^{-1} with respect to ξ we have to retain the first two terms in D since the tensor D_0 in (4.10.17) does not have an inverse, i.e. $\det D_0 = 0$. It is sufficient to show that all the components of the tensor D_0 containing the index 3 are equal to zero. In fact, as will be shown below, in the integral for D_0 the tensor $D(\varphi, 1)$, which corresponds to $S_0(n)$ for $n_1 = n_2 = 0$, $n_3 = 1$, occurs. But from the general expression (4.9.43) for $S_0(n)$ it follows that in the case of a rhombic structure, the corresponding components of $S_0(n)$ which contain the index 3 are equal to zero for $n_1 = n_2 = 0$. We should expect that this is true also for an arbitrary anisotropy.

For the computation of D_0 and D_1, we consider $f_2(t, \xi_1)$ as a generalized function on the segment $|t| \leq 1$ with identified points ± 1, i.e. on the circumference. This is possible by virtue of the fact that $D(\varphi, t)$ is even and continuous. We can verify that for $\xi_1 \to 0$

$$f_2(t, \xi_1) = \delta(t \pm 1) + \frac{\xi_1}{2(1 - t^2)^{3/2}} + O(\xi_1^2) \,.$$

Substitution into (4.10.16) gives

$$D_0 = \frac{\alpha}{2\pi} \int_0^{2\pi} \frac{D(\varphi, \pm 1)}{\cos^2 \varphi + \alpha^2 \sin^2 \varphi}\, d\varphi \,, \qquad (4.10.18)$$

$$D_1 = \frac{\alpha^2}{4\pi} \int_0^{2\pi} \frac{d\varphi}{(\cos^2 \varphi + \alpha^2 \sin^2 \varphi)^{2/2}} \int_{-1}^{1} \frac{D(\varphi, t) - D(\varphi, \pm 1)}{(1 - t^2)^{3/2}}\, dt$$

where the latter integral is written in regularized form. We note that the expressions obtained become considerably simpler for the circular ($\alpha = 1$) and for the narrow ($\alpha \ll 1$) crack. In the isotropic case the integrals (4.10.18) can be easily computed and the inversion of the tensor (4.10.17) allows us to obtain explicit expressions for the components of the principal term of the expansion of D^{-1}

$$(D^{-1})_{1313} = \frac{1}{2\kappa_0} \frac{1-\alpha^2}{(1-\nu_0-\alpha^2)\,E(\sqrt{1-\alpha^2}) + \alpha^2\nu_0\,K(\sqrt{1-\alpha^2})} \frac{1}{\xi},$$

$$(D^{-1})_{2323} = \frac{1}{2\kappa_0} \frac{1-\alpha^2}{(1-\alpha^2-\nu_0\alpha^2)\,E(\sqrt{1-\alpha^2}) - \nu_0\alpha^2 K(\sqrt{1-\alpha^2})} \frac{1}{\xi},$$

$$(D^{-1})_{3333} = \frac{2}{\kappa_0} \frac{1}{E(\sqrt{1-\alpha^2})} \frac{1}{\xi}. \tag{4.10.19}$$

Here $K(\alpha)$ and $E(\alpha)$ are complete elliptic integrals of the first and second kind, respectively. The remaining components can be considered equal to zero with the accuracy of $O(1)$. From this it follows that a contribution to the singular stress $\sigma(n)$ at the surface of the crack is given only by the components of the external field σ_0 which have index equal to 3.

Passing to the narrow crack $\alpha \to 0$ yields

$$(D^{-1})_{1313} = \frac{1}{4\mu_0}, \quad (D^{-1})_{2323} = \frac{1}{2\kappa_0} \frac{1}{\xi},$$

$$(D^{-1})_{3333} = \frac{2}{\kappa_0} \frac{1}{\xi}, \tag{4.10.20}$$

which coincides with (4.10.14) for $\xi \to 0$. In the case when the medium has a rhombic symmetry, the tensor D^{-1} has a similar structure and its components can be expressed in the general case in terms of elliptic integrals. They become considerably simpler for the narrow crack and have the form

$$(D^{-1})_{1313} = \frac{1}{4\sqrt{c_{55}c_{66}}} \frac{1}{\xi}, \quad (D^{-1})_{2323} = \frac{c_{33}}{4(c_{22}c_{33} - c_{23}{}^2)} \frac{\sqrt{|u_1|} + \sqrt{|u_2|}}{\xi}$$

$$(D^{-1})_{3333} = \frac{\sqrt{c_{22}c_{33}}}{c_{22}c_{33} - c_{23}{}^2} \frac{\sqrt{|u_1|} + \sqrt{|u_2|}}{\xi} \tag{4.10.21}$$

which, obviously, in the particular case of the isotropic medium coincide with (4.10.20).

We emphasize that, representing an independent interest, the case of the narrow crack not only makes obvious the mechanism of the variation of the stress concentration when passing from a needle to a crack, but also allows us to pass to the plane problem if we consider the median section of the crack.

We now proceed to the direct investigation of the stress at the surface of the crack. We start with a pure extension. Since the components σ_0^{11} and σ_0^{22}, according to (4.10.19), do not contribute to the singularity, we consider only the extension σ_0^{33} along the axis x^3 normal to the crack.

From (4.9.32, 37) and (4.10.19–21) we find

$$\sigma^{\alpha\beta}(n) = S_0^{\alpha\beta 33}(n)(D^{-1})_{3333}\sigma_0^{33}. \tag{4.10.22}$$

It is convenient to carry out the investigation of $\sigma^{\alpha\beta}(n)$ in a local system of coordinates $x^{\varkappa'}$, connected at each point of the surface of the ellipsoidal crack with the normal n. As a local basis we take

$$e_{3'} = n, \; e_{1'} = n^0 \times e_3, \; e_{2'} = n \times (n^0 \times e_3)$$

$$n^0 = \frac{1}{\sqrt{n_1^2 + n_2^2}} (n_1 e^1 + n_2 e^2) \,, \tag{4.10.23}$$

where e_α is the unit vector of the coordinate system connected with the axes of the ellipsoid, while the vector n^0 is normal to the boundary of the crack at the point $(n_1, n_2, 0)$. It is obvious that for the points of the surface near the edge, which are of primary interest, the axes $x^{1'}$ and $x^{2'}$ are, respectively, parallel and normal to the edge. With respect to the local axes, the tensor $\sigma^{\alpha'\beta'}(n)$ is planar since all components with index 3 are equal to zero. This follows from the equilibrium conditions.

First we consider the isotropic case. Transforming in (4.10.22) to local coordinates and taking into account (4.9.43) and (4.10.19, 20), we obtain

$$\sigma^{1'1'}(n) = \nu_0 \sigma_{\max}(1 - n_3^2), \; \sigma^{2'2'}(n) = \sigma_{\max}(1 - n_3^2), \; \sigma^{1'2'}(n) = 0 \tag{4.10.24}$$

Here for the finite and the narrow crack we have, respectively,

$$\sigma_{\max} = \frac{2}{E(\sqrt{1 - \alpha^2})} \frac{1}{\xi} \sigma_0^{33}, \quad \sigma_{\max} = \frac{2}{\xi} \sigma_0^{33} \,. \tag{4.10.25}$$

Thus, for the given load the selected system of coordinates coincides, to the accepted degree of accuracy, with the principal axes of the tensor $\sigma(n)$. Since the principal stresses $\sigma_1 = \sigma^{1'1'}$ and $\sigma_2 = \sigma^{2'2'}$ have the same sign, the maximum tangential stress is $\tau_{\max} = |\sigma_2|/2$. It is clear from (4.10.24) that in the case of an extension the largest value of the stress in any cross section normal to the edge is attained for $n_3 = 0$, i.e. on the edge itself.

In the anisotropic case the computations are similar but more cumbersome. We give the final expressions for the stresses $\sigma(n)$ at the surface of a narrow crack in an orthotropic medium in the cross sections $n_1 = 0$ and $n_2 = 0$. In the selected local basis, which in the present case coincides with the principle axes of the tensor $\sigma(n)$, we have for the cross section $n_2 = 0$

$$\begin{aligned} \sigma_1(n) = \; & \Delta^{-1} c_{55} [c_{11} c_{23} - c_{12} c_{13}) \, n_1^2 \\ & + (c_{12} c_{33} - c_{13} c_{23}) n_3^2] \, n_1^2 (D^{-1})_{3333} \, \sigma_0^{33} \,, \end{aligned} \tag{4.10.26}$$

$$\begin{aligned} \sigma_2(n) &= \Delta^{-1} c_{55} (c_{11} c_{33} - c^{13}_{2}) n_1^2 (D^{-1})_{3333} \, \sigma_0^{33} \\ \Delta &= c_{11} c_{55} n_1^4 + (c_{11} c_{33} - 2 c_{13} c_{55} - c_{13}^2) n_1^2 n_3^2 + c_{33} c_{55} n_3^4 \,. \end{aligned} \tag{4.10.27}$$

Here $(D^{-1})_{3333}$ is given by (4.10.21). The stresses in the cross section $n_1 = 0$ are obtained by replacing in the right-hand sides the indices 1 and 5 by 2 and 4, respectively.

We consider now a pure shear. The component σ_0^{12} does not contribute to the singularity and the situation for σ_0^{13} and σ_0^{23} is similar. Therefore we assume that the external field coincides with σ_0^{13}.

In the isotropic case, we find from (4.9.32, 37) and (4.10.19)

$$\sigma^{\alpha\beta}(n) = 4S_0^{\alpha\beta 13}(n)(D^{-1})_{1313}\sigma_0^{13} . \tag{4.10.28}$$

If in the above-introduced local coordinate system we perform a rotation around the normal n, i.e. the axis $x^{3'}$, through an angle

$$\varphi_0 = -\frac{1}{2}\tan^{-1}\frac{n_2}{n_1 n_3} , \tag{4.10.29}$$

then the new axes are principal for $\sigma^{\alpha'\beta'}(n)$. For the principal stresses we find

$$\begin{aligned} \sigma_1(n) &= \sigma^{1'1'}(n) = \sigma_{\max}\,[(1-\nu_0)\,\sqrt{(1-n_1^2)\,(1-n_3^2)} - (1+\nu_0)\,n_1 n_3] , \\ \sigma_2(n) &= \sigma^{2'2'}(n) = -\sigma_{\max}\,[(1-\nu_0)\,\sqrt{(1-n_1^2)\,(1-n_3^2)} + (1+\nu_0)\,n_1 n_3] . \end{aligned} \tag{4.10.30}$$

Here for the finite and the narrow crack we have, respectively,

$$\begin{aligned} \sigma_{\max} &= \frac{1-\alpha^2}{(1-\nu_0-\alpha^2)\,E(\sqrt{1-\alpha^2}) + \nu_0\alpha^2 K(\sqrt{1-\alpha^2})}\,\frac{1}{\xi}\,\sigma_0^{13} , \\ \sigma_{\max} &= \frac{1}{1-\nu_0}\,\frac{1}{\xi}\,\sigma_0^{13} . \end{aligned} \tag{4.10.31}$$

The largest absolute value of the principal stresses σ_1 and σ_2, equal to $\sigma_{\max}$, occurs at the cross section $n_2 = 0$ in the points

$$n^* = \left(\pm\frac{1}{\sqrt{2}}, 0, \mp\frac{1}{\sqrt{2}}\right), \quad n^* = \left(\pm\frac{1}{\sqrt{2}}, 0, \pm\frac{1}{\sqrt{2}}\right) \tag{4.10.32}$$

which are the maxima (minima) of the functions $\sigma_1(n)$ and $\sigma_2(n)$. In this cross section σ_1 and σ_2 are equal to zero at the edge. It is essential that the maximum point is situated at a very small distance from the edge of the crack, which according to (4.9.14, 30) is

$$\Delta x^1 = \frac{a_1}{2}\varepsilon^2 + O(\varepsilon^4) . \tag{4.10.33}$$

We can also show that the characteristic width of the peak is of order ε. By varying the cross section from $n_2 = 0$ to $n_1 = 0$ the magnitude of the peak decreases while the point of the maximum approaches the edge coinciding with it for $n_1 = 0$. For the cross sections in which σ_1 and σ_2 are of the same sign, the character of the variation of $\tau_{\max}$ is evident. In particular, in the cross section

$n_2 = 0$ the quantity τ_{max} has a sharp peak of height $\sigma_{max}/2$ at the point n^* and vanishes at the edge. In the case when σ_1 and σ_2 have different signs

$$\tau_{max} = (1 - \nu_0)\, \sigma_{max} \sqrt{(1 - n_1^2)(1 - n_3^2)}\, . \tag{4.10.34}$$

This quantity reaches its maximum value equal to $(1 - \nu_0)\, \sigma_{max}$ at the edge in the cross section $n_1 = 0$.

Thus, in the case of shear, a remarkable increase of stress takes place, which, if not taken into account, will lead, in numerical computations, to qualitatively irregular results on the stress concentration in the crack. This shows the necessity of investigating the stresses not only at the characteristic points of the contour of the crack, but also over its entire surface.

A similar phenomenon takes place in an anisotropic medium. The principal stresses at the surface of a narrow crack in an orthotropic medium in the cross section $n_2 = 0$ are

$$\begin{aligned}
\sigma_1(n) &= -\, 4\Delta^{-1} c_{55}\, [(c_{11}c_{23} - c_{12}c_{13})\, n_1^2 \\
&\quad + (c_{12}c_{33} - c_{13}c_{23}) n_3^2]\, n_1 n_3 (D^{-1})_{1313}\, \sigma_0^{13}\, , \\
\sigma_2(n) &= -\, 4\Delta^{-1} c_{55} (c_{11}c_{33} - c_{13}^2)\, n_1 n_3 (D^{-1})_{1313}\, \sigma_0^{13}\, .
\end{aligned}$$

Here Δ and $(D^{-1})_{1313}$ are given by (4.10.27 and 21), respectively. In the cross section $n_1 = 0$

$$\sigma_1(n) = -\sigma_2(n) = 4\, \frac{c_{55}c_{66}n_2}{c_{66}n_2^2 + c_{55}n_3^2} (D^{-1})_{1313}\, \sigma_0^{13}\, .$$

Obviously, as in the isotropic case, the increase effect takes place in the cross section $n_2 = 0$.

4.11 Crack in a Homogeneous Medium [6]

Let us first consider a flattened cavity, which occupies a finite simply-connected region V with a smooth boundary. Then, according to (4.8.42, 43), the corresponding integral equations have the form

$$\varepsilon(x) - \int K_0(x - x')\, C_0 \varepsilon^+(x') dx' = \varepsilon_0\, , \tag{4.11.1}$$

$$\sigma(x) + \int S_0(x - x')\, \varepsilon^+(x') dx' = \sigma_0\, . \tag{4.11.2}$$

From (4.11.1) one can find $\varepsilon^+(x)$; then (4.11.2) determines the field $\sigma(x)$ outside the cavity.

---

[6] Sections 4.11 and 4.12 were written in collaboration with S. K. Kanaun.

Let us consider the limiting transition from the cavity to a crack, i.e. to a cut along a smooth oriented surface Ω which is bounded by a closed contour Γ which lies on the boundary of V. Let us choose a local coordinate system at the point $x' \in \Omega$, such that its z-axis is directed along the normal $n(x')$. Let $h(x')$ be the transverse dimension of the cavity and $z_1(x', h)$, $z_2(x', h)$ be the coordinates of the points of intersection of the z-axis with the boundary of V, while $z_1, z_2 \to 0$, as $h \to 0$. For a fixed point $x \notin V$, the kernel $S_0(x - x')$ in (4.11.2) is a smooth, bounded function. Hence,

$$\int S_0(x - x')\, \varepsilon^+(x')\, dx' = \int_\Omega S_0(x - x')\, \varepsilon^+(x', h)\, d\Omega' + O(h)\,, \qquad (4.11.3)$$

where

$$\varepsilon^+(x, h) = \int_{z_1(x,h)}^{z_2(x,h)} \varepsilon^+(x + zn(x))dz, \quad x \in \Omega\,. \qquad (4.11.4)$$

Note that the quantity $\varepsilon^+(x, h)$ can be interpreted as the coefficient of the principal term of the expansion of the function $\varepsilon^+(x)V(x)$ in multipoles, which are concentrated on Ω,

$$\varepsilon^+(x)V(x) = \varepsilon^+(x, h)\delta(\Omega) + \cdots. \qquad (4.11.5)$$

Here $\delta(\Omega)$ is the δ-function concentrated on Ω (Sect. 11.1).

It is important that all other terms in this expansion vanish as $h \to 0$, i.e. the total strain field $\varepsilon(x)$ contains a singular component proportional to $\delta(\Omega)$.

On the other hand, for the limiting case of a crack, the stress $\sigma(x)$ and displacement $u(x)$ satisfy the conditions

$$n \cdot \sigma|_\Omega = 0, \quad [u]|_\Omega = b\,, \qquad (4.11.6)$$

where $b(x)$ $(x \in \Omega)$ is the vector of a discontinuity in displacement (or the opening of the crack), i.e. a quantity which is to be found from the solution of the crack problem. From here, it follows that the singular component of $\varepsilon(x)$ for the crack has a coefficient which is equal to the symmetrized product of the vectors b and n. A comparison with (4.11.5) yields

$$b_{(\alpha}(x)n_{\beta)}(x) = \bar{\varepsilon}^+_{\alpha\beta}(x) = \lim_{h\to 0} \varepsilon^+_{\alpha\beta}(x, h),\ x \in \Omega\,. \qquad (4.11.7)$$

From this, in particular, it is seen that, as $h \to 0$, the normal components of $\bar{\varepsilon}^+(x)$ tend to infinity, whereas the function $\varepsilon^+(x)$ is bounded.

Thus, with (4.11.2, 7) and the symmetry of the tensor $S_0(x)$ taken into account, the stress field outside the crack is connected with $b(x)$ by the relation

$$\sigma(x) = \sigma_0(x) - \int_\Omega S_0(x - x')n(x')b(x')\,d\Omega' . \tag{4.11.8}$$

From this and from the first condition (4.11.6) we find the equation for the vector field $b(x)$ on Ω

$$\int_\Omega T(x, x')b(x')d\Omega' = n(x)\sigma_0(x),\ x \in \Omega , \tag{4.11.9}$$

where

$$T(x, x') = n(x)S_0(x - x')n(x') . \tag{4.11.10}$$

The operator T, defined by these relations, admits the following regular representation on continuously differentiable on the surface Ω functions $b(x)$ (Appendix A.3):

$$(Tb)(x) = \fint_\Omega T(x, x')\,[b(x') - b(x)]d\Omega' - n(x)\Gamma(x)b(x) , \tag{4.11.11}$$

where $\Gamma(x)$ is a tensor-valued function, which depends on the contour Γ.

The operator T is continued to a wide enough class of generalized functions on Ω and belongs to the class of pseudo-differential operators. For functions which one has to deal with in the crack problem, the existence and uniqueness of a solution of an equation of the type $Tb = f$ is shown, for example, in [4.3]. For what follows, it is essential that the asymptoties of the solution of this equation near Γ for smooth $f(x)$ has the form

$$b(x) = \beta(x_0)r^{1/2} + O(r^{3/2}) , \tag{4.11.12}$$

where r is the distance from the point x along the normal to a point $x_0 \in \Gamma$, $\beta(x_0)$ is a smooth function on Γ.

Let ut consider the asymptotics of the stress field $\sigma(x)$ outside the crack in the neighborhood of the edge Γ of the crack. Let y_1, y_2, y_3 be the local cartesian coordinates at the point $x_0 \in \Gamma$, the y_2-axis being directed along the limiting normal to Ω at the point x_0, and the y_3-axis being tangent to Γ. Then the y_1-axis lies in the plane which is tangent to Ω at the point x_0. Taking into account (4.11.12), we have

$$b(y) = \beta(x_0)\sqrt{y_1} + O({y_1}^{3/2}) . \tag{4.11.13}$$

Using (4.11.8) and (4.11.3) we can write down the expression for σ at the point $z = (-\rho\cos\theta, -\rho\sin\theta, 0)$ $(\rho > 0)$

$$\sigma(z) = \sigma_0(z) - \frac{1}{\sqrt{\rho}}\int_\Omega S_0(\cos\theta + \xi_1', \sin\theta + \xi_2', \xi_3')\, n(\rho\xi')\beta(\rho\xi')\sqrt{\xi_1}\,d\Omega'$$
$$+ O(\sqrt{\rho}) . \tag{4.11.14}$$

Here $\xi = \rho^{-1}y_i$ and it is taken into account that $S_0(x)$ is an even homogeneous function of degree -3. It is easy to show that, as $\rho \to 0$, the integral tends to a finite limit, and hence, the stress has a singularity of the order of $\rho^{-1/2}$.

Let us introduce a tensor coefficient of stress intensity, which is of interest for applications and defined by the expression

$$J(\theta, x_0) = \lim_{\rho \to 0} \sqrt{\rho}\, \sigma(z)\,. \tag{4.11.15}$$

From (4.11.14) it follows that

$$J(\theta, x_0) = s(\theta)n(x_0)\beta(x_0)\,, \tag{4.11.16}$$

where

$$s(\theta) = -\int_0^\infty \sqrt{\xi_1}\, d\xi_1 \int_{-\infty}^\infty S_0(\cos\theta + \xi_1, \sin\theta, \xi_3)\, d\xi_3 \tag{4.11.17}$$

and $n(x_0)$ is the limiting normal to Ω at $x_0 \in \Gamma$. In a more detailed description,

$$J^{\alpha\beta}(\theta, x_0) = s^{\alpha\beta\lambda\mu}(\theta)\, n_\lambda(x_0)\beta_\mu(x_0)\,. \tag{4.11.18}$$

The intensity coefficient J_∞ admits an obvious interpretation, if one observes that $\int_{-\infty}^\infty S_0(\xi_1, \xi_2, \xi_3)\, d\xi_3$ coincides, in essence, with the Green's tensor for stress in the problem of plane deformation and complex shear (in dimensionless variables ξ) of a homogeneous medium with moduli C_0, the normal to the plane coinciding with the ξ_3-axis. Then J coincides with the stress at the point $\xi_1 = -\cos\theta$, $\xi_2 = -\sin\theta$, when a discontinuity of the displacement vector $\beta(x_0)\sqrt{\xi_1}$ is given along the positive semi-axis of ξ_1. In the isotropic case, the function $J(\theta)$ is easily found in an explicit form.

The tensor J can be represented as a sum of three tensors, which correspond to the three components of the vector β,

$$J = J_1 + J_2 + J_3\,, \tag{4.11.19}$$

where

$$J_i^{\alpha\beta}(\theta, x_0) = s^{\alpha\beta\lambda i}(\theta)n_\lambda(x_0)\, \beta_i(x_0) \tag{4.11.20}$$

(do not sum over i!). The tensors J_1 and J_2 are found from the solution of the corresponding planar problem, and J_3 is derived from the solution of an anti-planar problem (complex shear). All these problems can be solved by known methods for the case of an arbitrary anisotropy.

Note that in the case of linear fracture mechanics, the asymptotics of the stress field in the neighborhood of the edge of a crack is always characterized by the stress intensity factors $K_{\rm I}$, $K_{\rm II}$, $K_{\rm III}$. It can be shown that these factors coincide with the three components of $\beta(x_0)$ up to constant factors, the latter depending on the elastic constants.

Since the vector $\beta(x_0)$ depends linearly on the external field, it can be represented in the form

$$\beta(x_0) = B(x_0)\sigma_0 , \tag{4.11.21}$$

where $B(x_0)$ is the corresponding linear operator. Then J becomes

$$J(\theta, x_0) = F(\theta, x_0)\sigma_0 , \tag{4.11.22}$$

where

$$F(\theta, x_0) = s(\theta)\, n(x_0)B(x_0) . \tag{4.11.23}$$

This representation is convenient since $s(\theta)n(x_0)$ depends on a neighborbood of the point x_0 only, and the operator $B(x_0)$ depends on the whole surface Ω.

Thus, the calculation of the tensor intensity coefficient J is finally reduced to solving a certain planar problem and constructing the operator B. An explicit expression for the latter, for the case of an elliptic crack, will be obtained in the next section.

4.12 Elliptic Crack

An elliptic crack is the simplest case which is of interest for applications and admits an exact solution.

Let Ω be a plane cut with an elliptic boundary in an elastic medium, and $b(x)$ be the discontinuity of the displacement vector at the cut. We have the following analog of the theorem about polynomial conservation: for a polynomial external field $\sigma_0(x)$ of degree m, the vector $b(x)$ has the form

$$b(x_1, x_2) = B(x_1, x_2)\sqrt{1 - \frac{x_1^2}{a_1^2} - \frac{x_2^2}{a_2^2}} , \tag{4.12.1}$$

where $B(x_1, x_2)$ is a polynomial in x_1 and x_2 of a degree no higher than m (here the coordinate system x_1, x_2 is connected with the principal axes $2a_1$ and $2a_2$ of the ellipse). [7]

In order to prove this, let us consider the elliptic crack as the limit of an ellipsoidal cavity, as the semiaxis $a_3 = h$ of the ellipsoid tends to zero. According to (4.11.3), the vector $b(x)$ is determined by the limiting value of the quantity

$$\varepsilon^+(x, h) = \int_{-hz(x)}^{hz(x)} \varepsilon^+(x + \xi n)\, d\xi , \tag{4.12.2}$$

[7]This result was first obtained (in a very complicated way) in [4.4].

where $x(x_1, x_2) \in \Omega$, n = const is the normal to Ω, ξ is a coordinate along the normal and

$$z(x_1, x_2) = \sqrt{1 - \frac{x_1^2}{a_1^2} - \frac{x_2^2}{a_2^2}}\,. \tag{4.12.3}$$

From the polynomial conservation theorem, it follows that $\varepsilon^+(x + \xi n)$ is a polynomial of a degree not higher than m with respect to the coordinates x_1, x_2, ξ and, consequently, can be represented in the form

$$\varepsilon^+(x + \xi n) = \varepsilon^+_{(0)}(x) + \cdots + \varepsilon^+_{(m)}(x)\xi^m\,, \tag{4.12.4}$$

where $\varepsilon^+_{(k)}(x)$ is a polynomial of a degree not higher than $m - k$ with respect to the coordinates x_1, x_2. After this decomposition is substituted into (4.12.2), the integrals of odd powers of ξ vanish, and the integrals of terms with ξ^{2k} give expressions of the form

$$Q_{2k}(x_1, x_2)z(x_1, x_2)\,,$$

where Q_{2k} is a polynomial of a degree not higher than $2k$. Hence it follows that

$$\varepsilon^+(x, h) = P_m(x_1, x_2, h)\, z(x_1, x_2)\,, \tag{4.12.5}$$

where P_m is a polynomial of degree $\leq m$. Finally, setting $h = 0$, we obtain the required result.

Let us now consider the method of computing the polynomial $B(x_1, x_2)$. The vector $b(x)$ is a solution of (4.11.9), where $\sigma_0^+(x)$ is a polynomial of degree m, given on Ω. In the considered case, it is convenient to write down the equation for $b(x)$ in the equivalent form (Appendix A.3)

$$\int T(x - x')\, b(x')dx' = n\cdot\sigma_0^+(x),\ x \in \Omega\,, \tag{4.12.6}$$

where the integration is carried out over the whole plane, which contains Ω, the vector $b(x')$ is defined to be zero outside Ω, and $T(x)$ is a generalized function, which is generated by the generalized function $S_0(x)$ in three-space. Substituting (4.12.1) for $b(x)$ into (4.12.6) and taking into account (4.12.3), we have

$$\int T(x - x')z(x')B(x')dx' = n\cdot\sigma_0^+(x),\ x \in \Omega\,, \tag{4.12.7}$$

where $z(x)$ is continued by zero outside Ω.

According to the property of polynomial conservation, the operator T transforms the polynomial $B(x)$ (multiplied by $z(x)$) into the polynomial $n\cdot\sigma_0^+(x)$ of the same degree; in particular, a constant vector B is transformed into a constant tensor T_0B, which depends linearly on B. It is obvious that the constant coefficient T_0 is determined by the expression (the integration is carried out over the plane)

$$T_0 = \int T(x - x')\, z(x')\, dx'$$

$$= \int T(x)z(x)dx = \frac{1}{(2\pi)^2} \int T(k)z(k)\, dk\,, \tag{4.12.8}$$

where

$$T^{\alpha\beta}(k_1, k_2) = \frac{1}{2\pi} \int_{-\infty}^{\infty} S_0^{\alpha 3\beta 3}(k_1, k_2, k_3)\, dk_3\,, \tag{4.12.9}$$

$$z(k_1, k_2) = 2\pi a_1 a_2\, |ak|^{-2} \left(\frac{\sin|ak|}{|ak|} - \cos|ak|\right) \tag{4.12.10}$$

and $|ak|$ is the modulus of the vector (a_1k_1, a_2k_2).

Conversely, a constant external field $\sigma_0^{\alpha\beta}$ = const being given, we have

$$B = D^0\sigma_0,\ D^0 = T_0^{-1}\,. \tag{4.12.11}$$

For an arbitrary polynomial field, the solution can be constructed in the following way. The operator T transforms the homogeneous polynomial $B(x) = x^m$ of degree m (multiplied by $z(x)$) into a polynomial of the same degree. We find the coefficients of the latter, taking values at $x = 0$ of the successive derivatives in x from $T(x^m)$. Then, solving a system of linear equations, we obtain the coefficients of the polynomial $B_m(x)$ for the external field, which is proportional to x^m, and consequently, for an arbitrary polynomial $n \cdot \sigma_0^+(x)$.

Problem 4.12.1. Show that, if $n \cdot \sigma_0^+(x)$ is a linear homogeneous function $\sigma_i^\alpha x^i$ $(i = 1, 2)$, then $B_\alpha(x) = B_{\alpha i} x^i$, where

$$B_{\alpha i} = D_{\alpha\beta i}{}^{j}\sigma_j^\beta,\ D = T^{-1}\,, \tag{4.12.12}$$

$$T^{\alpha\beta i\cdot}{}_{j} = -\frac{1}{(2\pi)^2} \int \frac{\partial}{\partial k_i} [T^{\alpha\beta}(k)k_j]\, z(k)\, dk\,. \tag{4.12.13}$$

Problem 4.12.2. Verify that (4.12.1) for the vector $b(x)$ has, in a neighborhood of the point $x_0 \in \Gamma$, the following asymptotics in the local coordinate system y_1, y_2 (Sect. 4.11)

$$b(y_1, y_2) = \beta(x_0)\sqrt{y_1} + O(y_1^{3/2})\,, \tag{4.12.14}$$

where

$$\beta(x_0) = \sqrt{2}\, B(x_0)\, (x_0 a^{-4} x_0)^{1/4}, \qquad (4.12.15)$$

$$a = \begin{pmatrix} a_1 & 0 \\ 0 & a_2 \end{pmatrix},$$

and $B(x_0)$ is the value of $B(x_1, x_2)$ at the point $x_0 \in \Gamma$.

Thus, for a polynomial external field, the polynomial $B(x)$ being considered as a known, we have an explicit expression for $\beta(x_0)$, and consequently, for the tensor intensity coefficient

$$J = \sqrt{2}\, s(0) n B(x_0)(x_0 a^{-4} x_0)^{1/4}. \qquad (4.12.16)$$

Let us consider an isotropic medium and a constant external field as an example. An expression for $s(\theta)$ is easily found from (4.11.17); therefore it is sufficient to write down an expression for $B(x_0)$ = const. Taking into account (4.12.8, 11), after simple calculations, we find

$$D^0_{\alpha\beta} = \frac{2a_1^2(1-\nu)}{\mu a_2} d_\alpha \delta_{\alpha\beta} \quad \text{(do not sum!)} \qquad (4.12.17)$$

where

$$d_1 = \frac{1}{-c_3 + \nu(c_2 - 2c_3)}, \quad d_2 = \frac{1}{c_3 + \nu(c_1 - 2c_3)}, \quad d_3 = \frac{1}{c_3},$$

$$c_1 = -c_2 + 3c_3, \quad c_2 = c_3 - \frac{E(\alpha) - K(\alpha)}{\alpha^2}, \quad c_3 = \frac{E(\alpha)}{1 - \alpha^2},$$

$$\alpha^2 = 1 - \frac{a_2^2}{a_1^2} \quad (a_1 \geq a_2)$$

and $E(\alpha)$ and $K(\alpha)$ are the complete elliptic integrals of the first and second kinds.

4.13 Interaction Between Ellipsoidal Inhomogeneities

The results of Sects. 4.9–12 enable one, in particular, to deduce an explicit solution of the problem of interaction of inhomogeneities and cracks, which have ellipsoidal or elliptic form. For simplicity, let us consider only the principal asymptotics of perturbed fields. As is shown in Sect. 4.5, this means that, in the expansion with respect to multipoles (4.5.12) for the operator P of the interaction energy, we can restrict ourselves to the term P_{00}.

Omitting simple calculations, we present the final result for an ellipsoidal inhomogeneity

$$P_{00} = -v\mathrm{C}_1(\mathrm{C}_1 + \mathrm{C}_1 A \mathrm{C}_1)^{-1}\mathrm{C}_1 = -v(\mathrm{C}_1^{-1} + A)^{-1}. \tag{4.13.1}$$

Here v is the volume of the ellipsoid, C_1 is the tensor of perturbed elastic constants, and A is a tensor which depends on parameters of the ellipsoid and on the tensor of elastic constants C_0 of the external medium. In the general case, the latter is given by the (4.9.21, 23), and for an isotropic medium this is given by (4.9.39). The second equality in (4.13.1) is valid if C_1^{-1} exists.

For an ellipsoidal cavity, $C_1 = -C_0$ and

$$P_{00} = v\mathrm{C}_0 D^{-1}\mathrm{C}_0 = v(B_0 - A)^{-1}, \tag{4.13.2}$$

where $B_0 = C_0^{-1}$ is the tensor of elastic compliance of the medium, and D is given by (4.9.21, 23 and 40).

Using (4.5.7), we can now find an explicit expression for the asymptotics of the perturbed field. In the case when C_1 is an isotropic tensor, this expression coincides with the one obtained by ESHELBY [4.5, 7]. One can find a correspondence between the ellipsoidal inhomogeneity and the equivalent point defect (4.3.1) in the Debye's quasicontinuum with the characteristic

$$\tilde{C}_1 = (\mathrm{C}_1^{-1} + A - vg^0)^{-1}, \tag{4.13.3}$$

where g^0 is defined by (4.3.4). A perturbed field caused by this defect coincides with the asymptotics of the perturbed field of the ellipsoidal inhomogeneity. In order to show this, it is sufficient to compare (4.3.2) and (4.5.2), where $P(y, y')$ in the last expression is to be substituted by the dominant term of its expansion (4.5.12). Note that, in the case of a spherical inhomogeneity, $A = vg^0$ and, consequently, $\tilde{C}_1 = C_1$.

Let us consider two ellipsoidal inhomogeneities and assume

$$P_{00}^{i} = -v_i(\mathrm{C}_i^{-1} + A_i)^{-1}, \quad i = 1, 2 \tag{4.13.4}$$

to be given. According to (4.5.11), the interaction of ellipsoidal inhomogeneities is described by the matrix $P^{ij}(x_1 - x_2)$, which, in the approximation under consideration, has the form

$$P_{00}^{ij} = \begin{pmatrix} P_{00}^1 & P_{00}^1 \nabla\nabla G^0(x_1 - x_2) P_{00}^2 \\ P_{00}^2 \nabla\nabla G^0(x_2 - x_1) P_{00}^1 & P_{00}^2 \end{pmatrix}, \tag{4.13.5}$$

where x_1, x_2 are the position vectors of the centers of the ellipsoids. As is shown in Sect. 4.5, the central symmetry of the inclusions enables us to ignore the matrices P_{01}^{ij} u P_{10}^{ij}.

Substituting P_{00}^{ij} in (4.5.13), we find an expansion of the interaction energy Φ_{int} in powers of $r_{12} = |x_1 - x_2|$

$$\Phi_{\text{int}} = \Phi_0^1 + \Phi_0^2 + \Phi_1 r_{12}^{-3} + O(r_{12}^{-4}), \tag{4.13.6}$$

where

$$\Phi_0^i = \frac{1}{2}\,\varepsilon^0 P_{00}^i \varepsilon^0, \tag{4.13.7}$$

$$\Phi_1 r_{12}^{-3} = \varepsilon^0 P_{00}^1 \nabla\nabla G^0(x_1 - x_2) P_{00}^2 \varepsilon^0. \tag{4.13.8}$$

It is obvious, that here Φ_0^i is the self-energy of interaction of *i-th* defect with the external field ε^0, and Φ_1 is a quadratic function of the external field, which depends on C_0 and C_i ($i = 1, 2$), these depending on the defects and the unit vector of the straight line which connects the defects.

We present below the explicit expressions for Φ_0^i and Φ_1 in the case, when the external field is pure dilatation $\varepsilon_{\alpha\beta}^0 = \varepsilon^0 \delta_{\alpha\beta}$ and the inhomogeneities are two spheroids with a common rotation axis x^2 and with isotropic elastic constants

$$\Phi_0^i = -\frac{v_i \varepsilon_0^2}{2\kappa_0 \Delta_i}(3P_1^i + P_2^i), \tag{4.13.9}$$

$$\Phi_1 = \frac{4 v_1 v_2 \varepsilon_0^2}{\kappa_0 \Delta_1 \Delta_2}[(1 - 2\nu_0)(P_1^1 P_2^2 + P_2^1 P_2^2) + 2(1 - \nu_0) P_1^2 P_2^2],$$

where

$$P_1^i = (5 - 4\nu_0) I_2^i - 3 I_{12}^i + 8\pi(1 - \nu_0)\frac{\mu_0}{\mu_i}, \tag{4.13.10}$$

$$P_2^i = 2(1 - 2\nu_0)(I_1^i - I_2^i),$$

$$\Delta_i = [4(1 - \nu_0) I_2^i - 2 I_{12}^i - 8\pi(1 - \nu_0)\frac{\mu_0}{\mu_i}\cdot\frac{1}{(1 + \nu_i)}]$$

$$\times [(3 - 4\nu_0) I_1^i - I_{21}^i + 8\pi(1 - \nu_0)\frac{\mu_0}{\mu_i}\cdot\frac{(1 - \nu_i)}{(1 + \nu_i)}]$$

$$- 2[- I_2 + I_{12} - 8\pi(1 - \nu_0)\frac{\mu_0}{\mu_i}\cdot\frac{\nu_i}{1 + \nu_i}]^2.$$

The quantities I_p, I_{pq} contained in these expressions are given by (9.42), κ_0 is expressed by (4.9.41), the index i is the number of a defect, and the lower index of the quantities P_q^i is connected with the principal axes of the ellipsoids, these taking the values 1, 2 only (due to the rotation symmetry). A limiting case of a rigid inclusion is obtained as $\mu_i \to \infty$; the case of a cavity is obtained as, $\mu_i \to -\mu_0$, $\nu_i \to \nu_0$.

If one of the ellipsoids, for example the second one, is a sphere, then $P_2^2 = 0$ and the expression for Φ_1 is greatly simplified

$$\Phi_1 = \frac{3v_1v_2\varepsilon_0^2}{\pi\kappa_0} \cdot \frac{1-2\nu_0}{1+3\dfrac{(1-\nu_0)}{(1-2\nu_0)}\dfrac{\mu_0(1-2\nu_0)}{\mu_2(1+\nu_2)}} \cdot \frac{I_1^1 - I_2^1}{\Delta_1} . \tag{4.13.11}$$

Finally, if both ellipsoids are spheres, then $P_1^1 = P_2^2 = 0$ and $\Phi_1 = 0$. This is in agreement with the result of [4.8], where it is shown that the interaction energy of two isotropic spherical inclusions is of the order of r_{12}^{-6}.

In conclusions, a few words about the model of point defects in a classical elastic continuum. We have already mentioned earlier that the δ-functional model of an inhomogeneity is, strictly speaking, correct only for the quasi-continuum. This is evident, for example, from (4.13.3), where $g^0 \to \infty$ during limiting transition to the continuum. A point defect in a continuum has meaning only as a limiting image which corresponds to the asymptotics of the field of a finite inhomogeneity, and is completely characterized by the tensor P_{00}. In the particular case of an ellipsoidal inhomogeneity, P_{00} is given by (4.13.1).

The matrix P_{00}^{ij} describes the interaction of two point defects. It is essential that if one extrapolates this interaction to small distances, then the interaction energy appears to have a singularity at distances of the order of $v^{1/3}$, i.e. of the order of the dimension of the starting (finite) defect. This result can be obtained by substituting (4.13.5) into (4.5.13) for the interaction energy. Qualitatively, the existance of a characteristic scale parameter follows from dimensional reasonings.

Thus, the model of a point defect in an elastic continuum possesses an inner restriction at small distances. Later, these considerations will be taken into account, when considering a random field of point defects in Chap. 7.

4.14 Notes

The content of Sects, 4.1–4.5 is based on [B6.14, 20, 23]. As regards point defects, see [B6.27, 39]. The Green's tensor for displacement in an anisotropic medium was investigated in [B1.2] The Green's tensors for strain and stress were introduced in [B6.17, 24]. Elasticity in terms of projection operators was formulated in [B6.25, 26]. Discontinuities of stress and strain on the boundary of two media were obtained in [B6.24].

The polynomial conservation theorem for an ellipsoidal inclusion was proved for an isotropic medium and homogeneous field in [B6.5–7] and for an anisotropic medium and arbitrary polynomial field in [B6.22]. The complete investigation of stress on an ellipsoidal inhomogeneity and its limiting forms was carried out in [B6.21, 24], see also [B6.12, 31, 37, 42].

A survey on cracks can be found, for example, in [B6.8, 29]. As regards integral equations for cracks, see [B6.43, 7.13, 8.6].

5. Internal Stress and Point Defects

Vacancies, interstitial atoms and dislocations are well known examples of crystal lattice defects which are sources of internal stress. In this chapter we give a general outline of the theory of internal stress in nonlocal elasticity. Point defects, which are sources of internal stress as well as inhomogeneities, are considered in detail.

5.1 Internal Stress in the Nonlocal Theory

In all the preceeding discussions, it was the displacement vector or the field of the symmetric strain tensor of rank two, which is uniquely determined by the former, that described a state of a medium of simple structure. Conversely, if the strain field is given, satisfying the usual compatability conditions, then the displacement field is determined to within a rigid translation and a rotation of the medium. The deformation is connected with the stress tensor by Hooke's law (local or nonlocal), the stress tensor being connected by means of the equations of equilibrium, with the external body or boundary forces. In the absence of external forces, both the strain and the stress are equal to zero. According to such a scheme the classical as well as the nonlocal theory of elasticity are constructed.

However, another situation is possible, occurring in a number of important cases. In order to give an illustration, let us carry out a thought experiment.

Let us consider a medium which consists of particles connected by an arbitrary system of elastic bonds of the type of springs. Let us also assume that, in the initial state, the elastic energy of all the springs is equal to zero and consequently forces of interaction between particles of the medium also vanish. Let us take some of the springs, change their length (e.g. they are shortened) and return the springs to their original places. This, generally speaking, will result in a deformation of the medium, and the latter will proceed to a new state which corresponds to the minimum of the elastic energy. In this state of equilibrium, as distinct from the initial state, the elastic energy of bonds is no longer equal to zero. A convenient quantitative measure of interaction between particles of the medium is the stress tensor, which is not connected at all with the external forces. To emphasize this fact, let us call this tensor an internal stress tensor, while the usual stress, caused by external forces, will be called the external stress.

It is easily seen that the displacement of points of the medium cannot adequately describe the internal stressed state of the medium. In fact, the bonds could be changed in such a way that the internal forces would be self-compensated and the particles would not be subject to any displacements. This shows that it is necessary to introduce a new kinematic variable which determines the state of the medium.

Consider first a general case of a stressed state, caused by the action of the external forces as well as by sources of internal stress. The role of the latter in the crystal can be played by point defects of the type of foreign atoms and of vacancies, as well as by dislocations, grain boundaries and so forth. In macroscopic bodies internal stress can be caused, for example, by a heterogeneous temperature field or by purely mechanical causes of the type of prestress or prestrain. Hereafter, we confine ourselves to the static case and therefore we shall not introduce explicitly the internal degrees of freedom. According to the results of the Chap. 3, one can assume that they are eliminated from the equations and make an implicit contribution to the effective operators of elastic moduli.

Let us assume that the field variable, which uniquely determines the state of the medium, is a nonsymmetric tensor of rank two, of the elastic distortion $\chi_{\lambda\alpha}$, and that the elastic energy Φ in the harmonic approximation is a quadratic functional of the distortion:

$$\Phi = \frac{1}{2} \langle \chi_{\lambda\alpha} \mid \Phi^{\lambda\alpha\mu\beta} \mid \chi_{\mu\beta} \rangle \,. \tag{5.1.1}$$

Hence, it follows that the operator $\Phi^{\lambda\alpha\mu\beta}$ is self-adjoint, i.e. its kernel satisfies the following conditions

$$\Phi^{\lambda\alpha\mu\beta}(x, x') = \Phi^{\mu\beta\lambda\alpha}(x', x) \tag{5.1.2}$$

or in the k-representation

$$\Phi^{\lambda\alpha\mu\beta}(k, k') = \overline{\Phi^{\mu\beta\lambda\alpha}(k', k)} \,. \tag{5.1.3}$$

It is natural also to consider the form (5.1.1) as positive definite.

The distortion $\chi_{\lambda\alpha}$ can be represented in the form

$$\chi_{\lambda\alpha} = \zeta'_{\lambda\alpha} + \zeta_{\lambda\alpha} \,, \tag{5.1.4}$$

where $\zeta'_{\lambda\alpha}$ and $\zeta_{\lambda\alpha}$ are the external and internal distortions, respectively. The external distortion $\zeta'_{\lambda\alpha}$ is defined as a gradient of the displacement u_α, caused by external forces, i.e.

$$\zeta'_{\lambda\alpha}(x) \overset{\text{def}}{=} \partial_\lambda u_\alpha(x) \,. \tag{5.1.5}$$

Thus, we have

$$\mathrm{rot}^{\beta\lambda}\zeta'_{\lambda\alpha}(x) = 0 \quad (\mathrm{rot}^{\beta\lambda} \stackrel{\mathrm{def}}{=} \varepsilon^{\beta\mu\lambda}\partial_\mu)\,. \tag{5.1.6}$$

This condition is a characteristic one for the external distortion.

It is obvious that the external distortion is invariant with respect to translation, and under rotation, given by a constant anti-symmetric tensor $a_{\lambda\alpha}$, it is transformed according to the law

$$\zeta'_{\lambda\alpha}(x) \to \zeta'_{\lambda\alpha}(x) + a_{\lambda\alpha}\,. \tag{5.1.7}$$

Let us postulate that these transformational properties are possessed also by the total distortion $\chi_{\lambda\alpha}$ and consequently by the internal distortion $\zeta_{\lambda\alpha}$.

The requirement of invariance of the elastic energy with respect to rotation yields

$$\langle a_{\lambda\alpha} \mid \Phi^{\lambda\alpha\mu\beta} \mid \chi_{\beta\mu}\rangle = 0\,. \tag{5.1.8}$$

Hence, taking into account the arbitrariness of $a_{\lambda\alpha}$ and $\chi_{\mu\beta}$, we obtain a condition for $\Phi^{\lambda\alpha\mu\beta}$ [cf. 2.4.18)]

$$\Phi^{[\lambda\alpha]\mu\beta}(0, k') = 0\,. \tag{5.1.9}$$

The energy operator $\Phi^{\lambda\alpha\mu\beta}$ determines a scalar product in the functional space of distortions, transforming the latter into a Hilbert space. This Hilbert space can be decomposed into two orthogonal (with respect to energy) subspaces of external and internal distortions, and the total energy can be represented in the form of a sum of the external and internal energies.

We know from the results, obtained in the Chap. 2. that the most general expression for the external elastic energy can be represented in the form

$$\Phi = \frac{1}{2}\langle\zeta'_{\lambda\alpha} \mid \Phi^{\lambda\alpha\mu\beta} \mid \zeta'_{\mu\beta}\rangle = \frac{1}{2}\langle\varepsilon'_{\lambda\alpha} \mid c^{\lambda\alpha\mu\beta} \mid \varepsilon'_{\mu\beta}\rangle\,. \tag{5.1.10}$$

Here,

$$\varepsilon'_{\lambda\alpha}(x) = \partial_{(\lambda}u_{\alpha)}(x) = \zeta'_{(\lambda\alpha)} \tag{5.1.11}$$

is the symmetric strain tensor and the Hermitian operator of the elastic moduli $c^{\lambda\alpha\mu\beta}$ is symmetric with respect to the indices $\lambda\alpha$ and $\mu\beta$.

The external stress tensor $\sigma'^{\lambda\alpha}$ is connected with the external strain $\varepsilon'_{\lambda\alpha}$ by the nonlocal Hooke's law:

$$\sigma'^{\lambda\alpha}(x) = \int c^{\lambda\alpha\mu\beta}(x, x')\varepsilon'_{\mu\beta}(x')\,dx' \tag{5.1.12}$$

and it satisfies the conditions of equilibrium

$$\partial_\lambda\sigma'^{\lambda\alpha}(x) = -q^\alpha(x)\,. \tag{5.1.13}$$

The system of equations (5.1.11–13) (under the given conditions at infinity) uniquely determines the displacement and consequently the external distortion, strain and stress. In other words, there exists a Green's operator $G_{\alpha\beta}$, in terms of which the displacement is given by the relation

$$u_\alpha(x) = \int G_{\alpha\mu}(x, x')q^\beta(x')\, dx' . \tag{5.1.14}$$

We require, by definition, the fulfillment of the condition of orthogonality of external and internal distortions with respect to the energy, i.e.

$$\langle \zeta'_{\lambda\alpha} \mid \Phi^{\lambda\alpha\mu\beta} \mid \zeta_{\mu\beta} \rangle = 0 ; \tag{5.1.15}$$

this is equivalent to the representation of the total energy in the form of a sum of external and internal energies. The corresponding elastic fields, in this situation, do not interact and, hence, in the theory of internal stresses no external forces are needed. Let us emphasize that all this is valid only in the case of statics and, in particular, is not applicable to the case of moving dislocations.[1]

For the internal energy we now have

$$\Phi = \frac{1}{2} \langle \zeta_{\lambda\alpha} \mid \Phi^{\lambda\alpha\mu\beta} \mid \zeta_{\mu\beta} \rangle . \tag{5.1.16}$$

The quantity

$$\sigma^{\lambda\alpha}(x) \stackrel{\text{def}}{=} \int \Phi^{\lambda\alpha\mu\beta}(x, x')\, \zeta_{\mu\beta}(x')\, dx' \tag{5.1.17}$$

will be called the tensor of internal stress. From (5.1.15), taking into account (5.1.5) we find that $\sigma^{\lambda\alpha}$ satisfies the equation of equilibrium

$$\partial_\lambda \sigma^{\lambda\alpha}(x) = 0 . \tag{5.1.18}$$

Problem 5.1.1. Show that (5.1.15 and 18) are equivalent.

Problem 5.1.2. Show that if the internal distortion $\zeta_{\lambda\alpha}$ satisfies the condition (5.1.6), then $\zeta_{\lambda\alpha} = 0$. [Hint: Make use of the positive definiteness of the form (5.1.16)].

The right-hand side of the equation

$$\mathrm{rot}^{\mu\lambda}\zeta_{\lambda\nu}(x) = \alpha^{\mu}_{\cdot\nu}(x) \tag{5.1.19}$$

characterizes the density of sources of internal stress. If the physical source of internal stress are dislocations, then the tensor $\alpha^{\mu}_{\cdot\nu}(x)$ is called the density

[1]In the case of a moving dislocation, the external forces perform work.

of dislocations and, in the general case, the density of quasi-dislocations according to the terminology introduced by *Kröner* [5.1]. In what follows, unless otherwise stated, it is assumed that the sources of internal stress are localized in a bounded region.

Eq. (5.1.17 to 19) constitute a complete system which determines the internal distortion $\zeta_{\lambda\alpha}$ and the internal stress $\sigma^{\lambda\alpha}$ for a given dislocation density $\alpha^{\mu}_{\cdot\nu}$. The existence of a solution is proved in the usual manner and the uniqueness follows from Problem 5.1.2.

Let us consider the structure of the operator $\Phi^{\lambda\alpha\mu\beta}$, assuming that the action-at-a-distance is of the order of the scale parameter l. Let us represent $\Phi^{\lambda\alpha\mu\beta}$ in the form

$$\Phi^{\lambda\alpha\mu\beta} = c^{\lambda\alpha\mu\beta} + a^{\lambda\alpha\mu\beta}\,, \tag{5.1.20}$$

where $c^{\lambda\alpha\mu\beta}$ is the above-defined operator of elastic moduli which is symmetric with respect to the indices $\lambda\alpha$ and $\mu\beta$.

It is easily seen that the Hermitian operator $a^{\lambda\alpha\mu\beta}$ is to be localized in a region, where the sources of internal stress are present (more precisely, in an l-neighborhood of this region). In fact, outside the region, occupied by the sources of internal stress, the equations for the internal and external distortions must be identical, since there $\zeta_{\lambda\alpha}$ is locally representable in the form of the gradient of a displacement. This requirement leads to the condition

$$\partial_\lambda \partial'_\mu a^{\lambda\alpha\mu\beta}(x, x') = 0 \tag{5.1.21}$$

imposed on the kernel of the operator. Moreover, from the equations (5.1.9, 20), with the symmetry of $c^{\lambda\alpha\mu\beta}$ taken into account, it follows that

$$a^{[\lambda\alpha]\mu\beta}(0, k') = 0\,. \tag{5.1.22}$$

Problem 5.1.3. Show that the expansion of an entire analytic function $a^{\lambda\alpha\mu\beta}(k, k')$ in a series with respect to k, k', begins with terms of the type

$$a'^{\lambda\alpha\mu\beta}(k, k') = a_0^{\lambda\alpha\mu\beta\nu} k_\nu + a_0^{\mu\beta\lambda\alpha\nu} k'_\nu + \cdots\,, \tag{5.1.23}$$

where $a_0^{\lambda\alpha\mu\beta\nu}$ is a constant tensor, which is anti-symmetric with respect to the indices $\lambda\nu$ and symmetric with respect to the indices $\mu\beta$. In the presence of a central symmetry and in an isotropic medium, the expansion of $a^{\lambda\alpha\mu\beta}(k, k')$ begins with terms of the second order in k, k'.

If $l \to 0$, i.e. if a transition to a model of the ordinary elastic continuum is performed, then $a^{\lambda\alpha\mu\beta} \to 0$, and the operator $\Phi^{\lambda\alpha\mu\beta}$ coincides with the usual tensor of the elastic constants $c_0^{\lambda\alpha\mu\beta}$. In this case, the tensor of internal stress $\sigma^{\lambda\alpha}$ is symmetric. In the general case of a nonlocal model, $a^{\lambda\alpha\mu\beta} \neq 0$ and the tensor of internal stress, generally speaking, is not symmetric, but only in the region, where the sources of internal stress are present.

Here the following question can arise: why the tensor of internal stress cannot be symmetrized with the help of transformations, analogous to those carried out in Chap. 2. Recall, that there, besides conditions of invariance with respect to translation and rotation, an essential use was made of the representability of the external distortion in the form (5.1.5), which is ensured by the condition (5.1.6). However this condition is obviously invalid in the region, where the sources of internal stress are present.

It is assumed hereafter for the sake of simplicity that $\Phi^{\lambda\alpha\mu\beta} = c^{\lambda\alpha\mu\beta}$. The case $a^{\lambda\alpha\mu\beta} \neq 0$ can be considered in an analogous way but leads to more complicated formulae.

In view of our assumption, (5.1.16) for the internal energy can now be written in the form

$$\Phi = \frac{1}{2} \langle \varepsilon_{\lambda\alpha} \mid c^{\lambda\alpha\mu\beta} \mid \varepsilon_{\mu\beta} \rangle , \tag{5.1.24}$$

and (5.1.17) takes the form of the nonlocal Hooke's law:

$$\sigma^{\lambda\alpha}(x) = \int c^{\lambda\alpha\mu\beta}(x, x')\, \varepsilon_{\beta\mu}(x')\, dx' , \tag{5.1.25}$$

which connects the symmetric tensor of internal stress with the symmetric tensor of internal strain.

From the positive definiteness of the form (5.1.24), it follows that the operator $c^{\lambda\alpha\mu\beta}$ is nondegenerate, i.e. there exists an inverse operator $b_{\mu\beta\lambda\alpha}$, which will be called the operator of elastic compliance. This enables us to invert Hooke's law and write it down in the form

$$\varepsilon_{\mu\beta}(x) = \int b_{\mu\beta\lambda\alpha}(x, x')\, \sigma^{\lambda\alpha}(x')\, dx' . \tag{5.1.26}$$

In a short operator form

$$\sigma = C\varepsilon, \quad \varepsilon = B\sigma . \tag{5.1.27}$$

Correspondingly, instead of (5.1.24) we have

$$\Phi = \frac{1}{2} \langle \sigma \mid \varepsilon \rangle = \frac{1}{2} \langle \sigma \mid B \mid \sigma \rangle = \frac{1}{2} \langle \varepsilon \mid C \mid \varepsilon \rangle . \tag{5.1.28}$$

5.2 Geometry of a Medium with Sources of Internal Stress

In this section we briefly consider the geometrical meaning of the introduced measure of internal strain. A more complete geometrical theory of the internal stress can be found, for example, in [5.2–7].

Due to the assumption, concerning the equality of the operators $\Phi^{\lambda\alpha\mu\beta}$ and $c^{\lambda\alpha\mu\beta}$, the role of the main kinematic variable, which determines the internal energy (5.1.24), is now played not by the internal distortion $\zeta_{\lambda\alpha}$ but by the internal strain $\varepsilon_{\lambda\alpha}$. This enables us to simplify the geometric picture and to confine ourselves to purely metric considerations.

Two metrics can be connected with the elastic medium. The external one gives a distance ds' between points of the medium x^α and $x^\alpha + dx^\alpha$ in a deformed state, caused by the external forces. The corresponding expression has the form

$$ds'^2 = g'_{\alpha\beta}(x) dx^\alpha \, dx^\beta \,, \tag{5.2.1}$$

where $g'_{\alpha\beta}(x) = g'_{\beta\alpha}(x)$ is the external metric tensor, which determines the external metric in the given state. Since we are working in the harmonic approximation, i.e. we are considering small deformations and small displacements, the difference between Lagrangian and Eulerian systems of coordinates is not essential. Therefore, the tensor $g'_{\alpha\beta}(x)$ can be considered as a field quantity, which is related to a fixed coordinate system x^α in space, in which the deformed medium is embedded. Without loss of generality, it can be assumed that the coordinate system x^α is cartesian, with the constant metirc tensor $\delta_{\alpha\beta}$.

Let in the initial undeformed state the external metric tensor coincide with $\delta_{\alpha\beta}$. Then, as a measure of deformation we can take the difference

$$\varepsilon'_{\alpha\beta}(x) = g'_{\alpha\beta}(x) - \delta_{\alpha\beta} \tag{5.2.2}$$

called the tensor of external strain. But the deformed state can be also described in terms of the field of displacement $u_\alpha(x)$, with which the tensor $\varepsilon'_{\alpha\beta}(x)$ is connected by the relation (5.1.11). From the linear theory of elasticity it is well known that, for the representability of $\varepsilon'_{\alpha\beta}(x)$ in the form of (5.1.11), it is necessary and sufficient that the deformation $\varepsilon'_{\alpha\beta}(x)$ satisfies the Saint-Venant conditions of compatibility. The latter can be written in the form

$$\mathrm{Rot}^{\lambda\mu\alpha\beta} \varepsilon_{\alpha\beta}(x) = 0 \tag{5.2.3}$$

or briefly

$$\mathrm{Rot}\, \varepsilon = 0 \,, \tag{5.2.4}$$

where the operator Rot is defined by the expression [2]

[2]The operator Rot is sometimes denoted also by the symbol Ink. Attention should be paid to the correct order of indices: when (5.2.5) is substituted in (5.2.3), the operators rot are contracted with different indices.

$$\mathrm{Rot}^{\lambda\mu\alpha\beta} \overset{\mathrm{def}}{=} \mathrm{rot}^{\lambda\alpha}\, \mathrm{rot}^{\mu\beta} = \varepsilon^{\lambda\nu\alpha}\, \varepsilon^{\mu\rho\beta}\, \partial_\nu\, \partial_\rho\,. \tag{5.2.5}$$

The conditions of compatability and the operator Rot admit an interesting geometrical interpretation. The external metric of the medium is induced by the metric of the Euclidian space in which the medium is embedded before and after the deformation. Therefore, both in the initial and in the final deformed state, the metric tensor $g_{\alpha\beta}(x)$ must be Euclidean. But this means that the corresponding Riemann-Christoffel curvature tensor vanishes. For small changes of the metric, the curvature tensor is linearized, and the condition that it vanishes takes the form

$$\mathrm{Rot}\; g' = 0\,, \tag{5.2.6}$$

and this determines the geometrical meaning of the operator Rot.

Problem 5.2.1. Verify that the operator Rot satisfies the identities

$$\mathrm{div}\; \mathrm{Rot} = 0\,, \tag{5.2.7}$$

$$\mathrm{Rot}\; \mathrm{grad} = 0, \quad \mathrm{Rot}\; \mathrm{def} = 0\,, \tag{5.2.8}$$

where the operators grad and def act on vector fields (recall that the operator def is the symmetrized gradient).

Problem 5.2.2. Show that the operator Rot commutes with the operators of symmeterization and alternation, and hence, conserves the symmetry of a tensor of the second order.

Problem 5.2.3. Verify that analogously to the well-known identity for the Laplacian

$$\Delta = \mathrm{grad}\; \mathrm{div} - \mathrm{rot}\; \mathrm{rot} \tag{5.2.9}$$

there exists the following identity for the bi-Laplacian:

$$\Delta^2 = \mathrm{def}(2\Delta - \mathrm{grad}\; \mathrm{div})\; \mathrm{div} + \mathrm{Rot}\; \mathrm{Rot}. \tag{5.2.10}$$

Other properties of the operator Rot and its representations in curvilinear coordinate systems can be found, for example, in [5.8, 10].

The internal metric of the medium

$$ds^2 = g_{\alpha\beta}(x)\; dx^\alpha\; dx^\beta \tag{5.2.11}$$

as distinct from the external one, is invariant with respect to embeddings of the medium in the space and is connected with the defects of structure, which were discussed above. For the phenomenological definition of the internal metric, let us cut out an element of the medium which contains a point x and is small enough for the internal stress in it to be eliminated. Then distances between points of the element in this stress-free state are identified with the internal metric at the point x. For the nonlocal theory, this definition, strictly speaking, is valid only in the cases when internal stresses do not change appreciably over distances of the order of the scale parameter l.

Let, in the state without defects of structure, the internal metric tensor coincide with $\delta_{\alpha\beta}$. As a measure of internal deformation, we take the difference

$$\varepsilon_{\alpha\beta}(x) = g_{\alpha\beta}(x) - \delta_{\alpha\beta}\,, \tag{5.2.12}$$

i.e. the internal strain tensor.

It is easily seen that the internal metric can be Euclidean only in the absence of sources of internal stress, i.e. when $g_{\alpha\beta}(x) = \delta_{\alpha\beta}$. In fact, otherwise the internal strain $\varepsilon_{\alpha\beta}(x)$ could be compensated by an external strain $\varepsilon'_{\alpha\beta}(x) = \varepsilon_{\alpha\beta}(x)$, i.e. the internal stress could be removed (globally!) with the help of external forces. But, from the results obtained in the preceeding section, we know that this is impossible. Thus, the quantity

$$\eta^{\nu\rho}(x) \overset{\text{def}}{=} \mathrm{Rot}^{\nu\rho\alpha\beta}\,\varepsilon_{\alpha\beta}(x) \tag{5.2.13}$$

differs from zero only in the region, where there are sources of internal stress, and it characterizes their density. This quantity is called an incompatibility. From the properties of the operator Rot, it follows that $\eta^{\nu\rho}(x)$ is a symmetric tensor, the divergence of which is equal to zero

$$\partial_\nu \eta^{\nu\rho}(x) = 0\,. \tag{5.2.14}$$

Taking into account (5.1.18) and (5.1.27), let us write in abreviated notations, the complete system of equations for the internal stress

$$\mathrm{div}\,\sigma = 0, \quad \mathrm{Rot}\,B\sigma = \eta \quad (\mathrm{div}\,\eta = 0)\,. \tag{5.2.15}$$

Incompatibility η and natural conditions at the infinity being given, the existence and uniqueness of the solution follow from the above.

Problem 5.2.4. Show that the incompatibility and the dislocation (or quasi-dislocation) density are connected by the relation

$$\eta^{\nu\rho}(x) = \mathrm{rot}^{(\rho|\mu|}\alpha^{\nu)}_{\cdot\ \mu}(x)\,. \tag{5.2.16}$$

We see that a given dislocation density determines uniquely the incompatibility and hence the internal stress. However, the converse it not valid. If the stress σ is given then η is defined uniquely but α is defined with some arbitrariness.

Problem 5.2.5. Verify that a given internal strain ε determines α to within rot (def $u + \omega$), where u is an arbitrary rotational vector, and ω is an arbitrary rotational antisymmetric tensor (the latter is representable in the form $\omega^{\rho\nu} = \varepsilon^{\rho\mu\nu}\partial_\mu\varphi$, where φ is an arbitrary scalar).

While considering specific problems of the theory of dislocations, it will be shown that this indeterminacy is not essential and is fully compensated by the geometric clarity of the quantity α as a characteristic of dislocations.

For a number of problems it is convenient to introduce one more (the third) characteristic of the density of sources of stress, namely the density of moments of dislocations (or quasi-dislocations) $m^{\rho\nu}$, which determines uniquely the density of dislocations

$$\alpha^{\mu\nu}(x) = \mathrm{rot}^{\mu}{}_{\cdot\rho}\, m^{\rho\nu}(x)\,. \tag{5.2.17}$$

The three characteristics are connected by

$$\eta^{\lambda\mu} = \mathrm{rot}^{(\lambda}{}_{\cdot\nu}\alpha^{\mu)\nu} = \mathrm{rot}^{(\lambda}{}_{\cdot\nu}\mathrm{rot}^{\mu)}{}_{\cdot\rho}\, m^{\rho\nu} = \mathrm{Rot}^{(\lambda\mu)}{}_{\cdot\cdot\rho\nu} m^{\rho\nu}\,. \tag{5.2.18}$$

Thus, from the viewpoint of internal stress, the density of moments of dislocations is defined with even more indeterminacy than the density of dislocations itself. In particular, the anti-symmetric component $m^{[\rho\mu]}$ does not affect $\eta^{\lambda\mu}$ and, hence, $\sigma^{\alpha\beta}$ (but not $\alpha^{\mu\nu}$). We consider this situation in more detail in Section 6.1, with the example of a dislocation loop.

5.3 Green's Tensors for Internal Stress

Let us represent the solution of the system (5.2.15) in two forms

$$\sigma^{\alpha\beta}(x) = \int Z^{\alpha\beta\lambda\mu}(x, x')\,\eta_{\lambda\mu}(x')\,dx' = \int G^{\alpha\beta\lambda\mu}(x, x')\,m_{\lambda\mu}(x')\,dx' \tag{5.3.1}$$

or briefly,

$$\sigma = Z\eta = Gm\,. \tag{5.3.2}$$

The tensor operators Z and G obviously, play here the role of the Green's operators for internal stress. From (5.2.18) it follows that they are connected by the relation

$$G = Z\,\mathrm{Rot}\,. \tag{5.3.3}$$

Substituting (5.3.2) in (5.2.15), we find that the Green's operator G must satisfy the following equations

$$\operatorname{div} G = 0, \ \operatorname{Rot} BG = \operatorname{Rot}. \tag{5.3.4}$$

For G to be uniquely determined, let us require additionally that it be Hermitian: $G = G^{+}$. This is equivalent to the condition that its kernel $G^{\alpha\beta\lambda\mu}(x, x')$ possesses the symmetry of the kernel $c^{\alpha\beta\lambda\mu}(x, x')$ of the operator of elastic moduli. It is also necessary to require the fufillment of natural conditions at infinity.[3]

Taking into account the first equation of the system (5.3.4) and the condition of Hermiticity, let us represent the operator G in the form

$$G = \operatorname{Rot} H \operatorname{Rot}, \tag{5.3.5}$$

where the Hermitian operator H is determined not uniquely. Later we shall use this indeterminacy in order to obtain the simplest expression for H.

The internal elastic energy, which is defined by (5.1.24 or 28), can be now represented in the forms

$$\Phi = \frac{1}{2} \langle m \mid G \mid m \rangle = \frac{1}{2} \langle \alpha \mid \operatorname{rot} H \operatorname{rot} \mid \alpha \rangle = \frac{1}{2} \langle \eta \mid H \mid \eta \rangle, \tag{5.3.6}$$

the convenience of a choice of this or that being determined by a specific nature of sources of internal stress.

Note that (5.3.6) could be used as the basis for a definition of the operators G and H, if we consider the conditions which must be satisfied by the sources of internal stress as being given. Now, our concrete problem is an explicit construction of the Green's tensor $G(\mathbf{r})$ for the homogeneous isotropic medium with the kernel of the operator of elastic moduli (3.8.4), which, in the brief notation is written in the form [4]

$$c(k) = 2\mu(k)E_1 + \lambda(k)E_2. \tag{5.3.7}$$

Here E_1 and E_2 are the constant fourth order tensors:

$$E_1^{\alpha\beta\lambda\mu} = \frac{1}{2} (\delta^{\alpha\lambda}\delta^{\beta\mu} + \delta^{\alpha\mu}\delta^{\beta\lambda}), \ E_2^{\alpha\beta\lambda\mu} = \delta^{\alpha\beta}\delta^{\lambda\mu}. \tag{5.3.8}$$

The kernel $b(k)$ of the operator of the elastic compliance is obtained by inverting the tensor $c(k)$. In the Appendix A4, a table for multiplication of the fourth order tensors is given, from which it can be seen that E plays the role

[3] In the local limit, G coincides with the operator S of Chap. 4. Earlier, the symbol G denoted the Green's operator for displacements, but the latter is not used in Chap. 5 and 6, which excludes a possibility of confusion.

[4] Here, as in Sect. 2.8, k and r denote the moduli of the vectors $\mathbf{k}$ and $\mathbf{r}$.

of a unit in this tensor algebra. It can be easily proved that

$$b(k) = \frac{1}{2\mu(k)} E_1 - \frac{\lambda(k)}{2\mu(k)\,[3\lambda(k) + 2\mu(k)]} E_2\,. \tag{5.3.9}$$

Let us write the second equation of the system (5.3.4) in the k-representation

$$R^{\alpha\beta\nu\rho}(\mathbf{k}) b_{\nu\rho\kappa\tau}(k)\, G^{\kappa\tau\lambda\mu}(\mathbf{k}) = R^{\alpha\beta\lambda\mu}(\mathbf{k})\,, \tag{5.3.10}$$

where

$$R^{\alpha\beta\lambda\mu}(\mathbf{k}) = -(\varepsilon^{\alpha\rho\lambda}\,\varepsilon^{\beta\tau\mu}\, k_\rho k_\tau)_{(\lambda\mu)} \tag{5.3.11}$$

is the Fourier transform of the symmetrized operator Rot.

The solution of (5.3.10) must satisfy the first equation of the system (5.3.4), i.e.

$$k_\kappa G^{\kappa\tau\lambda\mu}(\mathbf{k}) - 0\,, \tag{5.3.12}$$

and have the required singularity, as $\mathbf{k} \to 0$, which ensures the fulfillment of the conditions for $G(\mathbf{r})$ at infinity. Omitting calculations, we present the final result

$$G(\mathbf{k}) = R(\mathbf{k}) H(k) R(\mathbf{k})\,, \tag{5.3.13}$$

where

$$H(k) = \frac{2\mu(k)}{k^4}\left(E_1 + \frac{\lambda(k)}{\lambda(k) + 2\mu(k)} E_2\right). \tag{5.3.14}$$

In order to pass to the $\mathbf{r}$-representation, it is necessary to give explicitly the functions $\mu(k)$ and $\lambda(k)$ or, equivalently $\omega_{\mathrm{l}}(k)$ and $\omega_{\mathrm{t}}(k)$, i.e. the frequencies of the longitudinal and transverse eigenvibrations, which are connected with $\mu(k)$ and $\lambda(k)$ by (2.8.10, 11).

In what follows we confine ourselves to the Debye model of quasi-continuum introduced in Sect. 2.9, i.e. we assume $k \leq \kappa = \pi/a$, where κ is Debye's truncation radius and a is the scale parameter. Let us take for $\omega_{\mathrm{i}}(k)$ ($\mathrm{i} = \mathrm{l}, \mathrm{t}$) the simplest polynomial approximation, which, as distinct from the case considered in Sect. 2.9, is conveniently represented in the form

$$\omega_{\mathrm{i}}^2(k) = v_{\mathrm{i}}^2 k^2 \left[- \gamma_{\mathrm{i}} \kappa^{-2} k^2 + \frac{1}{3}(2\gamma_{\mathrm{i}} - 1)\kappa^{-4} k^4 \right], \tag{5.3.15}$$

where v_{i} is the velocity of sound when $k = 0$, γ_{i} is an arbitrary parameter connected with the limiting value of the frequency $\omega_{\mathrm{i}}(\kappa)$,

$$\frac{2 - \gamma_{\mathrm{i}}}{3} = \frac{\omega_{\mathrm{i}}^2(\kappa)}{v_{\mathrm{i}}^2\, \kappa^2}\,. \tag{5.3.16}$$

Hence $-\infty < \gamma_i < 2$. If we assume additionally that the limiting frequencies are less than the Debye ones, then $-1 < \gamma_i < 2$. The value $\gamma_i = 0.8$ practically corresponds to the Born-Karman model. It is also easy to verify that (5.3.15) satisfies the conditions (2.9.7) for group velocity.

To simplify the formulae, let us assume that $\gamma_l = \gamma_t = \gamma$. This is equivalent to the assumption of independence of the Poisson coefficient $\nu = \nu_0$ of k. Then, a comparison of (5.3.15) with (2.8.10, 11) yields

$$\mu(k) = \mu_0 \tilde{\mu}(k) , \tag{5.3.17}$$

where $\mu_0 = \mu(0)$, and $\tilde{\mu}(k)$ is a dimensionless function:

$$\tilde{\mu}(k) = 1 - \gamma\kappa^{-2}k^2 + \frac{1}{3}(2\gamma - 1)\kappa^{-4}k^4 \tag{5.3.18}$$

Eq. (5.3.14) assumes the form

$$H(k) = 2k^{-4}\mu(k)\left(E_1 + \frac{\nu_0}{1 - \nu_0} E_2\right). \tag{5.3.19}$$

Thus, the problem is reduced to the determination of the inverse Fourier transform of the function $f(k) = k^{-4}\mu(k)$ under the condition $k \leq \kappa$. Calculations, analogous to those carried out in Sect. 2.9 yield

$$f(r) = \mu(\Delta)h(r) , \tag{5.3.20}$$

where $h(r)$ is given by (2.9.4) and

$$\begin{gathered} \mu(\Delta) = \mu_0 \tilde{\mu}(\Delta) , \\ \tilde{\mu}(\Delta) = 1 + \gamma\kappa^{-2}\Delta + \frac{1}{3}(2\gamma - 1)\kappa^{-4}\Delta^2 . \end{gathered} \tag{5.3.21}$$

For the Debye model we set $\tilde{\mu}(\Delta) = 1$.

The expression for $H(r)$ has now the form

$$H(r) = 2\left(E_1 + \frac{\nu_0}{1 - \nu_0} E_2\right)f(r) , \tag{5.3.22}$$

and taking into account (5.3.13), we obtain

$$G^{\alpha\beta\lambda\mu}(\mathbf{r}) = \mathrm{Rot}^{\alpha\beta\nu\rho}\,\mathrm{Rot}^{\lambda\mu\kappa\tau}\,H_{\nu\rho\kappa\tau}(r) . \tag{5.3.23}$$

This expression can also be represented in the form

$$G(\mathbf{r}) = \frac{2}{1 - \nu_0}\,\mathrm{Rot}\,\mathrm{Rot}\,\mu(\Delta)\,h(r) + \frac{2\nu_0}{1 - \nu_0}\,\mathrm{Rot}\,\mu(\Delta)\,g(r) , \tag{5.3.24}$$

where $g(r)$ is given by (2.9.3).

It is to be emphasized that, in the Debye quasicontinuum, the Green's tensors G and H are entire functions, and consequently, do not have singularities, as $r \to 0$. This enables us to write them in the form of well converging series. In particular, for $f(r)$, the expansion

$$f(r) = \frac{\mu_0}{3\pi^2\kappa} \sum_{n=0}^{\infty} \frac{(-1)^n[2(2n+1)(2n+5) - \gamma(2n-1)(2n+7)]}{(4n^2-1)(2n+3)(2n+1)!} (\kappa r)^{2n} \tag{5.3.25}$$

is valid. Using the expansion for $f(r)$, it is easy to obtain the value of $G(\mathbf{r})$ for $\mathbf{r} = 0$, which plays a significant role in the theory of point defects:

$$G(0) = \frac{\beta\mu_0\kappa^3}{45\pi^2(1-\nu_0)} [(7-5\nu_0)E_1 + (1+5\nu_0)E_2]. \tag{5.3.26}$$

The dimensionless multiplier β depends on the function $\omega(k)$ and is ultimately determined by the coefficient of r^4 in the expansion of $f(r)$. In the general case

$$\beta = \frac{3\rho}{\mu_0\kappa^3} \int_{x}^{0} \omega_t^2(k)\, dk = \frac{3}{\kappa^3} \int_{0}^{x} \tilde{\mu}(k)\, k^2\, dx. \tag{5.3.27}$$

For the Debye model and for the approximation (5.18) we have, respectively,

$$\beta = 1, \quad \beta = \frac{30 - 11\gamma}{35}. \tag{5.3.28}$$

The asymptotics of the Green's tensor, for large r or, equivalently, as $a \to 0$, is of interest. It is not difficult to show that the essential contribution to the asymptotics is made only by the first two terms of, the expansions of the type (5.3.15) of (5.3.18). The terms of higher orders give rapidly decaying oscillations with periods of the order of a, and must be excluded from the asymptotic expansion. The asymptotics of G is determined by the asymptotics of the function $f(r)$. The latter has the form

$$f(r) \simeq -\frac{\mu_0 r}{8\pi}\left(1 + \frac{2\gamma}{\kappa^2 r^2}\right). \tag{5.3.29}$$

The substitution into (5.3.23) yields

$$G(r) \simeq \frac{\mu_0}{r^3} f_1(n) + \frac{\mu_0\gamma a^2}{r^5} f_2(n), \tag{5.3.30}$$

where f_1, f_2 are dimensionless tensor functions of the unit vector $n^\alpha = x^\alpha/r$

and the Poisson coefficient ν_0. For the Debye model, the second term in (5.3.30) is absent and the asymptotics coincides with the zeroth approximation (with $a = 0$). In the general case, the second term in (5.3.30) accounts for the deviation of $\omega(k)$ from the linear law and contains approximate information about the presence of microstructure in the medium.

In conclusion, let us note that if the Green's tensor for an isotropic medium is known, an explicit expression for the Green's tensor of a medium with weak anisotropy can be constructed.

Let the operator of elastic compliance B have the form

$$B = B_0 - B_1 . \tag{5.3.31}$$

where B_0 is the operator of the isotropic medium, B_1 is the operator of perturbation. Then, taking into account (5.3.4), by standard methods of the perturbation theory, we obtain the expression for the Green's tensor in the first approximation

$$G = G_0 + G_0 B_1 G_0 , \tag{5.3.32}$$

5.4 Isolated Point Defect

The point defects like vacancies, interstitial atoms, substitutional impurity atoms, etc, are naturally considered both as sources of internal stress and as local inhomogeneities.

As a characteristic of a point defect as a source of internal stress, it is convenient to take a dimensionless symmetric tensor $M_{\lambda\mu}$ which is a moment of the defect. It characterizes the relative volume or shear distortion of an elementary cell, caused by the defect. In particular, in the case of a spherically symmetric defect (a center of dilatation) $M_{\lambda\mu} = M\delta_{\lambda\mu}$, where M is the ratio of the change of volume of the defect Δ, to the volume of an elementary cell v; more precisely $\Delta v/v = 3M$. To a single defect at the point x_0 there corresponds the density of moments

$$m_{\lambda\mu}(x) = vM_{\lambda\mu}\delta(x - x_0) . \tag{5.4.1}$$

The change of force constants in a neighborhood of the point defect, as distinct from Sect. 4.2, is now conveniently described with the help of the kernel $b(x, x')$ of the operator of elastic compliance B

$$b(x, x') = b_0(x - x') - vb_1\delta(x - x_0)\,\delta(x' - x_0) , \tag{5.4.2}$$

where $b_0(x - x')$ is the kernel of the operator B_0 of the homogeneous medium, b_1 is a constant tensor, which has the symmetry and dimension of the tensor of elastic compliance.

Let us denote by $G_0(x - x')$ the Green's tensor of the homogeneous medium, which corresponds to $b_0(x - x')$ and by $G(x, x')$ the Green's tensor of the medium with the defect, relating to $b(x, x')$. Omitting calculations, which are analogous to those, carried out in Sect. 4.2, let us write down the expression for $G(x, x')$ in terms of $G_0(x - x')$

$$G^{\alpha\beta\lambda\mu}(x, x') = G_0^{\alpha\beta\lambda\mu}(x - x') - G_0^{\alpha\beta\sigma\tau}(x - x_0)\, P_{\sigma\tau\kappa\nu}\, G_0^{\kappa\nu\lambda\mu}(x_0 - x'). \quad (5.4.3)$$

The constant tensor $P_{\sigma\tau\kappa\nu}$ has the symmetry of the tensor of elastic constants and is the inverse of the tensor

$$R^{\alpha\beta\lambda\mu} = G_0^{\alpha\beta\lambda\mu}(0) - v^{-1}\tilde{c}_1^{\alpha\beta\lambda\mu}, \quad (5.4.4)$$

where $\tilde{c}_1$ is the inverse of the tensor b_1.

The internal stress caused by the defect, according to (5.3.1), has the form

$$\begin{aligned} \sigma^{\alpha\beta}(x) &= vM_{\lambda\mu}G^{\alpha\beta\lambda\mu}(x, x_0) \\ &= vM_{\lambda\mu}[G_0^{\alpha\beta\lambda\mu}(x - x_0) - G_0^{\alpha\beta\nu\rho}(x - x_0)P_{\nu\rho\kappa\tau}G_0^{\kappa\tau\lambda\mu}(0)] \\ &= vM'_{\lambda\mu}G_0^{\alpha\beta\lambda\mu}(x - x_0), \end{aligned} \quad (5.4.5)$$

where $M'_{\lambda\mu}$ is the effective moment of the defect, which depends on the dispersion law and the ratio of elastic constant of the defect and that of the medium. Taking into account (5.4.4), after simple transformations one can obtain the expression for $M'_{\lambda\mu}$:

$$M'_{\lambda\mu} = [E'_{\lambda\mu\nu\rho} - vb'_{\lambda\mu\sigma\tau}G_0^{\sigma\tau}{}_{\cdot\cdot\nu\rho}(0)]^{-1}M^{\nu\rho}, \quad (5.4.6)$$

where E_1 is defined by (5.3.8).

From (5.3.6), we find the self-energy of the defect

$$\begin{aligned} 2\Phi &= v^2M_{\alpha\beta}G^{\alpha\beta\lambda\mu}(x_0, x_0)M_{\lambda\mu} = v^2M_{\alpha\beta}G_0^{\alpha\beta\lambda\mu}(0)M_{\lambda\mu} \\ &\quad - v^2M_{\alpha\beta}G_0^{\alpha\beta\nu\rho}(0)P_{\nu\rho\kappa\tau}G_0^{\kappa\tau\lambda\mu}(0)M_{\lambda\mu}. \end{aligned} \quad (5.4.7)$$

Taking into account (5.4.4), this expression can be transformed into

$$2\Phi = v^2M(g_0 - vb_1)^{-1}M, \quad (5.4.8)$$

where

$$g_0^{-1} = G_0(0). \quad (5.4.9)$$

Let us consider the interaction between the defect and another field of stress. We denote by $\tau(x)$ the unperturbed stress, i.e. the stress which would exist in the medium in the absence of the defect. If $\tau(x)$ is caused by external forces, then the defect behaves only as a local inhomogeneity. This case was considered in

Sect. 4.2. Therefore, let us assume that $\tau(x)$ is an internal stress, caused by some distant sources with the moment density $\tilde{m}(x)$. The resultant stress field will now consist of the defect field $\sigma(x)$, the field $\tau(x)$ and a perturbed field $\tilde{\tau}(x)$. For the latter, taking into account (4.3), it is easy to obtain

$$\tilde{\tau}^{\alpha\beta}(x) = \tilde{M}_{\lambda\mu} G_0^{\alpha\beta\lambda\mu}(x - x_0) , \tag{5.4.10}$$

where

$$\tilde{M}_{\lambda\mu} = -P_{\lambda\mu\nu\rho}\tau^{\nu\rho}(x_0) . \tag{5.4.11}$$

Thus, the perturbed field coincides with the field of a point source of internal stress with the moment $\tilde{M}$.

For the energy of interaction Φ_{int} of the field $\tau(x)$ with the field of the defect, we have

$$\Phi_{\mathrm{int}} = \langle m_{\alpha\beta} \mid G^{\alpha\beta\lambda\mu} \mid \tilde{m}_{\lambda\mu}\rangle = vM'_{\alpha\beta}\tau^{\alpha\beta}(x_0) , \tag{5.4.12}$$

where $M'_{\alpha\beta}$ is an effective moment of the defect. Hence, the force, which acts on the defect, is equal to

$$f_\lambda = -vM'_{\alpha\beta}\partial_\lambda\tau^{\alpha\beta}(x_0) . \tag{5.4.13}$$

Let us proceed to the case of an isotropic medium. The kernel of the operator B_0 is conveniently represented in the form [5]

$$b_0(\mathbf{r} - \mathbf{r}') = \left[\frac{1}{3}\chi_0^{-1}(\Delta)I_1 + \frac{1}{2}\mu_0^{-1}(\Delta)I_2\right]\delta(\mathbf{r} - \mathbf{r}') . \tag{5.4.14}$$

Here, I_1, I_2 is an orthonormalized basis for isotropic tensors

$$I_i \cdot I_k = \delta_{ik} I_k, \qquad I_1 = \frac{1}{3}E_1, \qquad I_2 = E_1 - \frac{1}{3}E_2 . \tag{5.4.15}$$

The operators $\chi_0(\Delta)$ and $\mu_0(\Delta)$ are the analogs of the modulus of volume compression χ_0 and the shear modulus μ_0. Assuming, as in Sect. 2.3, that the Poisson coefficient ν_0 is a constant, we have

$$\begin{aligned} \mu_0(\Delta) &= \mu_0\tilde{\mu}(\Delta) , \\ \chi_0(\Delta) &= \frac{2\mu_0(1 + \nu_0)}{3(1 - 2\nu_0)}\tilde{\mu}(\Delta) = \chi_0\tilde{\mu}(\Delta) , \end{aligned} \tag{5.4.16}$$

where $\tilde{\mu}(\Delta)$ is determined by the dispersion law. Recall (Sect. 2.3) that, for the

---

[5] As above, in the case of an isotropic medium we introduce the notations **r** and **k** for the position-vectors of points and r, k for their moduli.

Debye model, $\tilde{\mu}(\Delta) = 1$ and, for the model with a polynomial approximation of the dispersion curve, the operator $\mu(\Delta)$ is given by (5.3.21).

Set now $B = B_0 - B_1$, where B_1 is the operator of perturbed elastic compliance. For better physical clarity it is expedient to express the kernel of the operator B_1 in terms of the kernel of the operator of perturbed elastic moduli C_1. Confining ourselves to the case of a spherically symmetric defect, we obtain

$$b_1(\mathbf{r}, \mathbf{r}') = v\left(\frac{\chi_1}{3\chi_0(\chi_0 + \chi_1)} I_1 + \frac{\mu_1}{2\mu_0(\mu_0 + \mu_1)} I_2\right)\delta(\mathbf{r} - \mathbf{r}_0)\,\delta(\mathbf{r}' - \mathbf{r}_0)\,, \tag{5.4.17}$$

where χ_1, μ_1 is the perturbation of the elastic moduli in a neighborhood of the defect.

From (5.4.4), (5.3.26) and (5.4.17), in view of the properties of the basis (5.4.15), we obtain the expression for the tensor P

$$\begin{aligned} P &= p_1 I_1 + p_2 I_2\,,\\ p_1^{-1} &= \frac{2\mu_0(1+\nu_0)}{v(1-2\nu_0)}\left(\frac{2\beta(1-2\nu_0)}{3(1-\nu_0)} - \frac{\chi_0+\chi_1}{\chi_1}\right),\\ p_2^{-1} &= \frac{2\mu_0}{v}\left(\frac{\beta(7-5\nu_0)}{15(1-\nu_0)} - \frac{\mu_0+\mu_1}{\mu_1}\right). \end{aligned} \tag{5.4.18}$$

The formulae (5.4.3), (5.3.24) and (5.4.18) give an explicit expression for the Green's tensor $G(\mathbf{r}, \mathbf{r}')$ of an isotropic medium with a point defect.

Let the defect be a centre of dilatation, i.e. $M_{\lambda\mu} = M\delta_{\lambda\mu}$. Then, for the stress caused by the defect, we find from (4.15) (for convenience of notation we assume $\mathbf{r}_0 = 0$)

$$\sigma^{\alpha\beta}(\mathbf{r}) = \frac{v(1+\nu_0)M'}{\pi^2(1-\nu_0)}\,\mathrm{rot}^{\alpha\nu}\,\mathrm{rot}_{\nu}^{\cdot\beta}(\Delta)\,\frac{\mathrm{Si}\,\kappa r}{r}\,, \tag{5.4.19}$$

where the effective moment M', according to (5.4.6), is determined by the expression

$$M' = \left(1 - \frac{2\beta(1-2\nu_0)}{3(1-\nu_0)}\,\frac{\chi_1}{\chi_0+\chi_1}\right)^{-1} M\,. \tag{5.4.20}$$

Let us consider two limiting cases. When $\kappa r \gg 1$, we have

$$\sigma^{\alpha\beta}(\mathbf{r}) \approx \frac{v\mu_0(1+\nu_0)M'}{2\pi(1-\nu_0)}\,\partial^\alpha\partial^\beta\,\frac{1}{r}\,. \tag{5.4.21}$$

To the other limiting case there corresponds the value $r = 0$ and

$$\sigma^{\alpha\beta}(0) = \frac{4\beta(1+\nu_0)\mu_0 M'}{3(1-\nu_0)}\,\delta^{\alpha\beta}\,. \tag{5.4.22}$$

The expressions (5.4.21, 22) admit a simple interpretation. Suppose that in a homogeneous isotropic medium with the moduli μ_0, χ_0 there is a spherical cavity of volume v and let a sphere of volume $v + \Delta v$ with different elastic moduli $\mu' = \mu_0 + \mu_1$, $\chi' = \chi_0 + \chi_1$ be inserted into this cavity. Then the stresses inside and outside the sphere exactly coincide with (5.4.21, 22), if we set $\beta = 1$, which corresponds to the Debye model. Thus, (5.4.19) contains, as a particular limiting case, a solution of the classical problem of an elastic inclusion of the type of a center of dilatation [5.10].

In the case of rigid inclusion ($\chi_1 \to \infty$) the relation between M' and M, according to (5.4.20), is the following:

$$M' = \left(1 - \frac{2\beta(1 - 2\nu_0)}{3(1 - \nu_0)}\right)^{-1} M . \tag{5.4.23}$$

When $\chi_1 = 0$, the moments M' and M coincide. Finally, the case $\chi_1 = -\chi_0$ corresponds to a cavity for which the moment M and the stress σ vanish.

In connection with this, there arises a question about an adequate model of a vacancy. From physical reasonings, it is obvious that we should set $\mu_1 = -\mu_0$, $\chi_1 = -\chi_0$, i.e. a vacancy will behave as a cavity with respect to an external field. The components of the tensor P, according to (5.4.18), will have in this case the form

$$p_1 = \frac{3v(1 - \nu_0)}{4\beta\mu_0(1 + \nu_0)}, \quad p_2 = \frac{15v(1 - \nu_0)}{2\beta\mu_0(7 - 5\nu_0)} \cdot \tag{5.4.24}$$

However, at the same time, a vacancy is a source of internal stress, its asymptotics being of the form (5.4.21), and, as $r \to 0$, the condition $\sigma(0) = 0$ should also be satisfied. The corresponding model in an elastic medium is a hollow sphere, which is inserted with stress into a cavity.

It is easy to show that all the above conditions can be satisfied, if one assumes that to the vacancy there corresponds the moment density

$$m_{\lambda\mu}(\mathbf{r}) = M'\delta_{\lambda\mu}(1 + \alpha\kappa^{-2}\Delta)\,\delta(\mathbf{r}) . \tag{5.4.25}$$

Then the stress caused by the vacancy takes the form (for the Debye model)

$$\sigma^{\alpha\beta}(\mathbf{r}) = \frac{v(1 + \nu_0)\mu_0 M'}{\pi^2(1 - \nu_0)} \operatorname{rot}^{\alpha\nu} \operatorname{rot}^{\;\beta}_{\nu} \left[\frac{\operatorname{Si} \kappa r}{r} - 2\pi^2\alpha\kappa^{-2}\delta_\kappa(\mathbf{r})\right]. \tag{5.4.26}$$

The asymptotics of $\sigma(\mathbf{r})$ coincides with (5.4.21) and the coefficient α is found from the condition $\sigma(0) = 0$. Calculation yields $\alpha = 5/3$. It is natural that the effective moment M' of the vacancy is no longer determined by (5.4.20), it should be regarded as a known quantity.

Let us also present the expression for the self-energy Φ of the defect of the type of a dilatation center. From (5.4.8), for the Debye model, we obtain

$$\Phi = \frac{2v(1+\nu_0)\mu_0 M^2}{1-\nu_0}\left(1 - \frac{2(1-2\nu_0)}{3(1-\nu_0)}\,\frac{\chi_1}{\chi_0+\chi_1}\right)^{-1}. \qquad (5.4.27)$$

It can be shown that this expression coincides exactly with the energy of the field of a sphere inserted with stress into the elastic medium.

For interaction of the field of stress $\tau(\mathbf{r})$ with a defect of the type of a center of dilation, we obtain from (5.4.12, 13)

$$\Phi_{\text{int}} = vM' \operatorname{Tr}\{\tau(0)\}, \qquad (5.4.28)$$

$$\mathbf{f} = -vM' \operatorname{grad} \operatorname{Tr}\{\tau(0)\}, \qquad (5.4.29)$$

where $\operatorname{Tr}\{\tau\}=\tau^{\alpha}_{\alpha}$, and the moment M' is determined by (5.4.20).

5.5 System of Point Defects

Let us consider a system of point sources of internal stress with the moment density

$$m_{\lambda\mu}(x) = \sum_{\mu} M^{i}_{\lambda\mu}\delta(x - x_i). \qquad (5.5.1)$$

In the general case, with the sources of internal stress there is associated a local change of elastic constants, and hence, for the kernel of the operator B we should set

$$b_{\alpha\beta\lambda\mu}(x, x') = b^{0}_{\alpha\beta\lambda\mu}(x - x') - v\sum_{i} b^{i}_{\alpha\beta\lambda\mu}\delta(x - x_i)\delta(x' - x_i), \qquad (5.5.2)$$

where the constant tensors b^i have dimension and symmetry of the tensor of elastic compliance.

For the Green's tensor $G(x, x')$ of the medium with defects, it is possible to obtain the expression, see (4.4.2),

$$G^{\alpha\beta\lambda\mu\prime}(x, x') = G_0^{\alpha\beta\lambda\mu}(x - x') - \sum_{ij} G_0^{\alpha\beta\sigma\tau}(x - x_i)P^{ij}_{\sigma\tau\kappa\nu}G_0^{\kappa\nu\lambda\mu}(x_j - x'). \quad (5.5.3)$$

The matrix P^{ij} with tensor components is the inverse of the matrix

$$R_{ij} = G_0(x_i - x_j) - v^{-1}\tilde{c}_i\delta_{ij}, \qquad (5.5.4)$$

where $\tilde{c}_i = b_i^{-1}$ (note that $\tilde{c}_i$ do not coincide with c_i in (4.4.1)).

In the case of two defects, the components P^{ij} are expressed in terms of the components R_{ij} according to (9.4.5).

The internal stress caused by a system of defects with the moment density (5.4.1), according to (5.3.1), has the form

$$\sigma^{\alpha\beta}(x) = v \sum_i M^i_{\lambda\mu} G^{\alpha\beta\lambda\mu}(x, x_i) . \tag{5.5.5}$$

In the presence of local inhomogeneities, this expression cannot be, of course, obtained by superposing the fields of separate point sources of internal stress.

For the elastic energy of the system of defects from (5.3.6) we find

$$\Phi = \frac{v^2}{2} \sum_{ij} M^i_{\alpha\beta} M^j_{\lambda\mu} G^{\alpha\beta\lambda\mu}(x_i, x_j) . \tag{5.5.6}$$

The above expressions give a solution of the problem of elastic fields, self-energy, energy of interaction and consequently of the forces of interaction in a system of point defects.

Let us consider in more detail the case of two defects of the type of a centre of dilatation in an isotropic medium. Making use of a special tensor basis (Appendix A4) enables us, in this case, to write down the final results in an explicit form. Since the general formulae are rather involved, let us confine ourselves to the first terms of the expansion of the elastic energy in the inverse powers of the distance ρ between the defects

$$\Phi = \Phi_0 + \Phi_1 \rho^{-6} - \Phi_2 \rho^{-9} + O(\rho^{-11}) . \tag{5.5.7}$$

The first term Φ_0, as it should be, is equal to the sum of the self-energies of each defect, which are given by (5.4.27). For the coefficients Φ_1 and Φ_2, we have

$$\begin{aligned} \Phi_1 &= \frac{3\mu_0 v^3}{8\pi^2} \left(\frac{1+\nu_0}{1-\nu_0}\right)^2 \left(\frac{\mu_1'}{m_2 h_1^2} M_1^2 + \frac{\mu_2'}{m_1 h_2^2} M_2^2\right), \\ \Phi_2 &= \frac{3\mu_0 v^4}{8\pi^3} \frac{(1+\nu_0)^2 (2-\nu_0)}{(1-\nu_0)} \frac{3\mu_1' \mu_2'}{m_1 m_2 h_1 h_2} M_1 M_2 , \end{aligned} \tag{5.5.8}$$

where

$$\begin{aligned} m_i &= 1 - \beta \frac{7 - 5\nu_0}{15(1-\nu_0)} \mu_i', \quad h_i = 1 - \frac{2}{3} \beta \frac{1 - 2\nu_0}{1 - \nu_0} \chi_i', \\ \mu_i' &= \frac{\mu_i}{\mu_0 + \mu_i}, \quad \chi_i' = \frac{\chi_i}{\chi_0 + \chi_i} . \end{aligned} \tag{5.5.9}$$

The expression for the interaction force is obtained by differentiating (5.5.7) with respect to ρ. At sufficiently large distances, defects of the same sign repel and defects of different signs always attract each other. With decreasing distance between the defects, the interaction forces can change sign, provided the limiting value of the dispersion curve exceeds the Debye frequency. In this case, a generation of stable pairs of defects of different signs is possible, i.e. the so-called dipoles of dilatation.

In the paper by *Lifshitz* and *Tanatarov* [5.11], the first term of the expansion of the elastic energy of interaction for two identical centers of dilatation in an ordinary isotropic elastic medium was calculated. Comparison shows that it coincides with the term $\Phi_1 \rho^{-6}$ shown above for the case of identical defects in the Debye model.

5.6 Notes

Internal stress in the local theory of elasticity is discussed, for example, in [B6.5–7, 11, 15–18, 35]. Internal stress in the nonlocal theory of elasticity was considered in [B3.21]. Point defects as sources of internal stress in Debye's quasicontinuum were investigated by Brailsford [B6.2] and Kosilova, Kunin and Sosnina [B6.14].

6. Dislocations

There is an extensive literature devoted to crystal lattice dislocations and to the continuum theory of dislocations based on the local theory of elasticity. The purpose of this chapter is to consider the theory of dislocations on an intermediate level—in the scope of non-local elasticity. The main focus is on the specific non-local effects which cannot be described in principle in the local elasticity.

6.1 Elements of the Continuum Theory of Dislocations

We suppose that the reader is familiar with the basic physical notion for dislocations as elementary carriers of plastic deformation in crystals. At the present, elements of the theory of dislocations are standard tools for mechanical engineers and stress analysts. Descriptions of the theory of dislocations, which are intended for different groups of readers, can be found, for example, in [6.1, 2]. In this section we present some facts from the continuum theory of dislocations, which are necessary for what follows.

Recall that a dislocation in a crystal is a linear defect of the crystal lattice of a special type. In Fig. 6.1 an edge dislocation is schematically shown, its line

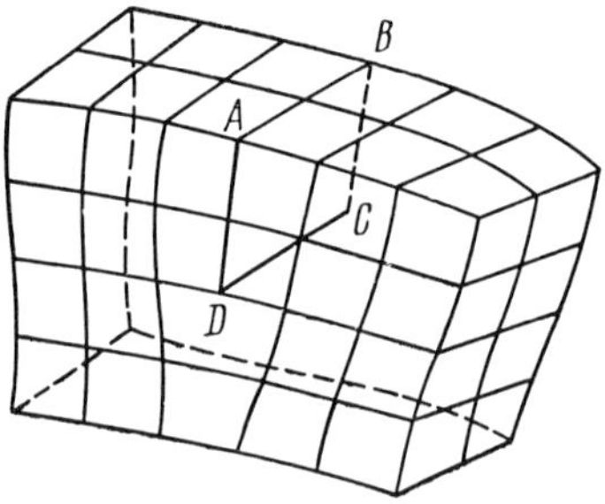

Fig. 6.1 Edge dislocation

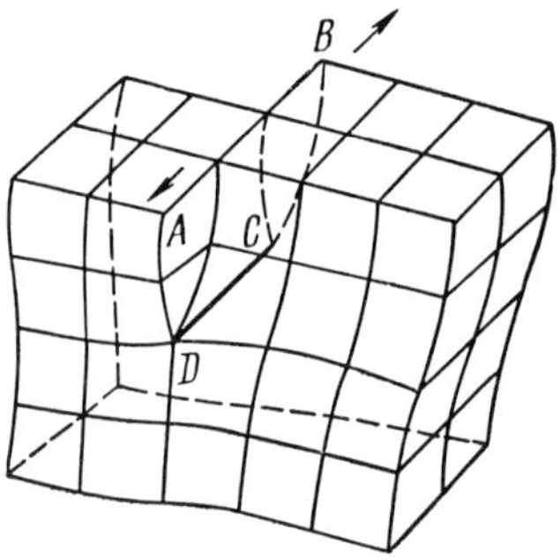

Fig. 6.2. Screw dislocation

coinciding with edge of an extra atomic semi-plane ABCD. The segment DC belongs to the line of dislocation. In Fig. 6.2, a screw dislocation is schematically shown, this being formed by the displacements of the surfaces of the cut ABCD in the direction AB. Here, the segment DC belongs to the line of the dislocation. The general case of a curvilinear dislocation can be considered as some sort of combination of edge and screw components.

In Fig. 6.3, the edge dislocation L is surrounded by a closed contour Γ which is called the Burgers circuit. Its image Γ' in a crystal without dislocation has a discontinuity. The vector b required to close the contour, is called the Burgers vector. For an edge dislocation the vector b is perpendicular to the dislocation line L, while for a screw dislocation it is parallel to the latter.

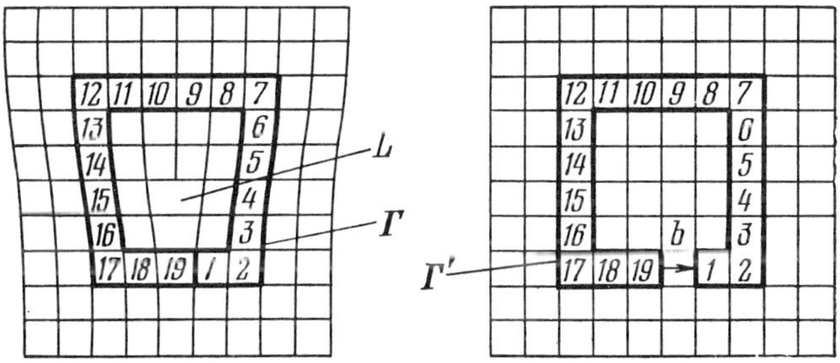

Fig. 6.3. Burgers circuits

In the general case, the Burgers vector has an arbitrary angle with respect to the dislocation line, this angle being able to change along the line for a linear dislocation, but the Burgers vector remains constant. It is the existence of a Burgers circuit with a non-zero constant vector b that is the definition of a dislocation in a crystal. From the definition, it follows that the dislocation line is always either closed or ends at the edge of the crystal (or goes to infinity in the case of an unbounded crystal).

The plane, which passes through the Burgers vector and is tangent to the line of dislocation, is called the slip plane. Dislocations move mostly along these planes.

In the continuum theory of dislocations a real crystal is replaced by a continuous elastic medium and the dislocation is modelled by a singular line L, which possesses the property that the integral

$$b_\mu = \int_\Gamma \partial_\lambda u_\mu d\Gamma^\lambda \tag{6.1.1}$$

has a constant value when taken over any closed contour Γ enclosing L. Here u_μ are displacements in the medium, and Γ and b are called the Burgers circuit and the Burgers vector, respectively.

Besides an individual dislocation, there are also dislocations, which are continuously distributed over a surface or a region. In such a case, the dislocations density $\alpha^{\mu\nu}(x)$ is defined as a total density of the Burgers-vector flux at the point x. It is convenient to introduce a formalism, which would enable us to consider in a unified manner arbitrary distributions of dislocations, in particular the singular ones [6.3].

Let L be a sufficiently smooth curve. The generalized function $\delta(L)$ on the space of indefinitely differentiable test functions $\varphi(x)$ with bounded support, is defined by the relation

$$\int \delta(L)\varphi(x)\,dx = \int_L \varphi(x_L)\,dL\,. \tag{6.1.2}$$

Analogously, for a surface S and region V we have $\delta(S)$ and $\delta(V)$ according to

$$\begin{aligned} \int \delta(S)\varphi(x)\,dx &= \int_S \varphi(x_S)\,dS\,, \\ \int \delta(V)\varphi(x)\,dx &= \int_V \varphi(x_V)\,dV\,. \end{aligned} \tag{6.1.3}$$

It is clear that $\delta(V)$ coincides with the characteristic function of the region. The above-introduced notation is due to the convenience in writing down the following formulae.

Concerning L, S, and V it is assumed only that the corresponding integrals in (6.1.2) and (6.1.3) exist.

The following representations are valid

$$\begin{aligned} &\delta(L) = \int_L \delta(x - x_L)\,dL,\ \delta(S) = \int_S \delta(x - x_S)\,dS\,, \\ &\delta(V) = \int_V \delta(x - x_V)\,dV\,, \end{aligned} \tag{6.1.4}$$

and they are to be understood in the sense that

$$\int dx\,\varphi(x) \int_L \delta(x - x_L)\,dL = \int_L dL \int \delta(x - x_L)\varphi(x)dx = \int_L \varphi(x_L)\,dL\,. \tag{6.1.5}$$

For bounded L, S, and V

$$\int \delta(L)\,dx = l,\ \int \delta(S)\,dx = s,\ \int \delta(V)\,dx = v\,, \tag{6.1.6}$$

where l, s and v are length, area and volume respectively.

Besides the scalar δ-functions, we shall also consider vector ones, which for oriented L and S are defined by

$$\int \delta(\mathbf{L})\varphi(x)\,dx = \int_L \varphi(x_L)\,d\mathbf{L}\,,$$
$$\int \delta(\mathbf{S})\varphi(x)\,dx = \int_S \varphi(x_S)\,d\mathbf{S}\,, \tag{6.1.7}$$

where $d\mathbf{L}$ and $d\mathbf{S}$ are the elements of the curve and the surface.

From Gauss' and Stockes' formulae, important relations for the vector δ-functions follow.

Let S be the boundary of a region V. Then

$$\delta(\mathbf{S}) = -\operatorname{grad}\delta(V)\,, \tag{6.1.8}$$

whence it follows that, for a closed surface S,

$$\operatorname{rot}\,\delta(\mathbf{S}) = 0\,. \tag{6.1.9}$$

Let S be now a surface and L be its boundary; then

$$\delta(\mathbf{L}) = \operatorname{rot}\,\delta(\mathbf{S}) \tag{6.1.10}$$

and consequently for a closed contour

$$\operatorname{div}\,\delta(\mathbf{L}) = 0\,. \tag{6.1.11}$$

In order to prove, for example, (6.1.8), let us write down the relations

$$\int \varphi(x)\partial_\lambda\delta(V)\,dx = -\int_V \partial_\lambda\varphi(x_V)\,dV =$$
$$= -\int_S \varphi_S(x_S)\,dS_\lambda = -\int \delta(S_\lambda)\varphi(x)\,dx\,, \tag{6.1.12}$$

where we have used the definitions of δ-functions, the usual rules of operation with derivatives of generalized functions and the analog of the Gauss theorem (recall that we are working in an affine coordinate system). Then (6.1.12) immediately implies (6.1.8).

Problem 6.1.1. Prove (6.1.10).

Let us now consider an oriented contour L and a scalar or tensor function $f(x_L)$ defined on it. Then the δ-function with weight $f(x_L)$ is defined by

$$\int [f(x_L)\delta(L)]\varphi(x)\,dx = \int_L f(x_L)\varphi(x_L)\,dL\,, \tag{6.1.13}$$

or

$$[f(x_L)\delta(L)] = \int_L \varphi(x - x_L) f(x_L)\, dL\,. \tag{6.1.14}$$

Analogously, other δ-functions with weight can also be defined. The derivatives of δ-functions with weight are defined, as usually, by a transposition of the operation to the test function. In the cases when no confusion can arise, we shall omit the brackets in the expressions for the δ-functions with weight and write, for example, $\mathrm{rot}^{\mu\lambda} M_\lambda(x_L)\delta(L)$, where the operator rot is applied to the entire δ-function with weight.

It can be easily seen that the following relations are valid:

$$\delta(L^\lambda) = l^\lambda(x_L)\delta(L), \qquad \delta(S^\lambda) = n^\lambda(x_S)\delta(S)\,, \tag{6.1.15}$$

where l^λ and n^λ are the unit tangent vector and the normal, respectively.

Note that formulae of the type (6.1.8) and (6.1.10) are applicable only to δ-functions with weight of the special form (6.1.15). More detailed information on δ-functions concentrated on lines and surfaces, may be found in [6.4, 5] .

Let us now proceed to the consideration of characteristics of a distribution of dislocations. Let L be a closed (or going to infinity) line, which is the boundary of the surface S, and let b^ν be a constant vector. The expression

$$\alpha^{\mu\nu}(x) = b^\nu \delta(L^\mu) \tag{6.1.16}$$

due to (6.1.11), satisfies the condition

$$\partial_\mu \alpha^{\mu\nu}(x) = 0 \tag{6.1.17}$$

and, according to (5.1.19), can be considered as a density of a distribution of dislocations. Let us draw an arbitrary surface F, which intersects the line L at one point and let us write the expression for the flux $\alpha^{\mu\nu}$ through this surface. Taking into account the properties of $\delta(L^\mu)$, we obtain

$$\int_F \alpha^{\mu\nu}(x_F)\, dF_\mu = b^\nu \int_F \delta(L^\mu)\, dF_\mu = \pm b^\nu\,, \tag{6.1.18}$$

where the signs + or − are taken depending on whether the orientations of L and F are in conformity or not.

Since this flux does not depend on the choice of the surface F which intersects the line L and is equal to zero if F does not interest L, we come to the conclusion that (6.1.16) is the density, which corresponds to an individual dislocation with the line L and Burgers vector b^ν. On the other hand, taking into account (6.1.10) we have

$$\alpha^{\mu\nu}(x) = \text{rot}^{\mu}_{\cdot\rho} b^{\nu}\delta(S^{\rho}) . \tag{6.1.19}$$

A comparison with (5.2.17) shows that this dislocation density can be associated with the density of moments

$$m^{\rho\nu}(x) = b^{\nu}\delta(S^{\rho}) = M^{\rho\nu}(x_S)\delta(S) , \tag{6.1.20}$$

where

$$M^{\rho\nu}(x_S) = b^{\nu}n^{\rho}(x_S) \tag{6.1.21}$$

is the surface density of moments.

Here, the degree of non-uniqueness of the moment density $m^{\rho\nu}(x)$ as a characteristic of a distribution of dislocation, becomes evident. The surface S in (6.1.20) can be replaced by any other surface S', which has the same line L as a boundary. The expression (6.1.20) can be also interpreted as a distribution of dislocations on the surface S with a constant Burgers vector b^{ν} or, equivalently, with the two-dimensional density of moments (6.1.21). According to (5.2.17), this distribution is equivalent to an individual dislocation with the line L and with the same Burgers vector.

We see that there is a far-reaching analogy between a density of dislocations and a density of currents (or vortices). In fact, if we ignore the additional index ν, then $\alpha^{\mu\nu}(x)$ can be considered either as a linear current with the line L, or as currents distributed with constant density over the surface S. Both distributions of currents are equivalent, since the currents on the surface S compensate each other, except for the boundary current. Generalizing this analogy, we note that the static continuum theory of dislocations can be compared with the magnetostatics, and to the usual static elasticity there corresponds the electrostatics. From a mathematical standpoint the only important distinction between these theories is the higher tensor dimension of quantities in the former theories in comparison with their analogs in the latter ones.

From the preceding, it becomes clear that a distribution of dislocations in a region can be prescribed in three equivalent ways

$$\begin{aligned} m^{\rho\nu}(x) &= M^{\rho\nu}(x_V)\delta(V) , \\ \alpha^{\mu\nu}(x) &= \text{rot}^{\mu}_{\cdot\rho}\, M^{\rho\nu}(x_V)\delta(V) , \\ \eta^{\lambda\mu}(x) &= \text{Rot}^{(\lambda\mu)}_{\cdot\cdot\rho\nu}\, M^{\rho\nu}(x_V)\delta(V) , \end{aligned} \tag{6.1.22}$$

where $M^{\rho\nu}(x_V)$ is the volume density of moments of dislociations. If the dislocations are distributed over a surface S or a contour L, then $\delta(V)$ is to be replaced by $\delta(S)$ or $\delta(L)$, and $M^{\rho\nu}(x_V)$ by a surface moment density $M^{\rho\nu}(x_S)$ or by a linear moment density $M^{\rho\nu}(x_L)$, respectively.

In the limiting case of an elementary dislocation loop, which can be obtained, for example, by contracting the surface S in (6.1.19) to a point x_0 or, equivalently, by moving away far enough the point of observation x, the expression for the density of moments takes the form

$$m^{\rho\nu}(x) = M^{\rho\nu}(x)\delta(x - x_0)\,, \tag{6.1.23}$$

where $M^{\rho\nu}(x_0) = b^\rho s^\nu$ is the moment of an elementary dislocation, s^ρ is the normal vector of the surface element S.

A comparison with (5.4.1) shows than an elementary dislocation loop is equivalent to a point source of internal stress.

Together with the results of the previous chapter, we have now all the necessary formalism to compute fields and interactions of dislocations.

6.2 Some Three-Dimensional Problems

The computation of static fields and the interaction of dislocations within the framework of the simplest model of an isotropic elastic continuum has been dealt with in many papers, their results are given in the monographs and reviews mentioned above. Computations, for specific crystals, carried out on computers, showed that, in many cases, formulae of the continuum theory of dislocations remain valid even in the field where they seem to be inapplicable, namely, for distances of the order of a few interatomic distances. However, at the same time, it was also found that a number of effects essentially depend on the discrete structure of the crystal, and for their theoretical description one has to refute the continuum model.

An intermediate place between the crystal lattice and the elastic continuum is occupied by the above investigated Debye's model of the quasi-continuum with the space dispersion taken into account or not. As we know, this model gives a qualitatively correct description of the discrete structure and it is not too complicated for theoretical analysis. In Sections 6.4–6 within the framework of this model, a few problems will be considered, in which typical effects of microstructure appear. In particular, taking into account a change of force constants in the core of the dislocation is one of such problems.

Let the operator of elastic compliance B for an isolated dislocation with the line L, has the form

$$B = B_0 - B_L\,, \tag{6.2.1}$$

where B_0 is the operator of a homogeneous medium, B_L is the operator, which describes the local change of elastic properties in the core of the dislocation. The simplest expression for the kernel of the operator B_L has the form ($s \sim a^2$)

$$B_L(x, x') = s\int_L B'(x_L)\delta(x - x_L)\delta(x' - x_L)\,dL$$

$$= s\delta(x - x')[B'(x_L)\delta(L)]\,, \tag{6.2.2}$$

where $B'(x_L)$ is a tensor, which has dimension and symmetry of the tensor of elastic compliance. It is obvious that $B'(x_L)$ must depend on the direction of the Burgers vector and on the tangent vector to the dislocation line.

The construction of the Green's tensor G, which corresponds to (6.2.1), the Green's tensor G_0 of the homogeneous medium being given, leads, in the general case, to the solution of an integral equation on the dislocation line L. If the perturbation of the force constants in the core of the dislocation is small, then the integral equation can be effectively solved by the method of successive approximations. In the first appoximation, we have

$$G = G_0 + G_0 B_L G_0 \tag{6.2.3}$$

or, in a more detailed description,

$$G(x, x') = G_0(x - x') + s\int_L G_0(x - x_L)B'(x_L)G_0(x_L - x')\,dL\,. \tag{6.2.4}$$

Similarly, for the Green's tensor H, we obtain

$$H(x, x')$$

$$= H_0(x - x') + s\int_L H_0(x - x_L)\mathrm{Rot}B'(x_L)\mathrm{Rot}H_0(x_L - x')\,dL\,. \tag{6.2.5}$$

The expression for the energy Φ of a dislocation with line L and a Burgers vector $\mathbf{b}$, taking into account (5.3.6) and (5.1.16), can be represented in the form

$$\Phi = \frac{1}{2}b_\alpha b_\beta \Phi^{\alpha\beta}(L)\,, \tag{6.2.6}$$

where the tensor $\Phi^{\alpha\beta}(L)$ depends only on the line of the dislocation

$$\Phi(L) = \iint_{LL} \mathrm{rot}\ \mathrm{rot}\ H(x_L, x_L')\,dL\,dL'\,. \tag{6.2.7}$$

The stress produced by the dislocation, according to (5.3.1) and (6.1.20), is given by

$$\sigma^{\alpha\beta}(x) = b_\nu \int_L \mathrm{rot}'_{\nu\mu}\, Z^{\alpha\beta\lambda\mu}(x, x_L')\,dL_\lambda\,. \tag{6.2.8}$$

Let us now consider the interaction of the dislocation with a stress field $\sigma(x)$, which is caused by sufficiently remote sources of stress. The total force $\mathbf{q}$, which acts on the dislocation, can be represented as the sum of two components

$$\mathbf{q} = \mathbf{q}_0 + \mathbf{q}' , \tag{6.2.9}$$

where $\mathbf{q}_0$ depends linearly on the Burgers vector $\mathbf{b}$ of the dislocation and on the external field σ, and $\mathbf{q}'$ is quadratic with respect to the external field and does not depend on $\mathbf{b}$. Omitting not difficult but cumbersome calculations, we present the final result.

The force $\mathbf{q}_0$ can be written in a form, which coincides with the known expression of Peach-Koeller [6.1]

$$q_0^\alpha = \varepsilon^{\alpha\beta\mu} b^\nu \int_L \tau_{\mu\nu}(x_L)\, dL_\beta , \tag{6.2.10}$$

but with effective stress $\tau_{\mu\nu}$. In the case of a small change of the force constants in the core of the dislocation,

$$\tau^{\alpha\beta}(x) = \sigma^{\alpha\beta}(x) + s \int_L G_0^{\alpha\beta\lambda\mu}(x - x_L)\, B'_{\lambda\mu\nu\rho}(x_L) \sigma^{\nu\rho}(x_L)\, dL . \tag{6.2.11}$$

Under the same assumption,

$$q'_\nu = - s \int_L \sigma^{\alpha\beta}(x_L)\, B'_{\alpha\beta\lambda\mu}(x_L)\, \partial_\nu \sigma^{\lambda\mu}(x_L)\, dL . \tag{6.2.12}$$

In an analogous manner, one can consider systems of dislocations or systems of dislocations and point defects.

6.3 Two-Dimensional Problems

Beginning with this section, we consider the straight line dislocations, which are parallel to the z-axis. In this case the stress and the strain depend only on the two-dimensional position-vector $\mathbf{r}(x, y)$.

Let us decompose the 3-dimensional stress tensor σ^{ik} ($i, k = 1, 2, 3$) into a two-dimensional tensor $\sigma^{\alpha\beta}$, vector σ^α and scalar σ

$$\sigma^{ik} = \left(\begin{array}{c|c} \sigma^{\alpha\beta} & \sigma^\alpha \\ \hline \sigma^\alpha & \sigma \end{array} \right), \quad \alpha, \beta = 1, 2 . \tag{6.3.1}$$

The equations of equilibrium imply the representability of $\sigma^{\alpha\beta}$ and σ^{α} in the form

$$\sigma^{\alpha\beta} = \mathrm{Rot}^{\alpha\beta}\,\varphi, \;\; \sigma^{\alpha} = \mathrm{rot}^{\alpha}\psi\,, \tag{6.3.2}$$

where φ, ψ are scalars, and the operators Rot and rot are defined by the expressions

$$\mathrm{Rot}^{\alpha\beta} = e^{\alpha\lambda}\, e^{\beta\mu}\,\partial_{\lambda}\partial_{\mu}, \;\; \mathrm{rot}^{\alpha} = e^{\alpha\lambda}\partial_{\lambda}\,, \tag{6.3.3}$$

where $e^{\alpha\beta}$ is the two-dimensional unit antisymmetric pseudotensor.

In an analogous manner, let us decompose the three-dimensional strain tensor ε_{ik} into two-dimensional tensor $\varepsilon_{\alpha\beta'}$, vector ε_{α} and a scalar ε.

The distribution of dislocations in the case under consideration is completely characterized by the density of Burgers vector $b_i(\mathbf{r})$. Let us represent b_i in the form of a two-dimensional vector b_{α} and a scalar $b = b_3$. It is obvious that the vector b_{α} describes the distribution of edge dislocations and the scalar b describes the distribution of screw dislocations.

Taking into account the relation (5.2.16) between the incompatability and the dislocations density, we find that (5.2.13) takes now the form

$$\mathrm{Rot}^{\alpha\beta}\varepsilon_{\alpha\beta} = -\,\mathrm{rot}^{\alpha}b_{\alpha}, \;\; \mathrm{rot}^{\alpha}\,\varepsilon_{\alpha} = -\,\frac{1}{2}\,b, \;\; \partial_{\alpha}\partial_{\beta}\,\varepsilon = 0\,. \tag{6.3.4}$$

From the last equation, taking into account the boundary conditions at infinity, it follows that $\varepsilon = 0$.

The operator form of the Hooke's law can be now written in the form

$$\begin{aligned} \varepsilon_{\alpha\beta} &= B_{\alpha\beta\lambda\mu}\sigma^{\lambda\mu} + 2B_{\alpha\beta\lambda}\sigma^{\lambda} + B_{\alpha\beta}\sigma\,, \\ \varepsilon_{\alpha} &= B^{+}_{\alpha\beta\lambda}\,\sigma^{\lambda\mu} + 2B'_{\alpha\lambda}\,\sigma^{\lambda} + B_{\alpha}\sigma\,, \\ \varepsilon &= B^{+}_{\lambda\mu}\,\sigma^{\lambda\mu} + 2B^{+}_{\lambda}\,\sigma^{\lambda} + B\sigma\,. \end{aligned} \tag{6.3.5}$$

The two-dimensional operators appearing here are connected with the three-dimensional operator B_{ijkl} in an obvious way, the cross denoting the Hermitian conjugation.

In view of (6.3.2, 4 and 6.5), taking into account that $\varepsilon = 0$, we find the equations for φ and ψ

$$\begin{aligned} &\mathrm{Rot}^{\alpha\beta}\,\mathscr{B}_{\alpha\beta\lambda\mu}\,\mathrm{Rot}^{\lambda\mu}\,\varphi + \mathrm{Rot}^{\alpha\beta}\,\mathscr{B}_{\alpha\beta\lambda}\,\mathrm{rot}^{\lambda}\,\psi = -\,\mathrm{rot}^{\alpha}b_{\alpha}\,, \\ &\mathrm{rot}^{\alpha}\mathscr{B}^{+}_{\alpha\lambda\mu}\,\mathrm{Rot}^{\lambda\mu}\,\varphi + \mathrm{rot}^{\alpha}\mathscr{B}_{\alpha\lambda}\,\mathrm{rot}^{\lambda}\psi = -\,b\,. \end{aligned} \tag{6.3.6}$$

Here

$$\begin{aligned} \mathscr{B}_{\alpha\beta\lambda\mu} = B_{\alpha\beta\lambda\mu} - B_{\alpha\beta}B^{-1}B^{+}_{\lambda\mu}, \;\; \mathscr{B}_{\alpha\beta\lambda} &= 2(B_{\alpha\beta\lambda} - B_{\alpha\beta}B^{-1}B^{+}_{\lambda})\,, \\ \mathscr{B}_{\alpha\lambda} &= 4(B'_{\alpha\lambda} - B_{\alpha}B^{-1}B^{+}_{\lambda})\,. \end{aligned} \tag{6.3.7}$$

If the operator $\mathscr{B}_{\alpha\beta\lambda}$ is equal to zero, the system (6.3.6) is uncoupled. This will take place in all cases, when, due to the symmetry conditions in the plane x, y, there do not exist two-dimensional tensorial material characteristics of a medium of an odd order. In particular, these conditions are satisfied in the presence of central symmetry, cubic symmetry or in the isotropic case.

Let us represent the solution of the system (6.3.6) in the form

$$\begin{aligned} \phi(\mathbf{r}) &= -\int G_{11}(\mathbf{r}, \mathbf{r}')\mathrm{rot}^{\alpha} b_{\alpha}(\mathbf{r}')\, d\mathbf{r}' + \int G_{12}(\mathbf{r}, \mathbf{r}') b(\mathbf{r}') d\mathbf{r}' , \\ \psi(\mathbf{r}) &= -\int G_{21}(\mathbf{r}, \mathbf{r}')\mathrm{rot}^{\alpha} b_{\alpha}(\mathbf{r}')\, d\mathbf{r}' + \int G_{22}(\mathbf{r}, \mathbf{r}') b(\mathbf{r}') d\mathbf{r}' , \end{aligned} \tag{6.3.8}$$

where G is the matrix Green's operator. Substitution in (6.3.2) yields an expression for the stress.

It can be shown that the elastic energy per unit of length of z-axis Φ is presented in terms of the Green's operator in the form

$$2\Phi = \langle b_{\alpha} | \mathrm{rot}^{\alpha}\, G_{11}\, \mathrm{rot}^{\beta} | b_{\beta} \rangle + \langle b | G_{22} | b \rangle + 2\langle b_{\alpha} | \mathrm{rot}^{\alpha}\, G_{12} | b \rangle . \tag{6.3.9}$$

Here, the two-dimenstional brackets are defined by

$$\langle u | G | v \rangle = \iint u(\mathbf{r}) G(\mathbf{r}, \mathbf{r}') v(\mathbf{r}')\, d\mathbf{r}\, d\mathbf{r}' = \langle v | G^{+} | u \rangle . \tag{6.3.10}$$

When the system (6.3.6) is uncoupled, the Green's operator is diagonal and hence, according to (6.3.9), in the case of a homogeneous medium, edge dislocations do not interact with the screw ones. If the change of force constants in the dislocation core is taken into account, then the above degeneration is removed, since the diagonal components of the Green's operator depend on the changes of force constants of all dislocations.[1]

Let us now proceed to a separate consideration of screw and edge dislocations.

6.4 Screw Dislocations

Suppose there are only screw dislocations in the medium, i.e. $b_{\alpha} = 0$. Let us further assume that the conditions of symmetry are satisfied, ensuring the uncoupling of the system (6.3.6). Then the elastic field of screw dislocations is described by a scalar function $\psi(\mathbf{r})$, which satisfies the equation

$$\mathrm{rot}^{\alpha} \mathscr{B}_{\alpha\beta}\, \mathrm{rot}^{\beta} \psi = -\, b . \tag{6.4.1}$$

[1]In connection with the divergence of the self-energy of a dislocation in a two-dimensional problem, as $\mathbf{r} \to \infty$, (6.3.9) determines the energy with logarithmic accuracy.

Let us introduce a Green's function $G(\mathbf{r}, \mathbf{r}')$ defined by

$$\int \mathrm{rot}^\alpha \mathscr{B}_{\alpha\beta}(\mathbf{r}, \mathbf{r}'') \, \mathrm{rot}^\beta G(\mathbf{r}'', \mathbf{r}') \, d\mathbf{r}'' = -\delta(\mathbf{r} - \mathbf{r}') . \tag{6.4.2}$$

In the case under consideration, only the components $\sigma^\alpha(\mathbf{r})$ of the stress differ from zero, and for them, according to (6.3.2), we have

$$\sigma^\alpha(\mathbf{r}) = \mathrm{rot}^\alpha \int G(\mathbf{r}, \mathbf{r}') b(\mathbf{r}') d\mathbf{r}' . \tag{6.4.3}$$

The energy Φ, as it follows from (6.3.9), has the form

$$2\Phi = \langle b|G|b\rangle . \tag{6.4.4}$$

Thus, the problem of determining elastic fields and of investigating the interaction between screw dislocations, is reduced to constructing the Green's function G. In this section we consider the case of a homogeneous isotropic medium without taking into account the change in the force constants in the core of the dislocation. The kernel of the operator $\mathscr{B}_{\alpha\beta}$ and the Green's function G are then of the difference type.

For an isotropic medium, taking into account (5.3.9), we obtain

$$\mathscr{B}_{\alpha\beta} = \mu^{-1}\delta_{\alpha\beta} , \tag{6.4.5}$$

where μ is an operator with the kernel $\mu(k)$. (6.4.2) takes the form

$$\mu^{-1}\Delta G(\mathbf{r}) = -\delta(\mathbf{r}) . \tag{6.4.6}$$

From the above discussion it follows that $G(\mathbf{r})$ is the inverse Fourier transform of $\mu(k)k^{-2}$, where $k = |\mathbf{k}|$. It is not difficult to prove that, in the two-dimensional case, the inverse Fourier transform of the function k^{-2} $(k \leq \kappa)$ is

$$g(r) = \frac{1}{2\pi}\left(\int_0^{\kappa r} \frac{J_0(\tau) - 1}{\tau} d\tau + c\right), \tag{6.4.7}$$

where $J_0(\tau)$ is the Bessel function of the zeroth order, $r = |\mathbf{r}|$. The constant c should be chosen from the condition that, at large distances ($\kappa r \gg 1$), the function $g(r)$ is to transform into the classical Green's function of the Laplace operator

$$g(r) \simeq -\frac{1}{2\pi} \ln \frac{r}{R} , \tag{6.4.8}$$

where the quantity $R \gg a$ has the meaning of a characteristic dimension of the body. Taking into account the asymptotic behaviour of (6.4.7), we find that $c = \ln \kappa R$. The solution of (6.4.6) may now be represented in the form

$$G(r) = \mu g(r) . \tag{6.4.9}$$

If for $\mu(k)$ the approximation (5.3.18) is taken, then

$$G(r) = \mu_0 \left[g(r) - \gamma \kappa^{-2} \delta_\kappa(\mathbf{r}) - \frac{2\gamma - 1}{3} \kappa^{-4} \Delta \delta_\kappa(\mathbf{r}) \right], \tag{6.4.10}$$

where

$$\delta_\kappa(\mathbf{r}) = - \Delta g(r) = \frac{\kappa}{2\pi r} J_1(\kappa r) \tag{6.4.11}$$

and $J_1(\kappa r)$ is the Bessel function of the first order. When $\kappa \to \infty$ $(a \to 0)$, the function $\delta_\kappa(r)$, becomes the ordinary two-dimensional δ-function.

For the Debye's model

$$G(r) = \mu_0 g(r) . \tag{6.4.12}$$

The asymptotics of $G(r)$, omitting the rapidly oscillating and decaying terms, has the form (independent of the model accepted)

$$G(r) \simeq - \frac{\mu_0}{2\pi} \ln \frac{r}{R} , \tag{6.4.13}$$

i.e., in the case of screw dislocations the space dispersion makes no additional contribution to the asymptotics of $G(r)$.

For a single screw dislocation at the point $\mathbf{r} = 0$ with the Burgers vector b, taking into account (6.4.3) we have

$$\sigma_x(\mathbf{r}) = b\partial_y G(r), \quad \sigma_y(\mathbf{r}) = - b\partial_x G(r) . \tag{6.4.14}$$

In particular, for the Debye's model,

$$\sigma_x(\mathbf{r}) = \frac{\mu_0 by}{2\pi} \cdot \frac{J_0(\kappa r) - 1}{r^2} , \qquad \sigma_y(\mathbf{r}) = - \frac{\mu_0 bx}{2\pi} \cdot \frac{J_0(\kappa r) - 1}{r^2} , \tag{6.4.15}$$

which coincides with the expressions obtained by Brailsford [6.6] When $\kappa r \gg 1$, (6.4.15) becomes the well-known solution of the continuum theory of dislocations [6.1].

The self-energy of the dislocation is given by

$$\Phi = \frac{1}{2} b^2 G(0) , \tag{6.4.16}$$

where the constant $G(0)$ for the model (5.3.18) and for the Debye's model is equal to

$$G(0) = \frac{\mu_0}{2\pi}\left(\ln \kappa R + \frac{1-4\gamma}{12}\right), \qquad G(0) = \frac{\mu_0}{2\pi} \ln \kappa R\,, \tag{6.4.17}$$

respectively.

Consider now a system of screw dislocations with the Burgers vectors b_j, the dislocations being located at points $\mathbf{r}_j$. Then, for the stress we have

$$\sigma_x(r) = \sum_j b_j \partial_y G(\mathbf{r} - \mathbf{r}_j), \quad \sigma_y(\mathbf{r}) = -\sum_j b_j \partial_x G(\mathbf{r} - \mathbf{r}_j)\,. \tag{6.4.18}$$

The elastic energy is equal to

$$\Phi = \frac{1}{2}\sum_{ij} b_i b_j G(\mathbf{r}_i - \mathbf{r}_j)\,. \tag{6.4.19}$$

Thus, in the case of two dislocations, we find for the energy of interaction

$$\Phi_{\text{int}} = b_1 b_2 G(\rho), \quad \rho = |\mathbf{r}_1 - \mathbf{r}_2|\,. \tag{6.4.20}$$

The force of interaction $\mathbf{q}$ is directed in this case along the vector $\mathbf{r} = \mathbf{r}_1 - \mathbf{r}_2$ and for the model (5.3.18) and for the Debye's model is equal to

$$q(\rho) = \mu_0 b_1 b_2 \left[\frac{1 - J_0(\kappa\rho)}{2\pi\rho} + \gamma\kappa^{-2}\frac{d}{d\rho}\delta_\kappa(\rho) + \frac{2\gamma - 1}{3}\kappa^{-4}\Delta\delta_\kappa(\rho)\right], \tag{6.4.21}$$

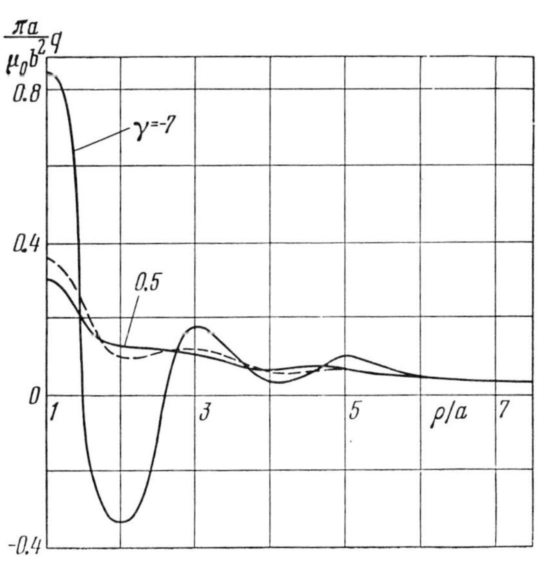

Fig. 6.4. Interaction force q as function of distance ρ

$$q(\rho) = \frac{\mu_0 b_1 b_2}{2\pi} \frac{1 - J_0(\kappa\rho)}{\rho}, \qquad (6.4.22)$$

respectively.

The asymptotics of (6.4.21, 22) coincide with the expression for the force of interaction, which is known in the continuum theory of dislocations.

Since $|J_0(\kappa\rho)| \leq 1$, it follows from (6.4.22) that, in the Debye's model, the sign of the force cannot change with decreasing distance between the dislocations. At the same time, an investigation of (6.4.21) shows that, when the space dispersion is taken into account, the sign of the force can change, provided the boundary value $\omega(\kappa)$ of the dispersion curve for transverse oscillations is greater than the corresponding Debye frequency. The graphs of $q(\rho)$ for different values of γ are presented in the Fig. 6.4. The dotted curve corresponds to the Debye model.

6.5 Influence of Change of the Force Constants in Cores of Screw Dislocations

In view of large relative displacements of atoms in the core of a dislocation, the force constants, generally speaking, differ from the force constants of a perfect crystal.

In the isotropic case, for screw dislocations, (6.4.5) is still satisfied, but the operator μ^{-1} is no longer a difference operator if the change in the elastic properties of the core are taken into account. Let us denote $B = \mu^{-1}$ and represent the kernel of the operator B for a system of screw dislocations in the form

$$B(\mathbf{r}, \mathbf{r}') = B_0(\mathbf{r} - \mathbf{r}') - s \sum_i B_i \delta(\mathbf{r} - \mathbf{r}_i)\delta(\mathbf{r}' - \mathbf{r}_i), \qquad (6.5.1)$$

where $B_0(\mathbf{r}, \mathbf{r}')$ is the kernel of the operator of the homogeneous medium, the B_i's are constants which have dimensions of μ_0^{-1}. For convenience of interpretation of the constants B_i, let us consider single screw dislocation within the framework of the Debye model. The kernel of the operator μ in this case can be represented in the form

$$\mu(\mathbf{r}, \mathbf{r}') = \mu_0 \delta(\mathbf{r} - \mathbf{r}') + s\mu_1 \delta(\mathbf{r} - \mathbf{r}_1)\delta(\mathbf{r}' - \mathbf{r}_1). \qquad (6.5.2)$$

The kernel of the inverse operator B has the form

$$B(\mathbf{r}, \mathbf{r}') = \mu_0^{-1} \delta(\mathbf{r} - \mathbf{r}') - sB_1 \delta(\mathbf{r} - \mathbf{r}_1)\delta(\mathbf{r}' - \mathbf{r}_1), \qquad (6.5.3)$$

where

$$B_1 = \frac{\mu_1}{\mu_0(\mu_0 + \mu_1)}. \qquad (6.5.4)$$

An analogous representation for B is conveniently preserved also in the general case of (6.5.1). From (6.5.4) it follows that to the values $B_i > 0$ there correspond more rigid elastic bonds in the dislocation core and to the values $B_i < 0$ there corresponds weakening of the elastic bonds.

Assuming that parallel screw dislocations differ from each other only in the sign of the Burgers vector **b** as in a crystal, it is natural to set $B_i = B_1$. Eq. (6.4.1) in the case under consideration takes the form

$$\operatorname{div} B \operatorname{grad} \psi = -b \,. \tag{6.5.5}$$

Let us denote by $G(\mathbf{r} - \mathbf{r}')$ the Green's function for the homogeneous medium, which was constructed in the preceeding section. Then, for the Green's function $G(\mathbf{r}, \mathbf{r}')$ for the operator (6.5.1), we can derive the expression

$$G(\mathbf{r}, \mathbf{r}') = G_0(\mathbf{r} - \mathbf{r}') - \sum_{ij} \partial_\alpha G_0(\mathbf{r} - \mathbf{r}_i) P_{ij}^{\alpha\beta} \partial_\beta G_0(\mathbf{r}_j - \mathbf{r}') \,, \tag{6.5.6}$$

where the matrix $P_{ij}^{\alpha\beta}$ is inverse to the matrix

$$R_{\alpha\beta}^{ij} = \partial_\alpha \partial_\beta G_0(\mathbf{r}_i - \mathbf{r}_j) + s^{-1} B^{-1} \delta_{ij} \delta_{\alpha\beta} \,. \tag{6.5.7}$$

For stress and energy of a system of dislocations we have

$$\sigma^\alpha(\mathbf{r}) = \sum_i b_i \operatorname{rot}^\alpha G(\mathbf{r}, \mathbf{r}_i) \,, \tag{6.5.8}$$

$$\Phi = \frac{1}{2} \sum_{ij} b_i b_j G(\mathbf{r}_i, \mathbf{r}_j) \,. \tag{6.5.9}$$

Taking into account that $\partial_\alpha G_0(0) = 0$, we find from (6.5.6, 8 and 9) that the change of the force constants in the core of an isolated screw dislocation does not contribute to the stress and the elastic self-energy.

Let us consider two screw dislocations with the Burgers vectors $b_1 = b$, $b_2 = \pm b$. Assume also that $B_1 = B_2 = B$. The elastic energy in this case has the form

$$\Phi = \frac{1}{2} b^2 [G(\mathbf{r}_1, \mathbf{r}_1) \pm 2G(\mathbf{r}_1, \mathbf{r}_2) + G(\mathbf{r}_2, \mathbf{r}_2)] \,. \tag{6.5.10}$$

The determination of the Green's function $G(\mathbf{r}, \mathbf{r}')$ is reduced to an inversion of the four-dimensional matrix R. Computations yield ($\rho = |\mathbf{r}_1 - \mathbf{r}_2|$)

$$\begin{aligned} G(\mathbf{r}_1, \mathbf{r}_1) &= G(\mathbf{r}_2, \mathbf{r}_2) = G_0(0) + \frac{\alpha [G_0'(\rho)]^2}{\alpha^2 - [G_0''(\rho)]^2} \,, \\ G(\mathbf{r}_1, \mathbf{r}_2) &= G_0(\rho) + \frac{[G_0'(\rho)]^2 G_0''(\rho)}{\alpha^2 - [G_0''(\rho)]^2} \,, \end{aligned} \tag{6.5.11}$$

where

$$\alpha = s^{-1}B^{-1} + \left[\frac{1}{\rho}G_0'(\rho)\right]_{\rho=0}. \tag{6.5.12}$$

The value $G_0(0)$ is given by (6.4.17), and the second term in (6.5.12), for the Debye's model and the model (4.3.18), respectively, has the form

$$\left[\frac{1}{\rho}G_0'(\rho)\right]_{\rho=0} = \begin{cases} -\dfrac{\mu_0\kappa^2}{8\pi}, \\ -\dfrac{\mu_0\kappa^2}{9\pi}\left(1 - \dfrac{5}{8}\gamma\right). \end{cases} \tag{6.5.13}$$

Let us write down the energy Φ in the form

$$\Phi = \Phi_0 + \Phi_{\text{int}}, \tag{6.5.14}$$

where Φ_0 is the self-energy of the dislocations, Φ_{int} is the interaction energy. For the latter, taking into account (6.5.10, 11), we obtain

$$\Phi_{\text{int}}(\rho) = b^2\left(\pm G_0(\rho) + \frac{[G_0'(\rho)^2]}{\alpha \mp G_0''(\rho)}\right). \tag{6.5.15}$$

The force of interaction is equal, by definition, to

$$q(\rho) = -\frac{d}{d\rho}\Phi_{\text{int}}(\rho). \tag{6.5.16}$$

Taking into account (6.4.13), for the asymptotics of $q(\rho)$, we obtain

$$q(\rho) \simeq \frac{\mu_0 b^2}{2\pi\rho}\left(\pm 1 + \frac{\mu_0}{\pi\alpha\rho^2}\right). \tag{6.5.17}$$

From this expression, it follows that the change of the sign of the force of interaction at large distances, is possible only under simultaneous fulfillment of two conditions: the values of the force constants in the core of the dislocation increase, and the limiting frequency $\omega(\kappa)$ is higher than the Debye's frequency.

An analysis of (6.5.16) in the general case, shows that change of the sign of the force at small distances (of the order of few interatomic distances) is possible only if the spatial dispersion is taken into account.

6.6 Edge Dislocations

Suppose that there are only edge dislocations in the medium, their lines being parallel to the z-axis and their distribution being given by a density of the vector $b_\alpha(\mathbf{r})$. As in the case of screw dislocations, we assume the fulfillment of symmetry conditions, which ensure the uncoupling of the system (6.3.6). Then the elastic field of the dislocation is described by a scalar function $\varphi(r)$, which satisfies the equation

$$\mathrm{Rot}^{\alpha\beta}\,\mathscr{B}_{\alpha\beta\lambda\mu}\,\mathrm{Rot}^{\lambda\mu}\,\varphi = -\mathrm{rot}^{\alpha}b_{\alpha}\,. \tag{6.6.1}$$

Let us introduce the Green's function $G(\mathbf{r}, \mathbf{r}')$, which is a solution of

$$\mathrm{Rot}^{\alpha\beta}\,\mathscr{B}_{\alpha\beta\lambda\mu}\,\mathrm{Rot}^{\lambda\mu}G = \delta(\mathbf{r} - \mathbf{r}')\,. \tag{6.6.2}$$

In this case, for non-zero components of stress, taking into account (6.3.2, 5), we have

$$\begin{aligned} \sigma^{\alpha\beta}(\mathbf{r}) &= -\int \mathrm{Rot}^{\alpha\beta}G(\mathbf{r}, \mathbf{r}')\mathrm{rot}^{\lambda}b_{\lambda}(\mathbf{r}')\,d\mathbf{r}'\,, \\ \sigma(\mathbf{r}) &= -\int B^{-1}B^{+}_{\alpha\beta}\upsilon^{\alpha\beta}\,. \end{aligned} \tag{6.6.3}$$

The energy Φ according to (6.3.9), can be written in the form

$$2\Phi = \langle b_{\alpha}|\,G^{\alpha\beta}\,|b_{\beta}\rangle\,, \tag{6.6.4}$$

where

$$G^{\alpha\beta}(\mathbf{r}, \mathbf{r}') = \mathrm{rot}^{\alpha}\mathrm{rot}'^{\beta}G(\mathbf{r}, \mathbf{r}')\,. \tag{6.6.5}$$

For a system of N dislocations with the Burgers vectors b^{i}_{α} $(i = 1, 2, \ldots, N)$

$$2\Phi = \sum_{ij} b^{i}_{\alpha}b^{j}_{\beta}G^{\alpha\beta}(\mathbf{r}_{i}, \mathbf{r}_{j})\,. \tag{6.6.6}$$

Let us first consider the case of a homogeneous isotropic medium without taking into account the change of force constants in the dislocation core. With (5.3.9) and assuming that the Poisson coefficient $\nu_0 = \mathrm{const}$, we transform (6.6.2) for the Green's function $G(\mathbf{r} - \mathbf{r}')$ into

$$\frac{1 - \nu_0}{2}\,\mu^{-1}\Delta^{2}G(\mathbf{r}) = \delta(\mathbf{r})\,. \tag{6.6.7}$$

Hence we have for $G(k)$

$$G(k) = \frac{2}{1 - \nu_0}\,\frac{\mu(k)}{k^4}\,. \tag{6.6.8}$$

It is not difficult to prove that the function $h(r)$, which is the inverse Fourier transform of k^{-4}, has the form

$$h(r) = \frac{1}{4\pi\kappa^2}\left[\pi\kappa^2 r^2 g(r) + J_0(\kappa r) - \frac{\kappa r}{2} J_1(\kappa r) + \frac{\kappa^2 r^2}{2}\right], \tag{6.6.9}$$

where $g(r)$ is defined by (6.4.7).

The Green's function $G(\mathbf{r}) = G(r)$ is then represented in the form

$$G(r) = \frac{2}{1-\nu_0}\mu h(r)\,. \tag{6.6.10}$$

For the Debye's model and for the model (5.3.18) we have

$$\begin{aligned} G(r) &= \frac{2\mu_0}{1-\nu_0} h(r)\,, \\ G(r) &= \frac{2\mu_0}{1-\nu_0}\left[h(r) - \gamma\kappa^{-2} g(r) + \frac{2\gamma - 1}{3}\kappa^{-4}\delta_\kappa(r)\right], \end{aligned} \tag{6.6.11}$$

respectively.

In the case of a single edge dislocation, the substitution of (6.6.11) into (6.6.3) enables us to obtain closed expressions for the stress. For the Debye model they coincide with the corresponding expressions obtained by *Brailsford* [6.6].

Consider now two parallel dislocations at the points $\mathbf{r}_1$ and $\mathbf{r}_2$ with the Burgers vectors $\mathbf{b}^1 = (b, 0)$, $\mathbf{b}^2 = (\pm b, 0)$ (the slip planes are parallel). Then, for the energy of interaction Φ_{int}, according to (6.6.6), we have

$$\Phi_{\text{int}} = \mp b^2 \partial_y^2 G(\rho),\ \rho = |\mathbf{r}_1 - \mathbf{r}_2|\,. \tag{6.6.12}$$

The projection of the force of interaction onto the slip plane for the model (5.3.18) is equal to

$$q_x = \pm\frac{2\mu_0 b^2}{1-\nu_0}\partial_x\partial_y^2\left[h(\rho) - \gamma\kappa^{-2} g(\rho) + \frac{2\gamma-1}{3}\kappa^{-4}\delta_\kappa(\rho)\right]. \tag{6.6.13}$$

For the Debye's model the last two terms in the above expression should be omitted.

In the polar coordinate system the asymptotics of q_x has the form

$$q_x \simeq \pm\frac{\mu_0 b^2}{2\pi(1-\nu_0)}\left(\frac{\cos\theta\cos 2\theta}{\rho} - 4\gamma\kappa^{-2}\frac{\cos\theta\,(1 - 4\sin^2\theta)}{\rho^3}\right). \tag{6.6.14}$$

The first term coincides with the known expression for q_x in the continuum theory of dislocations. As distinct from the case of a screw dislocation, the dispersion makes here the additional contribution of the order of ρ^{-3} to the asymptotics. For the Debye model the asymptotics is given by the first term only.

From (6.6.14) it follows that the equilibrium of a dislocation is possible either when $\theta = \pi/2$, or on some curve with the asymptotics $\theta = \pi/4$. In the continuum theory of dislocations, equilibria are possible only when $\theta = \pi/2$ and $\theta = \pi/4$. The qualitative correspondence between signs of the Burgers vectors and the stability of states of equilibrium are also preserved when the spatial dispersion is taken into account.

At small distances of the order of a few interatomic distances, the appearance of new equilibria is possible, but it happens only if the boundary frequency $\omega(\kappa)$ is larger, than the Debye frequency.

Let us now consider in the general form the influence of the change of the force constants in the cores of edge dislocations, on the Green's function.

From physical considerations it is clear that the change of the force constants in the core of an edge dislocation is different on different sides of the slip plane. On the side of the extra-semiplane, the atoms of the core come closer to each other and the rigidity of bonds increases. On the other side of the slip plane, the rigidity of the bonds decreases. Thus the perturbation of elastic bonds in the core has the nature of a dipole perpendicular to the Burgers vector.

For a system of parallel edge dislocations, let us represent the kernel of the operator (6.6.1) in the form

$$\mathscr{B}_{\alpha\beta\lambda\mu}(\mathbf{r}, \mathbf{r}') = \mathscr{B}^0_{\alpha\beta\lambda\mu}(\mathbf{r} - \mathbf{r}') - s^2 \sum_i \mathscr{B}^i_{\alpha\beta\lambda\mu} \nabla_i \delta(\mathbf{r} - \mathbf{r}_i) \nabla_i \delta(\mathbf{r}' - \mathbf{r}_i) . \qquad (6.6.15)$$

Here, $\nabla_i = n_i^\alpha \partial_\alpha$ is a scalar operator, n_i^α is the unit vector perpendicular to the Burgers vector $\mathbf{b}_i$ and directed to the side of the extra semiplane.

Let us denote by $G_0(\mathbf{r} - \mathbf{r}')$ the Green's function of the operator $\mathscr{B}^0$. Then, for the Green's function $G(\mathbf{r}, \mathbf{r}')$ of the operator $\mathscr{B}$, we have

$$G(\mathbf{r}, \mathbf{r}') = G_0(\mathbf{r} - \mathbf{r}') - \sum_{ij} \nabla_i G_0^{\alpha\beta}(\mathbf{r} - \mathbf{r}_i) P^{ij}_{\alpha\beta\lambda\mu} \nabla_j G_0^{\lambda\mu}(\mathbf{r}_j - \mathbf{r}') . \qquad (6.6.16)$$

The matrix P^{ij} is inverse to the matrix

$$R_{ij}^{\alpha\beta\lambda\mu} = \nabla_i \nabla_j G_0^{\alpha\beta\lambda\mu}(\mathbf{r}_i - \mathbf{r}_j) + s^{-2} C_i^{\alpha\beta\lambda\mu} \delta_{ij} , \qquad (6.6.17)$$

where

$$G_0^{\alpha\beta\lambda\mu}(\mathbf{r}) = \mathrm{Rot}^{\alpha\beta}\, \mathrm{Rot}^{\lambda\mu} G_i(\mathbf{r}),\ C_i = \mathscr{B}_i^{-1} . \qquad (6.6.18)$$

6.7 Notes

Basic physical notions of dislocation theory can be found, for example, in [B6.4, 9–11]. Mathematical aspects of the continuum theory of dislocations are considered in [B5.6, B6.1, 5–7, 15–20, 30, 32–35, 38, 41, 44].

Dislocations in the scope of models of elastic media with spatial dispersion and of the Debye's quasicontinuum have been investigated in works of Rogula [B6.36], Kosevich and Natsik [B6.13], Vdovin and Kunin [B6.40] and Brailsford [B6.2, 3].

7. Elastic Medium with Random Fields of Inhomogeneities [1]

In this chapter the method of the effective field is applied to solve problems for composites and cracked solids. Under the assumption of a random change of the effective field from one particle to another the formulae for the first and second moments of random stress-strain fields are presented.

7.1 Background

The problem of a random field of inhomogeneities in elastic medium has implications in a number of important applications for example, to composite materials in engineering. Such materials, which consist of essentially different components, facilitate a construction with extremely high strength and deformational properties. In calculating properties of such designs with respect to strength there is a problem of choice of an adequate mechanical model for the composites. In a number of cases their behavior is well described by the model of homogeneous elastic medium (matrix), which contains a random field of inclusions with elastic properties.

An other field of applications of the theory of inhomogeneous medium is the mechanics of fracture. The description of the process of microcracks, voids and other defects growing in solids, is connected with the problem of many-particle interactions of inhomogeneities in an elastic medium.

As a rule the sizes of inclusions and the distances between them are much smaller than size of a body and a typical scale for the change of an external field. That is why here an infinite homogeneous elastic medium containing a homogeneous random set of inclusions in a constant external field of stress is considered. The problem is to describe random stress and strain fields in such a medium.

It is well known [7.1] that the complete information about a random field is provided by its characteristic functional or an infinite set of its many points statistical moments. Unfortunately, the difficulties which arise in constructing all these objects for stress-strain fields in composits, can hardly be overcome. But for applications it is not necessary to have complete information about a random field. As a rule only several first statistical moments can be of interest.

---

[1]This chapter is written by S. K. Kanaun.

One of the most important characteristics of an inhomogeneous medium is its effective elastic moduli tensor. In order to determine this tensor we should have the first moment of the statistical solution or the mean values of the stress and strain fields. Most of the investigations in mechanics of an inhomogeneous medium are devoted to that very subject.

The fluctuations of the elastic fields are important for a description of phenomena depending on the fine structure of the composite, for example fracture and the beginning of the plastic shear. The numerical characteristics of the fluctuations (dispersions, correlation radii) can be found from the second statistical moments of the solution. The problem of the second moment is more complicated than that of calculating the overall moduli; the difficulties increase here substantially.

Most probably the first two statistical moments give us the most valuable information about the random stress-strain fields in composites. This information is necessary for the right description of the elastic-plastic deformation and fracture.

In many cases we can distinguish a typical element or a particle of the heterogeneous medium. An inclusion in a composite, a grain in a polycrystal, a crack in a cracked solid can be an example of a particle of this kind. The solution of the problem is reduced here to the description of many-particle interactions.

If the concentration of particles in a homogeneous medium is small and we can neglect their interactions, then the problem can be solved exactly, but the solution for an isolated particle must be known. In the cases of large concentrations, when the interaction is essential, as a rule, an approximate solution can only be obtained.

There are several methods to account for the interactions between particles in an elastic heterogeneous medium. A survey of these methods can be found in [7.2], for example.

Here we shall consider the method for evaluating the statistical moments of the random elastic fields in composities. The idea behind this method is closely related to the self-consistent method, well known in physics.

The self-consistent method (SCM) is one of the most powerful methods to solve the many-particle problem. In the quantum theory of the atom (the Hartree-Fock approach), or the description of phase transitions (Weiss method, Landau theory) SCM enables one to obtain good approximations in many important cases. These cases are well known. The external field for every particle must weakly depend on the concrete configuration of the surrounding particles and must be approximately equal to the combined field of all interacting particles.

In the theory of an inhomogeneous medium the self-consisted method is based on the following assumption. Every particle is considered isolated and contained in a homogeneous medium, the elastic moduli of which are equal to the unknown effective elastic moduli of the heterogeneous material. The external field for every particle, which is assumed equal to the external field, applies to the whole body. Let the solution of the last problem be known and therefore the function describing the state of every isolated particle can be expressed in terms of its properties, the effective parameters of the medium and an external field. After the summing of perturbations from all the particles an overall stress and strain may be found and an effective elastic moduli tensor of the composite is expressed in this way through the particle states functions.

As a result this way leads to a system of algebraic equations for the effective parameters of elasticity (Sect. 7.3). Such a modification of SCM will be called the method of an effective medium.

For the case of a composite medium another avenue for solving this problem can be suggested. Every particle may be considered isolated in a homogeneous medium—the matrix of the composite. The presence of surrounding particles is accounted for through an effective external field applying to every particle. The scheme which is based on such an assumption will be called the method of the effective field.

It is commonly accepted that the effective field is constant (and the same for all the particles). In this case the method of the effective field coincides with a modification of SCM, which is widely used in quantum mechanics, statistical physics and also in the mechanics of heterogeneous medium. The assumption about the constant effective field can be confirmed by calculating the effective parameters of an inhomogeneous medium. But in order to describe the fluctuations of elastic fields in a composite we have to account for the fluctuations of the effective field.

In this chapter the method of the effective field is applied to solve problems for composites and cracked solids. Under the assumption of a random change of the effective field from one particle to another the formulae for the first and second moments of random stress-strain fields are presented. The general scheme of the method is considered in Sect. 7.2–5. Sections 7.6–9 are devoted to the first approximation of the method, coinciding with a modification of SCM. Formulae for the effective constants of composites are obtained. The construction of the second moment is described in Sect. 7.11 for the case of point defects. The results are compared with the exact solutions for regular composites and experiments. Estimates for the accuracy of the method is given. Finally, the use of the results obtained for composite fracture is shown in Sect. 7.12.

7.2 Formulation of the Problem

Let us consider an infinite elastic medium the elastic moduli tensor of which is

$$C(x) = C_0 + C_1(x)\,, \tag{7.2.1}$$

where C_0 is the elastic moduli tensor of the homogeneous medium, $C_1(x)$ is a perturbation caused by the inhomogeneities. In the case of a composite medium $C_1(x)$ can be represented in the form

$$C_1(x) = \sum_i C_{1i} V_i(x) , \tag{7.2.2}$$

where $V_i(x)$ is the characteristic function of the region occupied by the i-th inclusion, the tensor C_{1i} is constant within the i-th inclusion but, generally speaking, varies from one inclusion to another.

The elastic compliance tensor of the inhomogeneous medium $B = C^{-1}$ is represented in the same way as (7.2.1, 2),

$$B(x) = B_0 + B_1(x), \qquad B_1(x) = \sum_i B_{1i} V_i(x) , \tag{7.2.3}$$

where $B_0 = C_0^{-1}$, B_{1i} are the constant tensors within each inclusion.

If the fields of stress $\sigma_i^+(x)$ and strain $\varepsilon_i^+(x)$ inside the inclusions are known, the tensors $\sigma(x)$ and $\varepsilon(x)$ at any point of the medium are represented in the form (Sect. 4.8)

$$\begin{aligned} \varepsilon(x) &= \varepsilon_0 - \int K_0(x - x') C_1 \varepsilon^+(x') \, dx' , \\ \sigma(x) &= \sigma_0 - \int S_0(x - x') B_1 \sigma^+(x') \, dx' , \end{aligned} \tag{7.2.4}$$

where ε_0 and σ_0 are the external strain and stress fields, respectively. Here we denote

$$\begin{aligned} C_1 \varepsilon^+(x) &= \sum_i C_{1i} \varepsilon_i^+(x) V_i(x) , \\ B_1 \sigma^+(x) &= \sum_i B_{1i} \sigma_i^+(x) V_i(x) . \end{aligned} \tag{7.2.5}$$

The kernels of the integral operators K_0 and S_0 in (7.2.4) are defined via the Green's tensor for displacements of the homogeneous medium (Sect. 4.6). It will be assumed further that the regions V_i occupied by inclusions constitute a homogeneous random field and all the tensors C_{1i} are independent random variables having the same known density functions. Such a model involves the most interesting composite materials. If an external field σ_0 (or ε_0) is constant, the functions $C_1 \varepsilon^+(x)$ and $B_1 \sigma^+(x)$ for each realization of the random distribution of the regions V_i in space may be represented at the sum of an almost periodic function and a function with bounded support. The operators K_0 and S_0 are defined upon such a class of functions in Sect. 4.8 and Appendix A3. It is essential that a unique definition of these operators is possible, if it is known what kind of external field is given: the stress field σ_0 or the strain field ε_0.

Further, for definiteness we assume that the external stress field σ_0 is always fixed. Then the operators K_0 and S_0 are defined on constants via (Sect. 4.8)

$$\int K_0(x - x')\, dx' = B_0, \qquad \int S_0(x - x')\, dx' = 0. \tag{7.2.6}$$

The problem for a fixed external strain field can be treated similarly.

Let us consider separately the important case of a homogeneous medium containing isolated cavities or cracks. For cavities $C_{1i} = -C_0$, and $B_{1i} = (C_0 + C_{1i})^{-1} - C_0^{-1}$ tends to infinity while $C_{1i} \to -C_0$. Since

$$B_{1i}\sigma_i^+ = [(C_0 + C_{1i})^{-1} - C_0^{-1}](C_0 + C_{1i})\varepsilon_i^+, \tag{7.2.7}$$

then $B_{1i}\sigma_i^+ \to \varepsilon^+$ while $C_{1i} \to -C_0$.

The relations (2.4) can be rewritten in this case as

$$\begin{aligned} \varepsilon(x) &= \varepsilon_0 + \int K_0(x - x')C_0\varepsilon^+(x')\, dx', \\ \sigma(x) &= \sigma_0 - \int S_0(x - x')\varepsilon^+(x')\, dx'. \end{aligned} \tag{7.2.8}$$

Note that the right-hand sides of (7.2.4 and 8) may be interpreted as the stress and strain fields in a homogeneous medium, containing, in the regions V_k, sources of dislocations moments with the density

$$m_k(x) = -B_{1k}\sigma_k^+(x) \tag{7.2.9}$$

in the case of inclusions and

$$m_k(x) = -\varepsilon_k^+(x) \tag{7.2.10}$$

in the case of cavities (cf. Sect. 5.3). It is essential that these dislocation moments are induced in the medium by applying the external field σ_0 because the fields $\sigma^+(x)$ and $\varepsilon^+(x)$ are linear functions of σ_0.

Let us consider now the case in which defects are cracks or cuts along the smooth, oriented surfaces Ω_k. If we pass to the limit from a cavity V_k to the crack Ω_k, as it was shown in Sect. 4.11, we have $\varepsilon_k^+(x) \to M_k(x)\delta(\Omega_k)$ where

$$M_k(x) = \frac{1}{2}[n_k(x)b_k(x) + b_k(x)n_k(x)], \tag{7.2.11}$$

$n_k(x)$ is the normal to the surface Ω_k, $b_k(x)$ is a vector field at Ω_k which can be interpreted as a jump of displacements across the crack, $\delta(\Omega_k)$ is a delta-function concentrated at Ω_k (Sect. 6.1).

As it follows from (7.2.6), the stress and strain fields in a cracked medium are represented in the form

$$\begin{aligned} \varepsilon(x) &= \varepsilon_0 + \int K_0(x - x')C_0M(x')\delta(\Omega)\, dx' , \\ \sigma(x) &= \sigma_0 - \int S_0(x - x')M(x')\delta(\Omega)\, dx' , \end{aligned} \tag{7.2.12}$$

where we denote

$$M(x)\delta(\Omega) = \sum_k n_k(x)b_k(x)\delta(\Omega_k) . \tag{7.2.13}$$

The equations for the vectors $b_k(x)$ follow from the boundary conditions on the surfaces of the cut and have the form

$$\sum_i \int n_k(x)S_0(x - x')n_i(x')b_i(x')\delta(\Omega_i)\, dx' = n_k(x)\,\sigma_0 , \quad \text{with } x \in \Omega_k,\ k = 1, 2, \ldots . \tag{7.2.14}$$

For the case of a single crack a similar equation is considered in Sect. 4.11.

7.3 The Effective Field

Suppose that isolated defects constitute a spatially homogeneous, random field. Let us fix one of its typical realizations and consider an arbitrary defect with number i. If an external field is applied to the medium, this defect is subject to the field $\bar{\sigma}_i(x)$; on the basis of previous considerations $\bar{\sigma}_i(x)$ can be represented in the form

$$\bar{\sigma}_i(x) = \sigma_0 + \sum_{k \neq i} \int S_0(x - x')m_k(x')\, dx', \qquad x \in V_i , \tag{7.3.1}$$

where $m_k(x)$ is defined by (7.2.9 or 10) for the case of inclusions or cavities and by (7.2.11) for the case of cracks. The field $\bar{\sigma}_i(x)$ defined by this relation in the region V_i may be interpreted as the external field of the i-th defect. In the field $\bar{\sigma}_i(x)$ this defect behaves as an isolated one.

Sometimes it is more convenient to consider an external strain field $\bar{\varepsilon}_i(x)$ of the i-th defect. The connection between $\bar{\varepsilon}_i(x)$ and $\bar{\sigma}_i(x)$ is represented by the obvious relation

$$\bar{\varepsilon}_i(x) = B_0\bar{\sigma}_i(x) . \tag{7.3.2}$$

Let the solution of the elastic problem for isolated defects in an arbitrary external field be known. This means that we have explicit expressions for the functions $m_k(x)$ in (7.3.1) henceforth denoted by $m_k(x, \bar{\sigma}_k)$. Thus the system of equations governing the fields $\bar{\sigma}_k(x)$ for each of the interacting defects, is obtained from (7.3.1) in the form

$$\bar{\sigma}_i(x) = \sigma_0 + \sum_{k \neq i} \int S_0(x - x') m_k(x', \bar{\sigma}_k)\, dx', \; x \in V_i \; (i = 1, 2 \ldots) . \tag{7.3.3}$$

If the solution of this system has been found, the stress and strain fields in an inhomogeneous medium are determined by

$$\begin{aligned} \sigma(x) &= \sigma_0 + \sum_k \int S_0(x - x') m_k(x', \bar{\sigma}_k)\, dx' , \\ \varepsilon(x) &= \varepsilon_0 - \sum_k \int K_0(x - x') C_0 m_k(x', \bar{\sigma}_k)\, dx' , \end{aligned} \tag{7.3.4}$$

which are the consequences of (7.2.4, 8). Thus the fields $\bar{\sigma}_k(x)$ can be considered as the main unknowns of the problem.

If inhomogeneities constitute a random field, $\bar{\sigma}_k(x)$ are random functions. For the purpose of constructing the statistical moments of $\bar{\sigma}_k(x)$ let us introduce the following simplifying assumptions about the structure of $\bar{\sigma}_k(x)$ (hypothesis of the effective field method):

H1) The field $\bar{\sigma}_k(x)$ is practically constant inside the region V_k (or Ω_k) but, generally speaking, these fields are different for the different defects.

H2) The random variables $\bar{\sigma}_k$ do not depend statistically on the geometric characteristics and elastic constants of the defect V_k.

Note that the constants $\bar{\sigma}_k$ may be regarded as the average of the external field $\bar{\sigma}_k(x)$ over the volume V_k.

The field $\bar{\sigma}_k(x)$ defined in the region $V = \bigcup V_k$ and equal to the constant $\bar{\sigma}_k$ in V_k will be called the effective field. Note that in the classical scheme of the self-consisted method $\bar{\sigma}$ is assumed constant and equal for each particle [7.3, 4]. In the present modification of the method this field is considered as a random one and as it will be shown below, it enables us to describe more precisely the interaction between defects in the various stochastic fields of inhomogeneities

The picture for the interactions corresponding to the hypotheses H1 and H2 may be described qualitatively in the following way. For the typical defect the external field $\bar{\sigma}_i(x)$ which is represented as a sum of the external field σ_0 and the fields caused by all the surrounding inhomogeneities, is approximately constant in the region V_i occupied by this defect. Furthermore the contribution of every individual defect to this field $(\bar{\sigma}_i)$ is negligible. Without considering the realm of application of these hypotheses, we proceed now to an analysis of the formal consequences.

Further we shall assume that defects are inclusions of ellipsoidal shape, or plane elliptical cuts (cracks). As follows from the solution for an isolated ellipsoidal inhomogeneity in a constant external field, the density of the dislocation moment $m_k(x, \bar{\sigma}_k)$ induced in the region V_k is represented in the form

$$m_k(x, \bar{\sigma}_k) = P_k \bar{\sigma}_k V_k(x) , \tag{7.3.5}$$

where

$$P_k = B_0 C_{1k}(I + A_k C_{1k})^{-1} B_0 . \tag{7.3.6}$$

Here A_k is the singular part of the operator K_0 which in the three-dimensional case has the form (Appendix A3)

$$A_k = \frac{1}{4\pi} \int_{\omega} K_0(a_k^{-1} k)\, d\omega . \tag{7.3.7}$$

The linear transformation a_k^{-1} transforms V_k into the unit sphere, $K_0(k)$ is the k-representation of the Green's function for strains, ω is the surface of the unit sphere in k-space.

In the case of elliptical cracks the density $m_k(x, \bar{\sigma}_k)$ in (7.3.3, 4) has the form

$$m_k(x, \bar{\sigma}_k) = P_k(x) \bar{\sigma}_k \delta(\Omega_k) , \tag{7.3.8}$$

where the expression for $P_k(x)$ is found from the solution of the problem for an elliptical crack in the constant external field $\bar{\sigma}_k$ (Sect. 4.12) and has the form

$$P_k(x) = P_k h_k(x). \tag{7.3.9}$$

Here the function $h_k(x)$ is defined on the surface Ω_k by the relation

$$h_k(x^1, x^2) = \frac{a_k^2}{b_k} \sqrt{1 - \left(\frac{x^1}{a_k}\right)^2 - \left(\frac{x^2}{b_k}\right)^2} , \tag{7.3.10}$$

where the axes x_1, x_2 are connected with the main axes of the crack. Here a_k and b_k are the half axes of the ellipse Ω_k, $a_k \geq b_k$. The constant tensor P_k in (7.3.9) is defined by

$$P_k = -n_k T_0^{-1} n_k , \tag{7.3.11}$$

where n_k is the unit normal to Ω_k and the tensor T_0 has the form of (4.12.8).

In the case of an isotropic medium T_0^{-1} is

$$T_{0\alpha\beta}^{-1} = \frac{2a_k^2(1 - \nu_0)}{b_k \mu_0} d_\alpha \delta_{\alpha\beta} \tag{7.3.12}$$

where the scalar coefficients d_α are defined by (4.12.8), if we substitute a_1 and a_2 into a_k and b_k, respectively.

The relations (7.3.4) for cracks may be rewritten as

$$\sigma(x) = \sigma_0 + \sum_k \int S_0(x - x')P_k(x')\bar{\sigma}_k\delta(\Omega_k)\, dx' , \tag{7.3.13}$$

$$\varepsilon(x) = \varepsilon_0 - \sum_k \int K_0(x - x')C_0P_k(x')\bar{\sigma}_k\delta(\Omega_k)\, dx' .$$

Let us fix now a point x_0 and introduce the region V_{x_0} by the relation

$$V_{x_0} = \begin{cases} V = \bigcup_i V_i & \text{if } x_0 \bar{\in} V \\ \bigcup_{i \neq j} V_i & \text{if } x_0 \in V_j \end{cases} \tag{7.3.14}$$

where V_j is the region occupied by j-th inclusion. $\bar{V}_{x_0}$ denotes the complement of V_{x_0} to V. Evidently $\bar{V}_{x_0}$ is the region V_j in which the point x_0 is situated. The characteristic functions (of argument x) of the regions V_{x_0} and $\bar{V}_{x_0}$ are denoted by $V(x_0; x)$ and $\bar{V}(x_0; x)$, respectively: hence

$$V(x) = V(x_0; x) + \bar{V}(x_0; x) . \tag{7.3.16}$$

Let $P(x)$ be an arbitrary continuous tensor field which coincides with P_i, see (7.3.6), in the regions V_i. Let the field $\bar{\sigma}(x)$ be defined in the region V by

$$\bar{\sigma}(x) = \sigma_0 + \int S_0(x - x')P(x')\bar{\sigma}(x')V(x; x')\, dx', \; x \in V . \tag{7.3.17}$$

If the hypothesis H1 is valid, the field $\bar{\sigma}(x)$ coincides with $\bar{\sigma}_i$ in the regions V_i and consequently with the effective field in V.

For the case of cracks let us introduce the region Ω of all surfaces of the cracks and the region Ω_{x_0} which is

$$\Omega_{x_0} = \begin{cases} \Omega = \bigcup_i \Omega_i & \text{if } x_0 \bar{\in} \Omega \\ \bigcup_{i \neq j} \Omega_i & \text{if } x_0 \in \Omega_j . \end{cases} \tag{7.3.18}$$

Thus the equation governing the effective field in the region Ω is represented in the form

$$\bar{\sigma}(x) = \sigma_0 + \int S_0(x - x')P(x')\bar{\sigma}(x')\delta(\Omega_x)\, dx', \qquad x \in \Omega \tag{7.3.19}$$

where $P(x)$ is an arbitrary continuous tensor field coinciding with $P_k(x)$, see (7.3.9), on the surfaces Ω_k.

Equations (7.3.17) and (7.3.19) represent the grounds for constructing the statistical moments of the effective field $\bar{\sigma}$. It will be necessary for the solution of this problem to calculate the various averages of the random fields which have to be considered here. In the following section we shall discuss in details the problem of construction of these averages.

7.4 Several Mean Values of Homogeneous Random Fields

In what follows, it will be assumed that $V(x)$ is the characteristic function of a random set of ellipsoidal regions, which are distributed homogeneously in the three-dimensional space. Fields of the type of the stress tensor $\sigma(x)$, the strain tensor $\varepsilon(x)$ as well as the effective field $\bar{\sigma}(x)$ are functionals of the random function $V(x)$. For a homogeneous external stress field, all these random fields are homogeneous, since their statistical characteristics must be invariant with respect to the group of translations.

Let $f(x, V)$ be one of the functions under consideration. The mean $\langle f(x)\rangle$ with respect to the ensemble of realizations is the integral in the functional space [7.5]

$$\langle f(x)\rangle = \int f(x, V)d\mu(V)\,, \tag{7.4.1}$$

where $\mu(V)$ is a measure in the functional space. Analogously, the correlation function of the field $f(x)$ is defined as

$$\langle f(x_1)f(x_2)\rangle = \int f(x_1, V)f(x_2, V)d\mu(V)\,, \tag{7.4.2}$$

and the higher moments of $f(x)$ are defined in the same manner.

In what follows, an important role will be played by the means of $f(x)$ calculated under the condition that the region V contains a fixed point x_1. For this conditional mean, we introduce the commonly accepted notation[2] $\langle f(x)|x_1\rangle$. By definition [7.5],

$$\langle f(x)|x_1\rangle = \langle V(x)\rangle^{-1}\int f(x)V(x_1)d\mu(V)\,. \tag{7.4.3}$$

Analogously, the mean of the field $f(x)$ under the condition that the region V contains the points $x_1, \ldots, x_n$ has the form

$$\langle f(x)|x_1, x_2, \ldots, x_n\rangle = \langle\prod_{i=1}^{n} V(x_i)\rangle^{-1}\int f(x_1)\prod_{i=1}^{n} V(x_i)d\mu(V)\,. \tag{7.4.4}$$

Let now $x \in V$, $x_1 \in V_x$. The mean of the field $f(x)$ under these conditions will be denoted by $\langle f(x)|x; x_1\rangle$, and the expression for this mean has the form

$$\langle f(x)|x; x_1\rangle = \langle V(x)V(x; x_1)\rangle^{-1}\int f(x)V(x)V(x; x_1)d\mu(V)\,. \tag{7.4.5}$$

---

[2]The symbol $\langle\,|\,\rangle$ is nowhere used in this chapter to denote the scalar product, therefore no confusion should arise.

Note that here the equality

$$\langle f(x)|x;\, x_1\rangle = \langle f(x)|x_1;\, x\rangle \tag{7.4.6}$$

is valid.

In the general case, the sign ";" separates variables which belong to the region V_x, from variables which belong to the region $\bar{V}_x$. Thus the mean of $f(x_1)f(x_2)$ under the condition $x_1 \in V$, $x_2 \in \bar{V}_{x_1}^{x_3, x_4} \in V_{x_1}$ is written in the form

$$\begin{aligned}\langle f(x_1)f(x_2)|x_1,\, x_2;\, x_3,\, x_4\rangle \\ = \langle V(x_1)\bar{V}(x_1;\, x_2)V(x_1;\, x_3)V(x_1;\, x_4)\rangle^{-1} \\ \int f(x_1)f(x_2)V(x_1)\bar{V}(x_1;\, x_2)V(x_1;\, x_3)V(x_1;\, x_4)d\mu(V)\,.\end{aligned} \tag{7.4.7}$$

Generalization of this formula to any number of functions and variables is obvious.

If the value of f at a point x does not depend statistically on values of the function $V(x)$ (though f can be a functional of V), then the following relations are valid:

$$\langle f(x)V(x)\rangle = \langle f(x)|x\rangle\,\langle V(x)\rangle\,, \tag{7.4.8}$$

$$\langle f(x)V(x_1;\, x)|x_1\rangle = \langle f(x)|x;\, x_1\rangle\,\langle V(x_1;\, x)|x_1\rangle \tag{7.4.9}$$

and so on.

Let us note the equality which is a consequence of (7.3.16)

$$\langle V(x)|x_1\rangle = \langle V(x_1;\, x)|x_1\rangle + \langle \bar{V}(x_1;\, x)|x_1\rangle\,. \tag{7.4.10}$$

For a homogeneous field of inclusions, both terms on the right-hand side are functions of the difference $x - x_1$. If the dimensions of inclusions are bound, the second term is essentially non-zero only in a finite neighborhood of the origin. The first term vanishes when $x = x_1$, since $V(x;\, x) \equiv 0$ in view of (7.3.14) (Appendix A4).

All homogeneous random fields we shall use will be assumed to be ergodic: the means with respect to the ensemble of realizations $V(x)$ coincide with the corresponding means with respect to the whole three-space of a fixed typical realization. Let to such a realization $V_*(x)$ correspond a field $f_*(x)$. Then

$$\langle f(x)\rangle = \lim_{v\to\infty} \frac{1}{v} \int_G f_*(x)\, dx\,, \tag{7.4.11}$$

$$\langle f(x_1)f(x_1 + x_2)\rangle = \lim_{v\to\infty} \frac{1}{v} \int_G f_*(x_1)f_*(x_1 + x_2)\, dx_1\,, \tag{7.4.12}$$

where on the left-hand sides we have the means with respect to the ensemble of realizations of the random function $V(x)$, G is a region of the volume v, which occupies the whole space at the limit.

Analogously, for conditional means, the ergodic property enables us to write

$$\langle f(x)|x\rangle = \langle V(x)\rangle^{-1} \lim_{v\to\infty} \frac{1}{v} \int_G f_*(x) V_*(x)\, dx\,, \tag{7.4.13}$$

where

$$\langle V(x)\rangle = \lim_{v\to\infty} \frac{1}{v} \int_G V_*(x)\, dx\,. \tag{7.4.14}$$

An example of a more complicated conditional mean is the following:

$$\begin{aligned}&\langle f(x_1) f(x_1 + x_2)|x_1, x_1 + x_2;\rangle \\ &\quad = \langle V(x_1)\bar{V}(x_1; x_1 + x_2)\rangle^{-1} \\ &\quad \lim_{v\to\infty} \frac{1}{v} \int_G f_*(x_1) f_*(x_1 + x_2) V_*(x_1)\, \bar{V}_*(x_1; x_1 + x_2)\, dx_1\,.\end{aligned} \tag{7.4.15}$$

Here

$$\langle V(x_1)\bar{V}(x_1; x_1 + x_2)\rangle = \lim_{v\to\infty} \frac{1}{v} \int_G V_*(x_1)\bar{V}_*(x_1, x_1 + x_2)\, dx_1\,. \tag{7.4.16}$$

The asterisk for a fixed realization will later be omitted.

When considering random fields of cracks, the role of the region V will be assumed by the set Ω of points of crack surfaces, and, instead of the characteristic function $V(x)$, the generalized function $\delta(\Omega)$ will be used (see the definition in Sect. 6.1). In order to simplify our formulae, let us introduce the notation

$$\delta(\Omega) \equiv \Omega(x)\,. \tag{7.4.17}$$

Then the mean of $f(x_1)$ under the condition that $x_1, x_2, \ldots, x_n \in \Omega$, becomes

$$\langle f(x_1)|x_1, \ldots, x_n\rangle = \langle \prod_{i=1}^{n} \Omega\,(x_i)\rangle^{-1} \int f(x_1) \prod_{i=1}^{n} \Omega(x_i) d\mu(\Omega)\,. \tag{7.4.18}$$

Here $\mu(\Omega)$ is a measure in the functional space, $\prod_{i=1}^{n}\Omega(x_i)$ is a generalized function on the direct product of Euclidean spaces.

Let now Ω_x be the surface, defined by (7.3.18), and $\bar{\Omega}_x$ be the complement to Ω_x in Ω. The generalized functions $\delta(\Omega_x)$ and $\delta(\bar{\Omega}_x)$ will be denoted as follows:

$$\delta(\Omega_x) \equiv \Omega(x;\, x'),\ \delta(\bar{\Omega}_x) \equiv \bar{\Omega}(x;\, x')\,. \tag{7.4.19}$$

The mean of the field $f(x)$ under the condition that $x \in \Omega$, $x_1, x_2, \ldots, x_n \in \bar{\Omega}_x$; $x_{n+1}, \ldots, x_m \in \Omega_x$ is represented in the form

$$\langle f(x)|x, x_1, \ldots x_n;\, x_{n+1}, \ldots x_m\rangle$$
$$= \langle \Omega(x) \prod_{i=1}^{n} \bar{\Omega}(x;\, x_i) \prod_{k=n+1}^{m} \Omega(x;\, x_k)\rangle^{-1} \int f(x_1)\Omega(x) \prod_{k=n+1}^{m} \Omega(x;\, x_k)\, d\mu(x)\,. \tag{7.4.20}$$

Note that it will always be clear from the context, which field (of cracks or of inclusions) is considered; therefore the left-hand sides of (7.4.4 and 18) have the same notations.

For ergodic fields of cracks, calculations of their means with respect to the ensemble of realization can be substituted by averaging throughout the space for a fixed typical realization $\Omega(x)$. In this case, the mean $\langle f(x_1)|x_1;\, x_1 + x_2\rangle$ for example, can be found from the relation, see (7.4.15),

$$\langle f(x_1)|x_1;\, x_1 + x_2\rangle$$
$$= \langle \Omega(x_1)\Omega(x_1;\, x_1 + x_2\rangle^{-1} \lim_{v\to\infty} \frac{1}{v} \int_G f(x_1)\Omega(x_1)\Omega(x_1;\, x_1 + x_2)\, dx_1\,, \tag{7.4.21}$$

where

$$\langle \Omega(x_1;\, x_1 + x_2)\rangle = \lim_{v\to\infty} \frac{1}{v} \int_G \Omega(x_1)\Omega(x_1;\, x_1 + x_2)\, dx_1\,. \tag{7.4.22}$$

In conclusion, let us note the formula for the mean $\langle \Omega(x)|x_1\rangle$, which is analogous to (7.4.10),

$$\langle \Omega(x)|x_1\rangle = \langle \Omega(x_1;\, x)|x_1\rangle + \langle \bar{\Omega}(x_1;\, x)|x_1\rangle\,. \tag{7.4.23}$$

For homogeneous fields of cracks, both terms on the right-hand side are functions of the difference $x - x_1$, the first term being equal to zero when $x = x_1$, and the support of the second one is concentrated in a finite neighborhood of the origin, if the dimensions of the cracks are bound.

The construction of means of the type (7.4.22, 23) for specific stochastic fields of cracks is considered in Appendix A4.

7.5 General Scheme for Constructing First Statistical Moments of the Solution

Let us begin with constructing the main statistical characteristics of the effective field $\bar{\sigma}(x)$. We denote by $\bar{\sigma}^n(x_1, \ldots, x_n)$ an n-tuple moment of the effective field, i.e. the mean of the tensor product $\bar{\sigma}(x_1) \cdot\ldots\cdot \bar{\sigma}(x_n)$ under the condition that the points $x_1, \ldots, x_n$ belong to V(or Ω). In particular, the expectation and the two-point moment of the effective field are conditional means of the form

$$\bar{\sigma}^1 = \langle\bar{\sigma}(x)|x\rangle, \ \bar{\sigma}^2(x_1 - x_2) = \langle\bar{\sigma}(x_1)\bar{\sigma}(x_2)|x_1, x_2\rangle . \tag{7.5.1}$$

In order to obtain an expression for $\bar{\sigma}'$, let us average both sides of (7.3.17) with respect to the ensemble of realizations of the random field of inhomogeneities under the condition $x \in V$

$$\langle\bar{\sigma}(x)|x\rangle = \sigma_0 + \int S_0(x - x') \langle P(x')\bar{\sigma}(x')V(x; x')|x\rangle \, dx' . \tag{7.5.2}$$

Since x is a fixed point here, $S_0(x - x')$ is a determined kernel. Using the hypothesis H2 (Sect. 7.3) about the statistical independence of the field $\bar{\sigma}(x)$ at the point x of the elastic properties, the form and the dimension of the inclusion located at this point, let us represent the mean in the integrand in (7.5.2) in the form of a product of means, analogously to (7.4.9). Then the expression for $\bar{\sigma}^1$ becomes

$$\bar{\sigma}^1 = \sigma_0 + \int S_0(x - x') \langle P(x')V(x; x')|x\rangle\langle\bar{\sigma}(x')|x'; x\rangle \, dx' , \tag{7.5.3}$$

where the mean $\langle\bar{\sigma}(x')|x'; x\rangle$ is calculated under the condition that $x' \in V$ and $x \in V_{x'}$.

In order to construct the mean $\langle\bar{\sigma}(x')|x'; x\rangle$ in the integrand of (7.5.3), let us again consider (7.3.17). Averaging both sides of the equation under the condition $x \in V$, $x_1 \in V_x$ and using the hypothesis H2 (Sect. 7.3) once more, we have

$$\begin{aligned}\langle\bar{\sigma}(x)|x; x_1\rangle &= \sigma_0 \\ &+ \int S_0(x - x')\langle P(x')V(x; x')|x; x'\rangle\langle\bar{\sigma}(x')|x; x_1, x'\rangle \, dx' .\end{aligned} \tag{7.5.4}$$

The equations (7.5.3, 4) are not closed, since their right-hand sides contain the mean values of the effective field, which are calculated under conditions, different from those in the left-hand sides. Thus, a chain of equations arises here, and to close it, one has to introduce additional assumptions about the means in (7.5.3, 4) or in analogous relations which connect more complicated conditional means.

As it was pointed out in Sect. 7.3, the classical scheme of the method of self-consistent field consists of assuming $\bar{\sigma}(x)$ to be constant and equal for all particles. Then $\bar{\sigma}'$ and the conditional mean $\langle\bar{\sigma}(x')|x'; x\rangle$ coincide with each other and with $\bar{\sigma}$, and the constant $\bar{\sigma}$ is given by (7.5.3).

In the scope of the modification (of the method) developed here, the equivalent assumption which enables one to obtain equations for expectation of the field $\bar{\sigma}^1$, is that [3]

$$\langle \bar{\sigma}(x')|x'; x \rangle = \langle \bar{\sigma}(x')|x' \rangle = \bar{\sigma}^1 . \tag{7.5.5}$$

Substituting this relation into (7.5.3) we arrive at the equation with respect to $\bar{\sigma}^1$

$$\bar{\sigma}^1 = \sigma_0 + \int S_0(x - x') \langle P(x')V(x; x')|x \rangle \, dx' \, \bar{\sigma}^1 . \tag{7.5.6}$$

Thus, the problem is reduced to constructing the mean by calculating the integral on the right-hand side of (7.5.6).

The next, more precise, approximation for $\bar{\sigma}^1$ can be obtained, if one assumes in (7.5.4) that

$$\langle \bar{\sigma}(x')|x; x_1, x' \rangle = \langle \bar{\sigma}(x')|x'; x_1 \rangle . \tag{7.5.7}$$

For a homogeneous random field of defects, the mean on the right-hand side depends on the difference $x' - x_1$ only:

$$\langle \bar{\sigma}(x')|x'; x_1 \rangle = \varphi(x' - x_1) . \tag{7.5.8}$$

The function $\varphi(x' - x_1)$ is the mean value of the effective field at the point x' under the condition that there is a defect at the point x_1, and this characterizes the pair interaction in the system of interacting defects. It is obvious that, as $|x' - x_1| \to \infty$, this function tends to a mean value of the effective field $\bar{\sigma}^1$. The equation for $\varphi(x)$ follows from (7.5.4, 7) and has the form

$$\varphi(x) = \sigma_0 + \int S_0(x - x') \langle P(x')V(x; x')|x; x', 0 \rangle \, \varphi(x') \, dx' . \tag{7.5.9}$$

To obtain the second moment $\bar{\sigma}^2(x_1 - x_2)$ of the effective field, let us take a product of values of the field $\bar{\sigma}(x)$ (the relation (7.3.17) at different points x_1 and x_2) and average the result under the condition $x_1, x_2 \in v$

$$\begin{aligned} \bar{\sigma}^2(x_1 - x_2) = {} & \sigma_0 \langle \bar{\sigma}(x_2)|x_1, x_2 \rangle \\ & + \int S_0(x_1 - x') \langle P(x')V(x_1; x')\bar{\sigma}(x')\bar{\sigma}(x_2)|x_1, x_2 \rangle \, dx' . \end{aligned} \tag{7.5.10}$$

The equation for the function $\bar{\sigma}^2(x_1 - x_2)$ will be obtained after splitting the mean in the integrand with the help of the hypothesis H2 (Sect. 7.3)

---

[3] Here, the influence of a defect at the point x on the mean value of the effective field at the point x' is not taken into account.

$$\begin{aligned}&\langle P(x')V(x_1;\,x')\bar{\sigma}(x')\bar{\sigma}(x_2)|x_1,\,x_2\rangle\\&\quad=\langle P(x')V(x_1;\,x')|x_1,\,x_2\rangle\,\langle\bar{\sigma}(x')\bar{\sigma}(x_2)|x',\,x_1,\,x_2\rangle\end{aligned}\tag{7.5.11}$$

and assuming, see (7.5.5, 7), that

$$\langle\bar{\sigma}(x')\bar{\sigma}(x_2)|x',\,x_1,\,x_2\rangle=\langle\bar{\sigma}(x')\bar{\sigma}(x_2)|x',\,x_2\rangle=\bar{\sigma}^2(x'-x_2)\,.\tag{7.5.12}$$

From here and from (7.5.10) we obtain

$$\begin{aligned}\bar{\sigma}^2(x_1-x_2)&=\sigma_0\,\varphi(x_2-x_1)\\&\quad+\int S_0(x_1-x')\,\langle P(x')V(x_1;\,x')|x_1,\,x_2\rangle\bar{\sigma}^2(x'-x_2)\,dx'\,,\end{aligned}\tag{7.5.13}$$

where $\varphi(x)$ is a solution of (7.5.9).

The way of constructing subsequent approximations for $\bar{\sigma}^2(x)$ and $\varphi(x)$ in the scope of the scheme given, is obvious. In what follows, we confine ourselves to the consideration of (7.5.6, 9 and 13). Solutions constructed on the basis of these equations apparently describe a number of experimental data and, in many cases interesting for application, give a good agreement with exact solutions (Sects. 7.7, 9).

Let us proceed to derive the first two statistical moments of the strain and stress fields in a composite medium. Let us substitute the expressions for $m_k(x,\,\bar{\sigma}^k)$ in the form of (7.3.5) into (7.3.4) for $\sigma(x)$ and $\varepsilon(x)$ and average the result with respect to the ensemble of realizations of the random field of inclusions. Using the hypotheses H1 and H2 (Sect. 7.3) about the structure of the effective field $\bar{\sigma}(x)$, we have

$$\begin{aligned}\langle\sigma(x)\rangle&=\sigma_0+\int S_0(x-x')\,\langle P(x')V(x')\rangle\,\langle\bar{\sigma}(x')|x'\rangle\,dx'\,,\\\langle\varepsilon(x)\rangle&=\varepsilon_0-\int K_0(x-x')C_0\,\langle P(x')V(x')\rangle\,\langle\bar{\sigma}(x')|x'\rangle\,dx'\,.\end{aligned}\tag{7.5.14}$$

For homogeneous random fields of inclusions, the means $\langle P(x)V(x)\rangle$ and $\langle\bar{\sigma}(x)|x\rangle=\bar{\sigma}^1$ are constants. Since the operators S_0 and K_0 act on constants in accordance with (7.2.6), then, from (7.5.14), we obtain

$$\langle\sigma(x)\rangle=\sigma_0,\;\langle\varepsilon(x)\rangle=\varepsilon_0-\langle P(x)V(x)\rangle\bar{\sigma}^1\,.\tag{7.5.15}$$

The tensor of effective elastic compliance B_* is introduced in a natural way

$$\langle\varepsilon\rangle=B_*\,\langle\sigma\rangle\,.\tag{7.5.16}$$

Then, from (7.5.15) and (7.5.6) it follows that

$$B_*=B_0-\langle P(x)V(x)\rangle\,\Lambda\,,\tag{7.5.17}$$

where the fourth rank tensor Λ connects the external field with the first moment $\bar{\sigma}^1$ of the effective field

$$\bar{\sigma}^1 = \Lambda\sigma_0 \tag{7.5.18}$$

and is determined by the relation

$$\Lambda = \left[1 - \int S_0(x - x')\,\langle P(x')V(x;\, x')|x\rangle\, dx'\right]^{-1}. \tag{7.5.19}$$

In order to construct the second statistical moments of the solution, let us consider the product of the right-hand side of the expressions (7.3.4) for stress (strain) taken at different points x_1 and x_2 and average the result with respect to the ensemble of realizations of fields of inclusions. Using the main hypothesis of the method of the self-consistent field and (7.2.6) which defines the action of the operators K_0 and S_0 on constants, we obtain

$$\begin{aligned}\langle\sigma(x_1)\sigma(x_2)\rangle &= \sigma_0^2\\ &+ \int S_0(x_1 - x')\, dx' \int S_0(x_2 - x'')\,\langle P(x')P(x'')V(x')V(x'')\rangle\bar{\sigma}^2(x' - x'')\, dx'',\\ \langle\varepsilon(x_1)\varepsilon(x_2)\rangle &= \varepsilon_0^2 + \varepsilon_0\langle P(x)V(x)\rangle\,\bar{\sigma}^1 + \langle P(x)V(x)\rangle\bar{\sigma}^1\varepsilon_0\\ &+ \int K_0(x_1 - x')C_0 dx' \int K_0(x_2\\ &- x'')C_0\langle P(x')P(x'')V(x')V(x'')\rangle\bar{\sigma}^2(x' - x'')\, dx''.\end{aligned} \tag{7.5.20}$$

Analogously, one can write down the expression for mean values of the energy density $\langle\Phi\rangle = \langle\sigma\varepsilon\rangle/2$, of a composite medium, of the density $\langle\Phi_{\text{int}}\rangle$ of defects interaction, etc.

When considering a random field of cracks, one has to substitute the functions $P(x')V(x')$ and $P(x')V(x;x')$ into the preceding relations for $P(X')\Omega(x')$ and $P(x')\Omega(x;\, x')$, respectively, where now an arbitrary continuous function $P(x)$ coincides with the functions $P_k(X)$, (7.3.9), on the surfaces Ω_k. Thus, to construct the first two moments of the random fields $\sigma(x)$ and $\varepsilon(x)$, one has to find the statistical moments $\bar{\sigma}^1$ and $\bar{\sigma}^2$ of the effective field and then to construct the means and to compute the integrals appearing in (7.5.17, 20).

In conclusion, let us compare the method of the self-consistent field with another self-consistent method of constructing effective elastic constants, namely, with the method of the effective medium. It was pointed out in Sect. 7.1 that the main assumption, on which the second is based, is that each inclusion in a composite behaves as an isolated one in a homogeneous medium, properties of the latter coinciding with the effective properties of the whole composite. The external field for each inclusion is assumed to be σ_0.

Using these assumptions, we can transform (7.3.4) for stress and strain fields, in the case of ellipsoidal inclusions, to the form

$$\sigma(x) = \sigma_0 + \int S_0(x - x') \sum_k P_{k*} V_k(x') \, dx' \sigma_0 ,$$
$$\varepsilon(x) = \varepsilon_0 - \int K_0(x - x_0) C_0 \sum_k P_{k*} V_k(x') \, dx' \sigma_0 \tag{7.5.21}$$

where the tensors P_{k*} are expressed through yet unknown effective elastic parameters B_* of the composite

$$P_{k*} = B_0 C_{1k*} (I + A_{k*} C_{1k*})^{-1} B_0 . \tag{7.5.22}$$

Here C_{1k*} is a perturbation of the elastic modulus for the k-th inclusion with respect to the effective tensor $C_* = B_*^{-1}$

$$C_{1k*} = C_0 + C_{1k} - C_* . \tag{7.5.23}$$

A tensor A_{k*} is defined by (7.3.7), where the function $K_0(a_k^{-1} k)$ is to be replaced by $K_*(k)$, being k-representation of the Green's tensor for the effective medium strain.

The equation for the unknown tensor B^* is obtained from the condition of self-consistency, which coincides with (7.5.16)

$$\langle \varepsilon \rangle = B_* \langle \sigma \rangle . \tag{7.5.16}$$

Having averaged (7.5.11) for the stress and strain tensors and acting then on the constant, which is the mean of the sum in the integrand, by the operators K_0 and S_0, we obtain from (7.5.16)

$$B_* = B_0 - \langle C_{1*} (I + A_* C_{1*})^{-1} V \rangle \, B_0 , \tag{7.5.24}$$

where

$$C_{1*} (I + A_* C_{1*})^{-1} V = \sum_k C_{1k*} (I + A_{k*} C_{1k*})^{-1} V_k(x) . \tag{7.5.25}$$

The relation (7.5.24) which connects the constants of the tensor C_* with each other, is to be considered an equation with respect to effective elastic constants of the composite. The latter are contained on the right-hand side through the tensor C_{1k*}, (7.5.23), and A_{k*}.

In the case of elliptic cracks, assumptions of the method of the effective medium lead to expressions for stress and strain tensors, which coincide with (7.3.13), if one sets $\bar{\sigma}_k = \sigma_0$ and substitutes $P_k(x)$ by $P_{k*}(x)$, while $P_{k*}(x)$ (an analog to $P_k(x)$ in (7.3.8)) is to be found from the solution of the problem about an elliptic crack in a homogeneous medium with the elastic moduli tensor C_*. Using (7.5.16), one can obtain equations for the effective tensor B_* for a medium with cracks in the form

$$B_* = B_0 - \langle P_* h(x) \Omega(x) \rangle , \tag{7.5.26}$$

where

$$P_* h(x) \Omega(x) = \sum_k P_{k*} h_k(x) \Omega_k(x) . \tag{7.5.27}$$

Solving (7.5.24 and 26) with respect to the components of B_* is possible if there are explicit expressions for the tensors A_* and P_{k*} in terms of the elastic constants of the effective medium. However, a construction of these expressions in the case of an arbitrary anisotropy of the tensor B_* is connected with serious technical difficulties (Sect. 4.8, 9). Besides, if the number of unknowns is large (21 in the general case), then one can obtain the solution only numerically, which is not always convenient for applications. Thus, the method of the effective medium is usually used when the composite is macroisotropic. A comparison of numerical values of effective elastic constants, calculated with the help of the method of the effective medium and of the method of the effective field, is carried out in the following sections.

7.6 Random Field of Ellipsoidal Inhomogeneities

Let us now consider in more detail the method of the effective field for a medium, which contains inclusions of the ellipsoidal form. In this section, the expectation $\bar{\sigma}^1$ of the effective field and the tensor B_* of the effective elastic compliance in the case of composite and polycrystal media will be derived.

From (7.5.17, 18), it follows that the problem of determining the tensors $\bar{\sigma}^1$ and B_* is reduced to constructing explicit expressions for the means

$$\langle P(x) V(x) \rangle = \langle \sum_k P_k V_k(x) \rangle , \tag{7.6.1}$$

$$\Psi(x - x') = \langle P(x') V(x; x') | x \rangle \tag{7.6.2}$$

and calculating the integral which represents the action of the generalized functions $S_0(x)$ on the function $\Psi(x)$ defined by the last equality. The tensor P_k in (7.6.1) has the form of (7.3.6).

Let us start with constructing the means (7.6.1, 2). Due to the ergodic property assumed, the mean (7.6.1) for homogeneous fields of nonintersecting inclusions, has the form

$$\langle P(x) V(x) \rangle = \lim_{v \to \infty} \frac{1}{v} \int_G \sum_{k=1}^{N} P_k V_k(x)\, dx = \lim_{v \to \infty} \frac{1}{v} \sum_{k=1}^{N} P_k v_k , \tag{7.6.3}$$

where v_k is the volume of the k-th inclusion, N is the number of inclusions in the region G. It is taken into account here that P_k is a constant tensor inside each of the inclusions.

If one considers the ensemble of realizations of the homogeneous random field $P(x)V(x)$, then the tensors $P_k v_k$ will be stochastic variables with one and the same distribution function for any k. Therefore, if both sides of (7.6.3) are averaged with respect to the ensemble of realizations once more, then

$$\langle P(x)V(x)\rangle = \lim_{v\to\infty}\frac{N}{v}\langle v(a)P(a)\rangle = \langle\frac{v(a)}{v_0}P(a)\rangle\,, \tag{7.6.4}$$

where v_0 is the mean volume per an inclusion, $v(a)$ is the volume of an ellipsoid with semiaxes a_1, a_2, a_3, $P(a)$ is a tensor determined by the form of the ellipsoid and by the elastic constants of the latter, see (7.3.6). On the right-hand side of (7.6.4), we have already a mean value of the random variable $v(a)P(a)$, the distribution function of which is determined by the statistical properties of the ensemble of inclusions and is assumed to be known.

The problem of constructing the mean (7.6.2) for a homogeneous random field of elliptic regions is considered in Appendix A4. Let us point out the main properties of the function $\Psi(x)$. This is a continuous, bound function, its values being tensors of rank four. If the dimensions of all inclusions are bound, then $\Psi(x)$ can be represented as a sum of a constant component $\langle\Psi\rangle$, an oscilating function with the zero mean value and a function with bounded support. The constant $\langle\Psi\rangle$ is easily calculated, if one takes into account that the constant components of the functions $\Psi(x)$ and $\langle P(x')V(x')|x\rangle$ coincide, since, for inclusions of bound dimensions, these functions differ from each other in a finite neighborhood of the origin only. Using the ergodic property and averaging later with respect to the ensemble, we obtain

$$\langle\Psi\rangle = \frac{1}{\langle V(x)\rangle}\langle\lim_{v\to\infty}\frac{1}{v^2}\int_G dx'\int_G P(x')V(x')V(x)\,dx\rangle = \langle\frac{v(a)}{v_0}P(a)\rangle\,, \tag{7.6.5}$$

where the definition of the conditional mean is taken into account (Sect. 7.4):

$$\langle P(x')V(x')|x\rangle = \frac{\langle P(x')V(x')V(x)\rangle}{\langle V(x)\rangle}\,. \tag{7.6.6}$$

The function $\Psi(x)$ characterizes the averaged density of defects, which surround a typical inclusion with its center at the origin. The form of $\Psi(x)$ is determined by a specific structure of the random field of inhomogeneities. If the latter possesses some symmetry (in the statistical sense), then that results in the corresponding symmetry of the function $\Psi(x)$. For example, if the regions V_k form an isotropic field, then the function $\Psi(x)$ is spherically symmetric, i.e. $\Psi(x) = \Psi(|x|)$.

A break of spherical symmetry of $\Psi(x)$ can result in the appearance of texture. By texture we understand the symmetry of the elastic constants tensor of the matrix being different from the symmetry of the effective constants tensor of the inhomogeneous medium. For many stochastic composites, the symmetry of the textuie connected with geometrical properties of the field of inhomogeneities is not too low and this can be described with the help of a second order tensor α. In such a simple case, α determines a linear transform, which transforms $\Psi(x)$ into a spherically symmetric function $[\Psi(\alpha x) = \Psi'(|x'|)]$. In the general case, such a linear transformation can not be found.

The function $\Psi(x)$ is contained in (7.5.17, 18) for $\bar{\sigma}^1$ and B_* due to the integral, which represents the action of the generalized function $S_0(x)$ on $\Psi(x)$. Using properties of this generalized function (Appendix A3 and Sect. 4.6) and the definition of the operator S_0 acting on constants (7.2.6), we have

$$\int S_0(x - x')\Psi(x - x')\,dx' = \int S_0(x)[\Psi(x) - \langle\Psi\rangle]\,dx$$
$$= -D\langle\Psi\rangle + \fint S_0(\alpha x)[\Psi(\alpha x) - \langle\Psi\rangle]\det\alpha\,dx\,, \tag{7.6.7}$$

where

$$D = \frac{1}{4\pi}\int_{\omega} S_0(\alpha^{-1}k)\,d\omega\,. \tag{7.6.8}$$

Here the tensor α determines an arbitrary nondegenerate transformation of the x-space, ω is the surface of the unit sphere in the k-space, and $\langle\Psi\rangle$ the mean value of $\Psi(x)$ of the form of (7.6.5). The last integral in (7.6.7) is understood in the sense of a principal value and converges at the origin and at infinity.

Let us consider an important particular case. If $\Psi(x)$ is a spherically symmetric function or there exists a linear transformation α, for which $\Psi(\alpha x)$ is spherically symmetric, then due to properties of integrals in the sense of a principal value, the last integral in (7.6.7) vanishes and the equality assumes the highly simple form

$$\int S_0(x - x')\Psi(x - x')\,dx' = -D\langle\Psi\rangle\,. \tag{7.6.9}$$

Substituting this relations into (7.5.17, 18) and taking into account (7.6.4), we obtain expressions for the tensors $\bar{\sigma}^1$ and B_* in the form

$$\bar{\sigma}^1 = \left[I + D\left\langle\frac{v(a)}{v_0}P(a)\right\rangle\right]^{-1}\sigma_0\,, \tag{7.6.10}$$

$$B_* = B_0 - \left\langle\frac{v(a)}{v_0}P(a)\right\rangle\left[I + D\left\langle\frac{v(a)}{v_0}P(a)\right\rangle\right]^{-1}, \tag{7.6.11}$$

where

$$P(a) = B_0 C_1 [I + A(a) C_1]^{-1} B_0 , \tag{7.6.12}$$

and the tensor $A(a)$ is defined by (7.3.7) with $a_k = a$.

Let us sum up. If, from a statistical investigation of a composite, one can obtain an expression for the function $\Psi(x)$ and find a mean value of the stochastic variable $v(a)P(a)$, then the problem of constructing the tensors $\bar{\sigma}^1$ and B_* in the scope of the first approximation in the method of the effective field is reduced to a calculation of the converging integral (7.6.7). In the case of spherical symmetry of the function or its simplest analog, this integral is calculated in a finite form and the required tensors, take the form of (7.6.10, 11). A more complicated case of a regular lattice of inclusions will be considered in Sect. 7.7, numerical methods being then necessary even for constructing the integral (7.6.7).

Note one important peculiarity of the solution obtained. Let all the inclusions have the form of a ball, the same elastic properties and be distributed in the space isotropically. Then the function $\Psi(x)$ is spherically symmetric and the expressions for the tensors B_* and $C_* = B_*^{-1}$ take the form

$$B_* = B_0 - pB_0 C_1 [C_0 + pC_1 + (1 - p) C_0 A C_1]^{-1} , \tag{7.6.13}$$

$$C_* = C_0 + pC_1 [I + (1 - p) A C_1]^{-1} , \tag{7.6.14}$$

where A is defined by (7.3.7) with $a_k^{\alpha\beta} = \delta^{\alpha\beta}$. It is easy to verify that the formal, limiting transition, as $p \to 1$, leads to a physically reasonable result: $B_* \to B_0 + B_1$. Thus, the results obtained are noncontradictory in the entire region of the concentration of inclusions, though, while p is close to 1, the hypotheses of the method of an effective field lose sense, since one can not speak of isolated inhomogeneities in this case.

Note one more important case. In the scope of the method presented, one can consider not only effective elastic constants, but also more delicate characteristics of a stress state in a composite. Indeed, by virtue of the hypothesis H1 (Sect. 7.3), each inclusion in the composite behaves as an isolated one in a homogeneous matrix under a constant field $\bar{\sigma}_k$, k being the number of the inclusion. From the solution for an isolated ellipsoidal inhomogeneity (Sect. 4.8), an expression follows for the stress tensor σ_k^+ inside each inclusion

$$\sigma_k^+ = (C_0 + C_{1k})(I + A_k C_{1k})^{-1} B_0 \bar{\sigma}_k . \tag{7.6.15}$$

In the first approximation, one can replace the tensor by its mean value $\bar{\sigma}^1$, namely (7.6.10). From here, it is easy to evaluate the stress concentration in the matrix on the boundaries of inclusions, making use of (4.7.16) [7.6].

Let us apply the method of an effective field to consider the most important class of inhomogeneous media, namely to one-phase polycrystals. In comparison with composites, polycrystals possess a pecularity: one cannot distinguish explicitly a basic medium (matrix). Let us take as B_0 the mean value of $B(x)$ with respect to the ensemble

$$B(x) = B_0 + B_1(x), \qquad B_0 = \langle B(x)\rangle . \tag{7.6.16}$$

The tensor $B_1(x)$ has a constant value inside each crystallite, but changes from one crystallite to another in accordance with a change of crystallographic axes.

Following the scope of the method of an effective field, let us assume that each crystallite k behaves as an isolated ellipsoidal inhomogeneity with the elastic properties $B_0 + B_{1k}$ in a homogeneous medium B_0 and an external field $\bar{\sigma}_k$, the latter being the sum of an external field $\bar{\sigma}_k$ and a perturbation caused by all other cristallites. In the first approximation, expressions for $\bar{\sigma}^1$ and the tensor of effective compliance B_* still have the form of (7.5.18, 17). Since now the inclusions (crystallites) fill the whole space,

$$V(x) = \sum_k V_k(x) \equiv 1 , \tag{7.6.17}$$

where $V_k(x)$ is the characteristic function of the region occupied by the k-th crystallite. From here, there follow the equalities

$$\langle P(x)V(x)\rangle = \langle P(x')V(x')V(x)\rangle = \langle P(a)\rangle , \tag{7.6.18}$$

where, on the right-hand side we have the mean of the stochastic variable $P(a)$, the latter still being of the form (7.6.12). It is determined by the distribution function of the orientation of the crystallites anisotropy axes and by the form of the crystallites.

The properties of the function $\Psi(x - x') = \langle P(x')V(x, x')|x\rangle$ coincide, in the case under consideration, with all the properties of the same function for composites, and its symmetry depends on the geometry of the random field of crystallites, which form the polycrystal.

If there exists a linear transformation α, which converts the function $\Psi(x)$ into a symmetric one $\Psi(\alpha x) = \Psi'(|x|)$, then, as before (7.5.17, 18), emply the equalities

$$\bar{\sigma}^1 = [I + D\langle P(a)\rangle]^{-1}\sigma_0 , \tag{7.6.19}$$

$$B_* = B_0 - \langle P(a)\rangle[I + D\langle P(a)\rangle]^{-1} , \tag{7.6.20}$$

where the tensor D has the form of (7.6.8).

The relation (7.6.15), if one uses $\bar{\sigma}^1$ in the form of (7.6.19) instead of $\bar{\sigma}_k$, enables one to evaluate the microstress σ_k^+ in the individual crystallites. It is easily proved that the values of σ_k^+ and of the tensor of the effective elastic compliance B_* in the form of (7.6.20) do not depend on the way of decomposing the tensor $B(x)$ into the constant B_0 and the perturbation $B_1(x)$, see (7.6.16); however, the value $\bar{\sigma}^1$ depends on this decomposition.

In conclusion, let us dwell on (7.5.24) for the effective elastic compliance, obtained with the help of the method of an effective medium. We consider, as above, the case of composite and polycrystalline media. Carrying out the averaging in (7.5.24) according to the scheme presented in this section, we have

$$B_* = B_0 - \left\langle \frac{v(a)}{v_0} [I + A_*(a)C_{1*}]^{-1} \right\rangle B_0 \,. \tag{7.6.21}$$

As it was already pointed out, the last relation can be considered an equation with respect to effective elastic parameters of the medium. The right-hand side contains components of the tensor B_* through the tensors C_{1*} and $A_*(a)$ (Sect. 7.5). In the case of polycrystals, one has to assume additionally, that $v(a)/v_0 = 1$.

The equation (7.6.21) was investigated in [7.7–10]. Values of the effective elastic compliance, obtained from its solution, satisfie rather delicate variational extimates of Hashin-Shtrikman [7.11], and, in some cases, coincide well with experimental data. However, for media, which contain cavities (pores), the method of an effective medium leads to negative values of the effective elastic moduli, when the concentration of pores is larger than in the case of close packing. The method of an effective field has an explicit advantage in this case: as it was pointed out, (7.6.11) for B_* is physically non-contradictorary up to a concentration of inclusions, which is equal to unity.

7.7 Regular Structures

Let us apply the method of an effective field for calculating the elastic macroconstants of composites. This problem is of particular interest, since in the two-dimensional case, for a number of regular structures there are explicit solutions, which enable us to estimate the error of the method.

Let identical ellipsoidal inclusions form a regular lattice in a homogeneous matrix. The tensor of effective compliance of the composite is defined as before, by

$$\langle \varepsilon(x) \rangle = B_* \langle \sigma(x) \rangle \,, \tag{7.7.1}$$

while averaging of the fields $\varepsilon(x)$ and $\sigma(x)$ is performed here with respect to an elementary cell of the lattice, that is obviously equivalent to averaging with respect to the whole space. The means of $\varepsilon(x)$ and $\sigma(x)$ in turn coincide with the means at a fixed point x under all possible translations of the regular lattice. This means that averaging in (7.7.1) can be understood as averaging with respect to the ensemble of stochastic functions $V(x)$, which is given by

$$V(x) = V_f(x + r), \tag{7.7.2}$$

where $V_f(x)$ is the characteristic function of the region, which is occupied by an arbitrary fixed lattice of inclusions, r is a stochastic vector, homogeneously distributed throughout the space. Note that, when averaging with respect to the ensemble (7.7.2), the ergodic property is fulfilled automatically. Thus, the general scheme of the method of an effective field, which was developed in preceeding sections, can be applied to the case of regular structures without any changes.

Let us consider (7.6.1, 2). Since all the inclusions are now identical, then $P_k = P(a) = \text{const}$, where $P(a)$ has the form of (7.6.12), and the tensor a^{-1} determines a linear transformation, which transforms each inclusion into a unit ball. The means (7.6.1, 2) become

$$\langle P(x)V(x)\rangle = pP(a), \tag{7.7.3}$$

$$\Psi(x - x') = \langle P(x')V(x; x')|x\rangle = P(a)\langle V(x; x')|x\rangle, \tag{7.7.4}$$

where p is still the inclusions density.

As it is shown in Appendix A4 for regular lattices of ellipsoidal regions in the space,

$$\langle V(x; x')|x\rangle = \sum_m{}' J(x - x' - m), \tag{7.7.5}$$

where $m = m_\alpha e^\alpha$ is the vector of the lattice formed by the centers of inclusions (Sect. 2.1), the prime attached to the summation sign means excluding the item with $m = 0$, and the function $J(x)$ is defined by the relation

$$J(x) = \begin{cases} \left(1 - \dfrac{|a^{-1}x|}{2}\right)^2\left(1 + \dfrac{|a^{-1}x|}{4}\right) & \text{with } |a^{-1}x| \le 2 \\ 0 & \text{with } |a^{-1}x| > 2\,. \end{cases} \tag{7.7.6}$$

From (7.7.4, 5), it follows that the function $\Psi(x)$ is the sum of a periodic function with the mean value $pP(a)$ and a function $-J(x)P(a)$ with bounded support. The action of the generalized function on such functions is defined by (7.6.7), from where it follows that

$$\int S_0(x - x')\Psi(x - x')dx' = pQP(a)\,. \tag{7.7.7}$$

The tensor Q is represented in the form of a converging series

$$Q = -D + \sum_m{}' \int S_0(x)\left[\frac{1}{p} J(x-m) - 1\right]dx\,. \tag{7.7.8}$$

where the tensor D is defined by (7.6.8).

Substituting this result into (7.5.17, 18), for the mean value $\bar{\sigma}^1$ of the effective field and the tensor of effective compliance B_*, we obtain

$$\bar{\sigma}^1 = [I - pQP(a)]^{-1}\sigma_0\,, \tag{7.7.9}$$

$$B_* = B_0 - pP(a)[I - pQP(a)]^{-1}\,. \tag{7.7.10}$$

Let us consider the example of an isotropic medium, in which spherical isotropic inclusions form a cubic lattice. The tensor B_* becomes

$$B_* = B_0 - pB_0C_1[I + pB_0C_1 + (1-p)AC_1 + p\alpha(p)\Gamma C_1]^{-1}B_0\,, \tag{7.7.11}$$

where the tensors A and Γ are defined by

$$A = \frac{1}{3\mu_0}\left[E^2 - \frac{1}{10(1-\nu_0)}(2E^1 + E^2)\right], \tag{7.7.12}$$

$$\Gamma_{\alpha\beta\lambda\mu} = \frac{1}{\mu_0(1-\nu_0)}\left[2E^1_{\alpha\beta\lambda\mu} + E^2_{\alpha\beta\lambda\mu} - 5\sum_{i=1}^{3}\delta_{i\alpha}\delta_{i\beta}\delta_{i\lambda}\delta_{i\mu}\right]. \tag{7.7.13}$$

Here μ_0 and ν_0 are the shear modulus and the Poisson ratio of the matrix, the components of the tensor Γ are written in the basis e^α of the lattice vectors, and the tensors E^i are defined in Appendix A1. The scalar function $\alpha(p)$ in (7.7.11) is illustrated by its graph if Fig. 7.1, which was obtained by numerical summation of a converging series of integrals.

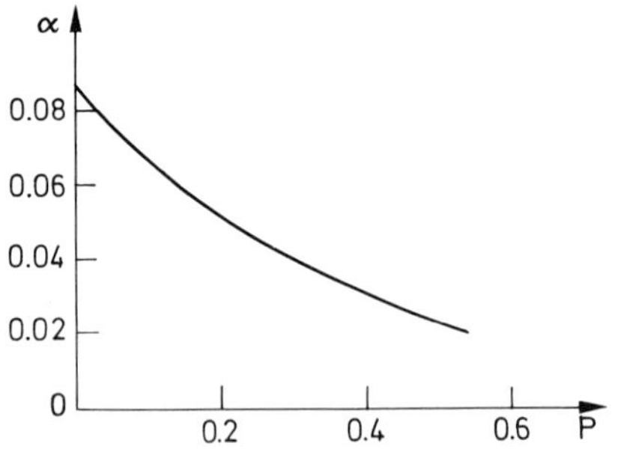

Fig. 7.1. Function $\alpha(p)$

Let us proceed to the two-dimensional problem. In this case, all constructions are carried out analogously to those of the three-dimensional case. The expression for the tensor of effective compliance has still the form of

(7.6.11) in the case of stochastic fields of inclusions and of (7.7.10) for regular structures. All tensors in these relations are to be replaced by their two-dimensional analogs. The two-dimensional problem was considered in more details in [7.12].

Let isotropic circular inhomogeneities form a regular lattice in an isotropic (two-dimensational) matrix. The following function is the two-dimensional analog of the function $J(x)$, (7.7.6),

$$J(x) = \begin{cases} \tan^{-1}\left(\frac{2r_0}{|x|}\sqrt{1-\frac{|x|}{2r_0}}\right) - \frac{|x|}{2r_0}\sqrt{1-\frac{|x|^2}{(2r_0)^2}} & \text{with } |x| \leq 2r_0 \\ 0 & \text{with } |x| > 2r_0 \end{cases} \tag{7.7.14}$$

where r_0 is the radius of an inclusion.

Let us consider a regular triangular lattice of inclusions. One can show [7.12] that, in this case, B_* has the form

$$B_* = B_0 - pB_0C_1[I + pB_0C_1 + (1-p)AC_1]^{-1}B_0\,, \tag{7.7.15}$$

which coincides with the expression for B_* in the case of an isotropic field of inclusions, see (7.6.13). Then the tensor A for a two-dimensional stress state if defined by

$$A = \frac{1}{16\mu_0}[8E^1 - (1+\nu_0)(E^2 + 2E^1)]\,, \tag{7.7.16}$$

where E^1, E^2 are the basis tensors, defined in Appendix A4, but for the two-dimensional case.

If inclusions form a square lattice, then the tensor B_* is defined by a relation of the form of (7.7.11), where A is the tensor (7.7.16), and

$$\Gamma_{\alpha\beta\lambda\mu} = \frac{1+\nu_0}{\mu_0}\left(2E^1_{\alpha\beta\lambda\mu} + E^2_{\alpha\beta\lambda\mu} - 4\sum_{i=1}^{2}\delta_{i\alpha}\delta_{i\beta}\delta_{i\lambda}\delta_{i\mu}\right). \tag{7.7.17}$$

Here the Greek indexes take values 1, 2.

The function $\alpha(p)$ in (7.7.11), for the two-dimensional case, was also found numerically, and a graph has been presented in [7.12].

In [7.13] the problem of the stress state of plane lattices of inclusions in a homogeneous external field was solved with the help of a series expansion in bi-periodic functions. The values of the effective elastice constants (of different lattices) presented there can be considered exact ones, since in the corresponding series a sufficient number of terms was conserved.

In Figs. 7.2 and 7.3, a comparison is given of exact values of effective elastic moduli with those obtained with the help of the method of the effective field for a regular triangular lattice. Here E_0 and μ_0 are the Young modulus and the shear modulus of the matrix, $E_0 + E_1$ and $\mu_0 + \mu_1$ are the same quan-

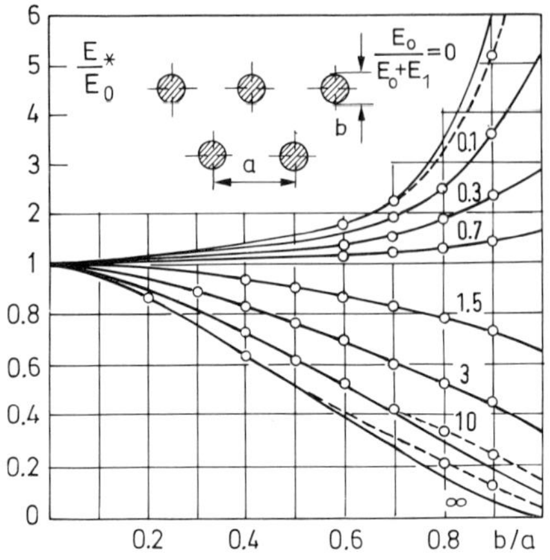

Fig. 7.2. Comparison of exact and effective Young's moduli

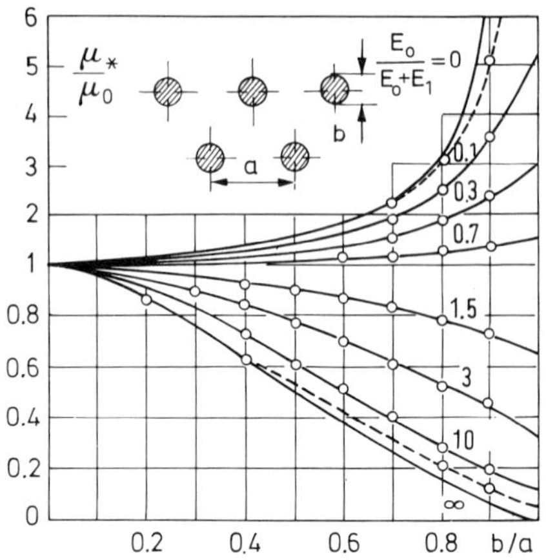

Fig. 7.3. Comparison of exact and effective shear moduli

tities for inclusions, E_*, and μ_* are the values of the effective elastic modulus. The solid curves represent the exact solutions, the dotted line with small circles corresponds to (7.7.15)

For this case, let us consider the error Δ which arises when evaluating effective constants with the help of the method of the effective field

$$\Delta = \frac{E_* - E_*^T}{E_*^T}. \tag{7.7.18}$$

Here E_*^T and E_* are the exact values of the effective Young modulus and the one calculated from (7.7.15), respectively. Calculations show that Δ is a function of the dimensionless parameter

$$\chi = P \frac{C_1}{C_0 + (1 - P)\gamma C_1}, \tag{7.7.19}$$

where c_0 is a characteristic value of the elastic modulus of the matrix, c_1 is the same for the perturbation of the modulus in inclusions, γ is a dimensionless scalar parameter, which depends on the form of inclusions and is equal to a characteristic value of components of the tensor AC_0. For circular inclusions, $\gamma = 0.66$. The dependence $\Delta(\chi)$ is exhibited in Fig. 7.4, where the scale on the absciss axis is $\ln(2 + \chi)$.

The parameter χ can vary between the limits $-1 \le \chi \le \infty$, while to negative values of χ there correspond inclusions more compliant than the matrix, and to positive χ the opposite. As it is seen from Fig. 7.4, an error higher than 10% arises in the regions $-1 < \chi < -0.8$ and $4 < \chi < 6$. The continuous part of the cuive in the Fig. 7.4 is obtained with the help of the results of [7.13].

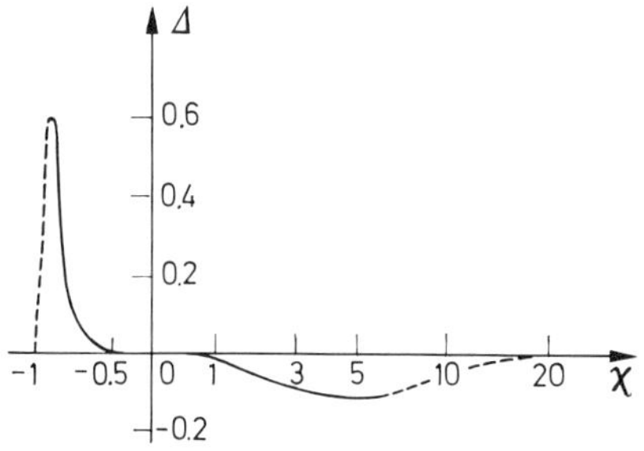

Fig. 7.4. Function $\Delta(\chi)$

The dotted line is an extrapolated value of the error. The fact that, as $p \to 1$, (7.7.15) gives a correct result for all C_1, is taken into account.

The relative error for the effective shear modulus practically coincides with $\Delta(\chi)$ for all values of the parameter χ.

A comparison of the effective field approximation with exact solutions for a square lattice of circular inclusions was carried out in [7.12]. The form of the error of calculating the effective elastic constants is here the same as in the case of a triangular lattice.

7.8 Fields of Elliptic Cracks

Let us proceed to a consideration of a random field of elliptic cracks in a homogeneous elastic medium. In what follows, it is assumed, for simplicity, that all cracks are opened, when the medium is loaded. This imposes certain restrictions on the external field σ_0.

The expressions (7.5.17, 18) for the tensors $\bar{\sigma}^1$ and B_*, in the case under consideration, have the form

$$\bar{\sigma}^1 = \Lambda\sigma_0\,, \tag{7.8.1}$$

$$B_* = B_0 - \langle P(x)\Omega(x)\rangle\Lambda\,, \tag{7.8.2}$$

where

$$\Lambda = [I - \int S_0(x-x')\,\langle P(x')\Omega(x;\,x')|x\rangle dx']^{-1}\,. \tag{7.8.3}$$

Here the functions $\Omega(x)$ and $\Omega(x;\,x')$ are defined by (7.4.17, 19), the continuous tensor field $P(x)$ coincides with $P_k(x)$, (7.3.9), on the surfaces of the cracks.

To construct the means in (7.8.2, 8) let us use the same procedure as in the determination of the means (7.6.1, 2) in the case of ellipsoidal inclusions. Using the ergodic property with the subsequent averaging with respect to the ensemble, we obtain

$$\langle P(x)\Omega(x)\rangle = \left\langle \lim_{v\to\infty} \frac{1}{v} \int_G P_k h_k(x)\Omega_k(x)dx \right\rangle = \frac{2\pi}{3}\left\langle \frac{a^3}{v_0} P(a,\, b)\right\rangle, \tag{7.8.4}$$

where $\Omega_k(x)$ is a generalized function concentrated on the crack surface Ω_k, $h_k(x)$ has the form of (7.3.10), the tensors P_k are given by (7.3.11), v_0 is as before the mean volume per crack. On the right-hand side of this relation, there appears the mean of the stochastic tensor $a^3P(a, b)$, where $P(a, b)$ has the same form as P_k, see (7.3.11), and is determined by the stochastic values a and $b (a \geq b)$ of the semiaxes of an elliptic crack and by the orientation n of the plane of the latter.

Let us consider the conditional mean in (7.8.3)

$$\Psi(x - x') = \langle P(x')\Omega(x;\, x')|x\rangle\,. \tag{7.8.5}$$

The constant components $\langle\Psi\rangle$ of the functions $\Psi(x)$ and $\Psi(x - x') = \langle P(x')\Omega(x')|x\rangle$ coincide, since their difference is a function, with bounded support, if the dimensions of the cracks are bound (Appendix A4). By analogy with (7.6.5), we have

$$\begin{aligned}\langle\Psi\rangle &= \frac{1}{\langle\Omega(x)\rangle}\left\langle \lim_{v\to\infty}\frac{1}{v^2}\int_G dx' \int_G P(x')\Omega(x')\Omega(x)dx\right\rangle \\ &= \frac{2\pi}{3\langle ab\rangle}\left\langle \frac{a^4 b}{v_0} P(a,\, b)\right\rangle.\end{aligned} \tag{7.8.6}$$

Here, the relations

$$\langle P(x')\Omega(x')|x\rangle = \frac{\langle P(x')\Omega(x')\Omega(x)\rangle}{\langle\Omega(x)\rangle}, \tag{7.8.7}$$

$$\langle\Omega(x)\rangle = \frac{\pi\langle ab\rangle}{v_0}, \tag{7.8.8}$$

are taken into account, the latter being easily obtained, as (7.8.4) was. For cracks of identical dimensions

$$\langle\Psi\rangle = \frac{2\pi}{3}\frac{a^3}{v_0}\langle P(a,\, b)\rangle\,, \tag{7.8.9}$$

where the mean is calculated only with respect to cracks orientations.

As in the case of ellipsoidal inclusions, the function $\Psi(x)$ for homogeneous fields of cracks is represented in the form of a sum of a periodic (generally speaking, generalized) or continuous oscillating function with the mean value equal to $\langle\Psi\rangle$ and of a function with bounded support. At the same time the integral in (7.8.3) admits regularization, which is analogous to (7.6.7)

$$\int S_0(x - x')\Psi(x - x')\,dx' = \int S_0(x)[\Psi(x) - \langle\Psi\rangle]\,dx$$
$$= D[\Psi(0) - \langle\Psi\rangle] + \textstyle\oint S_0(\alpha x)[\Psi(\alpha x) - \langle\Psi\rangle] \det\alpha\,dx\,, \tag{7.8.10}$$

where D is defined by (7.6.8), the tensor α being an arbitrary nondegenerate linear transformation of coordinates. The function $\Psi(x)$ is assumed to be continuous at the origin; this is true for most of the interesting distributions of cracks.

Let us consider some specific examples of stochastic fields of cracks.

1) *The Possion field of cracks.* Let, in a bounded volume v of the three-space, there be N cracks, dimensions and orientations of the latter being stochastic quantities with known distribution functions. The position of each crack does not depend on positions of others and is homogeneously distributed in v. Letting v and N tend to infinity so that $(v/N) = v_0 < \infty$, we obtain a homogeneous field of cracks, which will be called the Poisson field. In this case (Appendix A4)

$$\Psi(x) = \langle P(x')\Omega(x;\,x')|x\rangle = \langle\Psi\rangle\,, \tag{7.8.11}$$

i.e. $\Psi(x) = \text{const}$. By virtue of the definition (7.2.6) of the action of the operator S_0 on constants

$$\int S_0(x - x')\langle\Psi\rangle dx' = 0\,, \tag{7.8.12}$$

the expression (7.8.2) for B_* becomes

$$B_* = B_0 - \frac{2\pi}{3}\left\langle \frac{a^3}{v_0} P(a,\,b)\right\rangle. \tag{7.8.13}$$

For circular cracks of the stochastic radius a in an isotropic medium we then have

$$B_* = B_0 - \frac{8}{3}\,\frac{(1 - \nu_0)}{\mu_0(2 - \nu_0)}\,\frac{\langle a^3\rangle}{v_0}\langle 2E^5(n) - \nu_0 E^6(n)\rangle\,, \tag{7.8.14}$$

where the tensors E^i are defined by (A1) and the averaging of these tensors is carried out with respect to all possible orientations of cracks. If the distribution with respect to orientations is homogeneous, then

$$B_* = B_0 - \frac{8}{45}\,\frac{(1 - \nu_0)}{\mu_0(2 - \nu_0)}\,\frac{\langle a^3\rangle}{v_0}[\nu_0 E^2 - 2(5 - \nu_0)E^1]\,, \tag{7.8.15}$$

and, in particular, for the effective shear modulus μ_* and the Poisson coefficient ν_*, we have

$$\mu_* = \mu_0 \left(1 + \frac{32}{45} \frac{\langle a^3 \rangle}{v_0} \frac{(1-\nu_0)(5-\nu_0)}{(2-\nu_0)}\right)^{-1}, \tag{7.8.16}$$

$$\frac{\nu_*}{1+\nu_*} = \frac{\nu_0}{1+\nu_0} \frac{\mu_*}{\mu_0} \left(1 + \frac{16}{45} \frac{\langle a^3 \rangle}{v_0} \frac{(1-\nu_0^2)}{(2-\nu_0)}\right). \tag{7.8.17}$$

2) *The model with a restriction, concerning the intersections of cracks.* Let, around each crack, there be a neighborhood with a small probability of containing other cracks. This means that $\Psi(x) \approx 0$ in a neighborhood of the origin analogously to the case of ellipsoidal inclusions. If, moreover, $\Psi(x)$ is spherically symmetric, then the integral (7.8.10) takes the extremely simple form

$$\int S_0(x-x')\Psi(x-x')\,dx' = -D\langle \Psi \rangle. \tag{7.8.18}$$

At the same time, the tensor B_*, due to (7.8. 2–4), is

$$B_* = B_0 - \frac{2\pi}{3} \left\langle \frac{a^3}{v_0} P(a, b) \right\rangle (I + D\langle \Psi \rangle)^{-1}. \tag{7.8.19}$$

Note that the expression for B_* will have the same form, if there exists a linear transformation α, which makes $\Psi(x)$ a spherically symmetric function $\Psi'(|x|) = \Psi(\alpha x)$. Then the tensor D depends on α and is defined by the relation (7.6.8).

In the case of a spherically symmetric Ψ and of a homogeneous distribution with respect to orientations, B_* is an isotropic tensor

$$B_* = B_0 + \frac{8}{45} \frac{(1-\nu_0)}{\mu_0(2-\nu_0)} \cdot \frac{1}{(1-\beta_1)} \frac{\langle a^3 \rangle}{v_0} \times \left(\frac{2\beta_2(5-\nu_0) - \nu_0(1-\beta_1)}{1-\beta_1-3\beta_2} E^2 + 2(5-\nu_0) E^1 \right), \tag{7.8.20}$$

where

$$\beta_1 = \frac{32}{675} \frac{\langle a^3 \rangle}{v_0} \frac{(7-5\nu_0)(5-\nu_0)}{(2-\nu_0)}, \quad \beta_2 = \frac{32}{675} \frac{\langle a^3 \rangle}{v_0} \left(\frac{5-\nu_0}{2-\nu_0} + 10\nu_0 \right). \tag{7.8.21}$$

Here all cracks are circular and the radius a is stochastic.

3) *Regular lattices of cracks in the three-space.* Let us first consider a simple lattice, in which all cracks have identical dimensions and orientations. In this case the means (7.8.4, 5) become

$$\langle P(x)\Omega(x) \rangle = \langle \Psi \rangle = \frac{2\pi}{3} \frac{a^3}{v_0} P(a, b), \tag{7.8.22}$$

$$\Psi(x - x') = P(a, b) \langle h(x')\Omega(x; x')|x\rangle , \tag{7.8.23}$$

where $P(a, b)$ is a constant tensor of the form of (7.3.11) and the function $h(x)$ has the form of (7.3.10) at each crack. In these relations, averaging is carried out with respect to all translations, i.e. analogously to the case of regular lattices of inclusions (Section 7.7).

As it is shown in Appendix A4, the mean (7.8.23) for regular lattices of elliptic cracks is represented by

$$\Psi(x) = \frac{1}{\pi ab} \langle \Psi \rangle \sum_m{}' J(x - m)\Omega'_m(x) , \tag{7.8.24}$$

where m is a vector of the lattice formed by the centers of the cracks, the primed summation sign means excluding the term with $m = 0$, $\Omega'_m(x)$ is the δ-function concentrated on a flat surface with its center at the point m and parallel to all cracks. The form of the surface is an ellips with semiaxes, which are equal to doubled semiaxes of the crack under consideration. The function $J(x)$ is defined by the relation (A 4.47)

$$J(x) = \frac{3v_0}{2\pi} \begin{cases} \int_{\frac{1}{2}\xi}^{1} \frac{(\zeta - \xi)\sqrt{\xi(2\zeta - \xi)} + \zeta^2\left[\sin^{-1}\left(1 - \frac{\zeta}{\xi}\right) + \frac{\pi}{2}\right]}{\sqrt{1 - \zeta^2}} \zeta d\zeta & \text{with } \xi \leq 2 \\ 0 & \text{with } \xi > 2 \end{cases} \tag{7.8.25}$$

where

$$\xi = \sqrt{\frac{(x^1)^2}{a^2} + \frac{(x^2)^2}{b^2}} , \tag{7.8.26}$$

and the x^1- and x^2-axes are directed along the principal axes of the elliptic cracks.

From (7.8.10) we can now find the result of the action of the generalized function $S_0(x)$ on $\Psi(x)$ in the form of (7.8.24)

$$\int S_0(x - x')\Psi(x - x')\, dx' = Q\langle \Psi \rangle . \tag{7.8.27}$$

Here, the tensor Q has the form of a converging series of integrals

$$Q = -D + \sum_m{}' \oint S_0(x)\left[\frac{1}{\pi ab} J(x - m)\, \Omega'_m(x) - 1\right] dx , \tag{7.8.28}$$

where D is defined by (7.6.8) with $\alpha^{\alpha\beta} = \delta^{\alpha\beta}$. Substituting (7.8.27) into (7.8.1, 2) for the tensors $\bar{\sigma}^1$ and B^*, we obtain

$$\bar{\sigma}^1 = (I - Q\langle \Psi \rangle)^{-1} \sigma_0 , \tag{7.8.29}$$

$$B_* = B_0 - \langle \Psi \rangle (I - Q \langle \Psi \rangle)^{-1}, \tag{7.8.30}$$

where $\langle \Psi \rangle$ has the form of (7.8.22). Note that, for simple lattices, $\bar{\sigma}^1$ is the real external field $\sigma(x)$ averaged over the surface of any crack.

If cracks form a complex lattice, then the expression for $\Psi(x)$ consists of sums of the type of (7.8.24) over each sublattices. In this case, one can make the method of solution more precise by assuming that the value $\bar{\sigma}^1$ of the effective field is different for different lattices. Equations for the set of quantities $\bar{\sigma}_k^1$ (the values of $\bar{\sigma}^1$ on the k-th sublattice) follow from an analog of (7.5.6) for the case of cracks and become

$$\bar{\sigma}_k^1 + \sum_{i=1}^{n} Q_i^{(k)} \langle \Psi_i \rangle \bar{\sigma}_i = \sigma_0, \; k = 1, 2, \ldots, n, \tag{7.8.31}$$

where

$$Q_i^{(k)} = -D + \sum_{m_i} \int S_0(x) \left[\frac{1}{\pi a_i b_i} J(x - m_i - \xi_{ik}) \, \Omega'_{m_i}(x - \xi_{ik})^{-1} \right] dx, \tag{7.8.32}$$

and, with $i = k$, the expression for $Q_k^{(k)}$ has the form of (7.8.28), where the quantities a, b and m concern the k-th sublattice only. In these relations, m_i is a vector of the i-th sublattice, a_i, b_i are semiaxes of elliptic cracks in the i-th sublattice, ξ_{ik} is a vector, which connects knots of i-th and k-th sublattices in an elementary cell, the tensor $\langle \Psi_i \rangle$ is determined by the orientation and the dimensions of cracks in the i-th sublattice and has the form, analogous to (7.8.22).

Solving the system (7.8.31) with respect to the tensors $\bar{\sigma}_i^1$ and substituting the result into the expressions for the mean values of stress and strain, analogous to (7.5.14), using then the definitions (7.5.16) for the tensor B_* we obtain

$$B_* = B_0 - \sum_{i=1}^{n} \langle \Psi_i \rangle \Lambda_i, \tag{7.8.33}$$

where the tensors Λ_i connect $\bar{\sigma}_i^1$ with the external field σ_0

$$\bar{\sigma}_i^1 = \Lambda_i \sigma_0 \tag{7.8.34}$$

and are determined from the system (7.8.31).

Since we have assumed that each crack behaves as an isolated one in the field $\bar{\sigma}^1$, then, in the scope of our method, one can consider a value of the averaged stress intensity factor on the crack contour (for the definition of the stress intensity factors see Sect. 7.8.11). To achieve this, one has to use the solution of the problem for an isolated crack in a constant external stress field $\bar{\sigma}^1$ (Sect. 7.8.12).

Let us consider the example of an isotropic medium, which contains circular cracks of identical orientations, the cracks forming a lattice with three orthogonal planes of symmetry. In the case of cracks with identical orientations, the openings of the cracks are uniquely determined by the convolution of the tensor $\bar{\sigma}^1$ with the normal to the plane of the cracks. It can be shown that for the considered lattices $\bar{\sigma}^1 n$ has the form

$$\bar{\sigma}^{1\alpha\beta} n_\beta = \frac{1}{(1-\alpha)}\left(\delta^\alpha_\beta + \frac{\beta}{1-\alpha-\beta} n^\alpha n_\beta\right)\sigma_0^{\beta\nu} n_\nu\,, \tag{7.8.35}$$

the tensor B_* is determined by the relation

$$B_* = B_0 + \frac{8}{3}\frac{(1-\nu_0)}{\mu_0(2-\nu_0)}\frac{1}{(1-\alpha)}\frac{a^3}{v_0}\left[2E^5(n) + \left(\frac{2\beta(2-\nu_0)}{1-\alpha-\beta} - \nu_0\right)E^6(n)\right], \tag{7.8.36}$$

where

$$\alpha = \frac{4}{(2-\nu_0)}[\beta_0 - (2-\nu_0)\,\alpha_0],\ \beta = \frac{4(\nu_0-3)}{(2-\nu_0)}\beta_0\,, \tag{7.8.37}$$

and the radius of all cracks is a. The dimensionless scalar coefficients α_0 and β_0 depend only on the geometry of the lattice and are presented in the form of converging series of the type of (7.8.28). Fig. 7.5, displays the graphs of

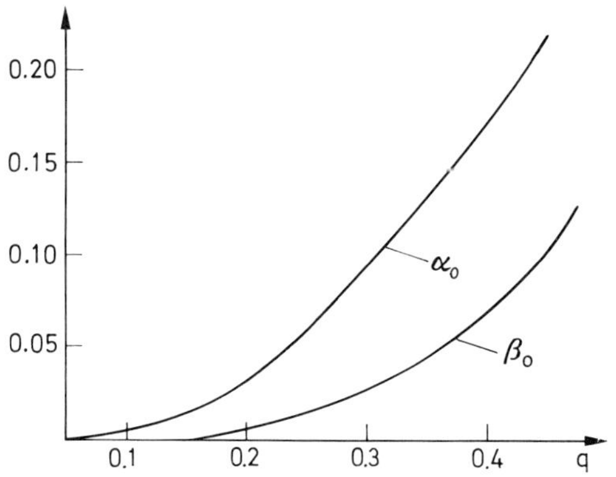

Fig. 7.5. Functions $\alpha_0(q)$ and $\beta_0(q)$

α_0 and β_0 as functions of the parameter $q = \langle a^3\rangle/v_0$ for a regular cubic lattice of cracks, all planes of the cracks being parallel to a fixed face of the cube.

The stress intensity factors on the contour of a crack are found from the solution of the problem for a circular isolated crack in the constant field $\bar{\sigma}^1$ of the form of (7.8.35). In the case of pure tension in the direction of the normal to the plane of cracks, $\sigma_0^{\alpha\beta} = \sigma_0 n^\alpha n^\beta$, and the stress intensity factor K_I has the form [7.8]

$$K_I = \frac{n\,\bar{\sigma}^1 n}{\sqrt{2a}} = \frac{1}{(1-\alpha-\beta)}\,\frac{\sigma_0}{\sqrt{2a}}\,. \tag{7.8.38}$$

This expression is to be considered as the mean stress intensity factor of the crack.

Let us consider a lattice of cracks located in one plane. This case can be obtained from the spatial one, if one of the parameters of the lattice tends to infinity. The problem of a square lattice of circular cracks on a plane was solved in [7.14]. In the Fig. 7.6, the value of the parameter

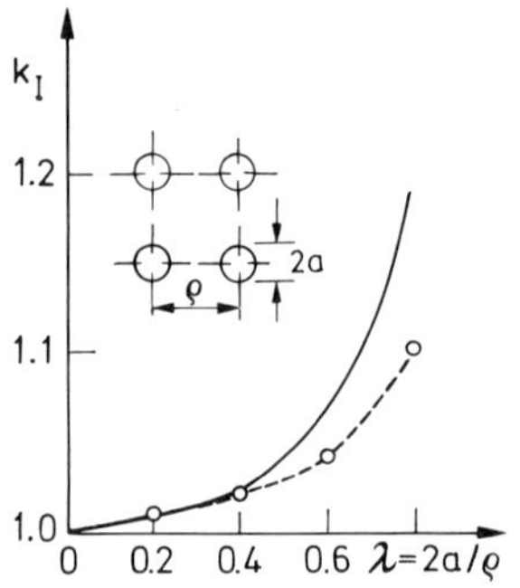

Fig. 7.6. Function $k_I(\lambda)$

$$k_I = \frac{\sqrt{2a}}{\sigma_0}\,K_I\,, \tag{7.8.39}$$

calculated with the help of the method of effective field (dotted curve) is compared with the maximun value of k_I on the edge of the cracks from [7.14] (continuous curve).

In conclusion, let us consider the expression for the tensor of the effective elastic compliance of a medium with cracks, to which the method of effective medium leads. Eq. (7.5.26) for the effective compliance is represented by

$$B_* = B_0 - \frac{2\pi}{3}\left\langle \frac{a^3}{v_0}\,P_*(a, b)\right\rangle, \tag{7.8.40}$$

where P_* is determined from the solution for an elliptical crack in the medium with the tensor B_* of elastic compliance and has the form, which is analogous to that of P_k in (7.3.8) (for more details, see the end of Sect. 7.5).

In the case of an isotropic medium, which contains a homogeneous field of circular cracks with homogeneously distributed orientations, the effective medium is also macroisotropic, and (7.8.40) becomes

$$B_* = B_0 - \frac{2\pi}{3}\left\langle \frac{a^3}{v_0} \right\rangle P_*\,, \text{ where} \tag{7.8.41}$$

$$P_* = -\frac{4}{15\pi}\,\frac{(1-\nu^*)}{\mu_*(2-\nu_*)}\,[\nu_*\,E^2 - 2(5-\nu_*)\,E^1]\,. \tag{7.8.42}$$

From here, the relations follow

$$\frac{\mu_*}{\mu_0} = -1\,\frac{32}{45}\,\frac{\langle a^3\rangle}{v_0}\,\frac{(1-\nu_*)\,(5-\nu_*)}{(2-\nu_*)}\,, \tag{7.8.43}$$

$$\frac{\langle a^3\rangle}{v_0} = \frac{45}{16}\,\frac{(\nu_0-\nu_*)\,(2-\nu_*)}{(1-\nu_*^2)[10\nu_0-\nu_*(1+3\nu_0)]}\,. \tag{7.8.44}$$

The latter can be considered as an equation with respect to ν_*.

The authors of [7.15] who obtained the result, pointed out that, with $\langle a^3\rangle/v_0 \geq 9/16$, the value of the shear modulus μ_* calculated according to (7.8.43, 44) becomes nonpositive and the formulae lose their physical sense. For comparison, note that (7.8.16, 17), to which the method of an effective field leads in the case of a Poisson set of cracks, are free of this defect: here the effective moduli are always positive. For the model with a restriction on cracks intersections [Eq. (7.8.20) for B_*], the effective moduli vanish for $q = \langle a^3\rangle/v_0$ much larger than 9/16 (for $q \approx 2$).

In the next section a comparison will be carried out of expressions for effective constants with the results of experiments and the exact solutions in the two-dimensional case; this will enable us to estimate the limits of applicability of both methods of self-consistency.

7.9 Two-Dimensional Systems of Rectilinear Cuts

In the two-dimensional case, a rectilinear cut is the analog of an elliptic crack. The formalism of the method of effective field is transfered without any important changes to the two-dimensional case. In this section, we shall present mostly the final results, outlining only the derivation. The planar problem is interesting due to the fact that there exist a number of exact solutions and of experimental observations, which enable one to find out the limits of applicability of different approximate approaches.

Let us start with an individual rectilinear cut. The δ-function, concentrated on L and defined by (6.1.2) will be denoted by $L(x)$

$$\delta(L) = L(x) . \tag{7.9.1}$$

When applying the stress $\bar{\sigma}$, one can interpret the cut at infinity as a line on which dislocation moments of some density $M(x^1)$ are induced. In the case of a constant external field, from the solution of the problem about an isolated rectilinear cut, it follows that

$$M(x^1) = P(n)\bar{\sigma}\sqrt{l^2 - (x^1)^2} , \tag{7.9.2}$$

where the x^1-axis of the cartesian coordinate system with the origin at the center of the cut is directed along L; l is the half length of the cut. The constant fourth order tensor P has the form

$$P_{\alpha\beta\lambda\mu} = -n_{(\alpha}T^{-1}_{\beta)(\lambda}n_{\mu)} , \tag{7.9.3}$$

$$\text{where } T^{\beta\lambda} = \frac{1}{4\pi}\int_{-\infty}^{\infty}\frac{J_1(|k_1|)}{|k_1|}dk_1\int_{-\infty}^{\infty}n_\alpha S_0^{\alpha\beta\lambda\mu}(k_1, k_2)n_\mu\, dk_2 . \tag{7.9.4}$$

Here $S_0(k)$ is the k-representation of the Green's function for stress in the two-dimensional problem, n is the normal to L, $J_1(k_1)$ is the Bessel function of the first order, the Greek subscripts in this section take values 1, 2. For an isotropic medium and for a plane state of stress

$$P_{\alpha\beta\lambda\mu} = -\frac{2}{\mu_0(1 + \nu_0)}n_{(\alpha}\delta_{\beta)(\lambda}n_{\mu)} . \tag{7.9.5}$$

Let us proceed to the case of a set L of rectilinear cuts on a plane and introduce the following notations. Let $L(x)$ be now the δ-function, concentrated on all cuts, and the function $L(x;\, x')$ be defined by the relation

$$L(x;\, x') = \begin{cases} L(x') = \sum_k L_k(x') & \text{with } x \bar{\in} L \\ \sum_{k \neq i} L_k(x') & \text{with } x \in L_i , \end{cases} \tag{7.9.6}$$

where $L_k(x)$ is the δ-function, concentrated on the k-th cut L_k.

In the case of a system of cracks in a plane, an analog of (7.3.17) for the effective field $\bar{\sigma}(x)$ has the form

$$\bar{\sigma}(x) = \sigma_0 + \int S_0(x - x')P(x')\bar{\sigma}(x')L(x;\, x')\, dx',\ x \in L , \tag{7.9.7}$$

where the continuous function $P(x)\bar{\sigma}(x) = M(x)$ has the form of (7.9.2) on each of the cuts L_k. The expressions for the tensors $\sigma(x)$ and $\varepsilon(x)$ are represented by

$$\sigma(x) = \sigma_0 + \int S_0(x - x')P(x')\bar{\sigma}(x')L(x')\,dx' , \tag{7.9.8}$$

$$\varepsilon(x) = \varepsilon_0 - \int K_0(x - x')C_0P(x')\bar{\sigma}(x')L(x')\,dx' . \tag{7.9.9}$$

From (7.9.7), using hypotheses H1 and H2 of the method of an effective field (Sect. 7.3), analogously to (7.5.17, 18), we have

$$\bar{\sigma}^1 = \Lambda\sigma_0 , \tag{7.9.10}$$

$$B_* = B_0 - \langle P(x)L(x)\rangle\Lambda , \tag{7.9.11}$$

where

$$\Lambda = \left[I - \int S_0(x - x')\langle P(x')L(x;\, x')|x\rangle\right]^{-1} . \tag{7.9.12}$$

The mean in (7.9.11) is represented by

$$\langle P(x)L(x)\rangle = \frac{1}{2}\left\langle \frac{\pi l^2}{\omega_0} P\right\rangle , \tag{7.9.13}$$

where ω_0 is an average area per crack, the tensor P has the form of (7.9.3) and depends on a stochastic orientation n of the crack. The mean in the integrand in (7.9.12)

$$\Psi(x - x') = \langle P(x')L(x;\, x')|x\rangle \tag{7.9.14}$$

possesses all the properties of the mean (7.8.5) of the three-dimensional case. The constant component of the function $\Psi(x)$ is equal to

$$\langle\Psi\rangle = \frac{\pi}{2\omega_0}\,\frac{\langle l^3P\rangle}{\langle l\rangle} \approx \frac{1}{2}\left\langle\frac{\pi\, l^2}{\omega_0}\right\rangle\langle P\rangle , \tag{7.9.15}$$

while the last equality occurs if the dimensions of all the cracks are approximately identical.

A regularization of the integral in (7.9.12) is defined by a relation analogous to (7.8.10). In the two-dimensional case, the tensor D becomes (Appendix A3)

$$D = \frac{1}{2\pi}\oint S_0(k)\,dL , \tag{7.9.16}$$

where the integration is carried out over the unit circle in the two-dimensional k-space. For an isotropic medium, the tensor D takes the form

$$D = -\frac{\mu_0(1+\nu_0)}{4}(E^2 + 2E^1). \tag{7.9.17}$$

Here E^1 and E^2 are the two-dimensional analogs of the basis tensors, defined by (A1).

Let us proceed to consider specific random fields of cracks on the plane.

1) *A Poisson field of rectilinear cuts* is defined analogously to a Poisson field of elliptic cracks in the three-dimensional case (Sect. 7.8), and here

$$\langle \Psi(x) \rangle = \langle \Psi \rangle = \text{const.} \tag{7.9.18}$$

Then the integral in (7.9.12) vanishes, and (7.9.10, 11) become

$$\bar{\sigma}^1 = \sigma_0 , \tag{7.9.19}$$

$$B_* = B_0 - \frac{1}{2}\left\langle \frac{\pi l^2}{\omega_0} P \right\rangle. \tag{7.9.20}$$

In the case of an isotropic medium

$$B_{*\alpha\beta\lambda\mu} = B_{0\alpha\beta\lambda\mu} + \frac{1}{\mu_0(1+\nu_0)}\left\langle \frac{\pi l^2}{\omega_0} n_{(\alpha}\delta_{\beta)(\lambda}n_{\mu)} \right\rangle. \tag{7.9.21}$$

If the distribution of the cracks in orientations is homogeneous and the distribution of dimensions does not depend on orientation, then

$$B_* = B_0 + \frac{q}{2\mu_0(1+\nu_0)} E^1 , \tag{7.9.22}$$

where q is a dimensionless parameter of the type of a concentration

$$q = \frac{\langle \pi l^2 \rangle}{\omega_0} . \tag{7.9.23}$$

This implies

$$\frac{E_*}{E_0} = \frac{\nu_*}{\nu_0} = \frac{1}{1+q} , \tag{7.9.24}$$

where E_0 and E_* are the Young moduli of the real and effective media, respectively.

2) *A model with a restriction on the intersections of cracks* (Sect. 7.8). In this case, the mean of $\Psi(x)$ defined by (7.9.14) is a continuous function, which is equal to zero at $x = 0$. If $\Psi(x)$ is also spherically symmetric, thus

$$\int S_0(x - x')\Psi(x - x')dx' = -D\langle \Psi \rangle , \tag{7.9.25}$$

where D has the form of (7.9.16), and $\langle \Psi \rangle$ has the form of (7.9.15). In this case, the tensors $\bar{\sigma}^1$ and B_* are represented by

$$\bar{\sigma}^1 = (I - D\langle \Psi \rangle)^{-1} \sigma_0 , \tag{7.9.26}$$

$$B_* = B_0 - \langle \Psi \rangle (I - D\langle \Psi \rangle)^{-1} . \tag{7.9.27}$$

Let all the cracks be circular, their distribution in orientations be homogeneous, and the matrix be isotropic. The preceeding relations imply the equalities

$$\bar{\sigma}^1 = \left(1 - \frac{1}{2} q\right)^{-1} \left(1 - \frac{1}{4} q\right)^{-1} \left[E^1 + \frac{1}{8} q (E^2 - E^1)\right] \sigma_0 , \tag{7.9.28}$$

$$B_* = B_0 + \frac{1}{2\mu_0(1 + \nu_0)} \cdot \frac{q}{\left(1 - \frac{1}{2} q\right)\left(1 - \frac{1}{4} q\right)} \left[E^1 + \frac{1}{8} q(E^2 - E^1)\right]. \tag{7.9.29}$$

Hence, for the relative elastic modulus E_*/E_0 and the Poisson coefficient ν_*/ν_0, we have

$$\frac{E_*}{E_0} = \left[1 + \frac{q\left(1 - \frac{3}{8} q\right)}{\left(1 - \frac{1}{2} q\right)\left(1 - \frac{1}{4} q\right)}\right]^{-1} \tag{7.9.30}$$

$$\frac{\nu_*}{\nu_0} = \frac{E_*}{E_0}\left[1 - \frac{q^2}{8\nu_0\left(1 - \frac{1}{2} q\right)\left(1 - \frac{1}{4} q\right)}\right], \tag{7.9.31}$$

where q has the form of (7.9.20).

In Fig. 7.7, the dependences (7.9.24) (curve 1), (7.9.30) (curve 2) and (7.9.31)

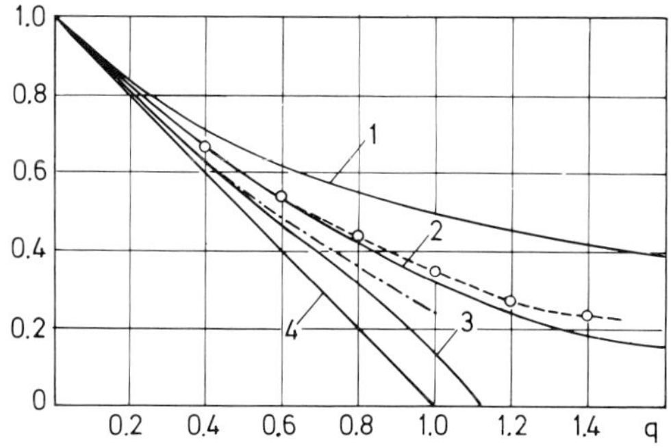

Fig. 7.7. Comparison of experimental and calculated elastic moduli

(curve 3) are compared with experimental data [7.16]. The experiments were carried out using thin rubber sheets, containing random fields of rectilinear cuts ($\nu_0 = 0.5$). Results of the experiments are exhibited in Fig. 7.7, by the dot line with small circles (E_*/E_0) and by the dot-and-dash line (ν_*/ν_0).

The statistical analysis shows that, to the field of cracks investigated in [7.16], there corresponds the model with the restriction on cracks intersection. The experimental data are described by (7.9.30, 31), obtained for this model, in the best way. Therefore one can state that the approximation of an effective field enables one to take into account correctly interaction between cracks in the situation under consideration.

Let us present here expressions for E_* and ν_*, which are given by the method of an effective medium in the case of an isotropic tensor of effective elastic compliance. The analog of (7.8.41) in the two-dimensional case has the form

$$B_* = B_0 + \frac{q}{E_*} E^1 . \tag{7.9.32}$$

Thus, we have

$$\frac{E_*}{E_0} = \frac{\nu_*}{\nu_0} = 1 - q . \tag{7.9.33}$$

To this dependence, there corresponds the straight line 4 in Fig. 7.7.

3) *Regular lattices of cracks in the plane.* In the case of cracks with identical orientations, the expression for $\Psi(x)$ becomes

$$\Psi(x) = \frac{1}{2l} \sum_m{}' J(x - m) L'_m(x) , \tag{7.9.34}$$

where m is a vector of the lattice formed by the centers of cracks, the primed summation sign means the exclusion of the term with $m = 0$, l is half the length of each crack, $L'_m(x)$ is the δ-function concentrated on the segment of the length $4l$ with its center in the m-th knot and parallel to all the cracks. The function $J(x)$ has the form

$$J(x) = \frac{2\omega_0}{\pi} \begin{cases} \left(1 - \frac{|x|}{l}\right)\sqrt{\frac{|x|}{l}\left(2 - \frac{|x|}{l}\right)} + \sin^{-1}\left(1 - \frac{|x|}{l}\right) + \frac{\pi}{2} & \text{if } |x| \le 2l \\ 0 \quad \text{if} \quad |x| > 2l \end{cases} \tag{7.9.35}$$

The action of the generalized function $S_0(x)$ on $\Psi(x)$ in the form of (7.9.34) is represented as in (7.8.27),

$$\int S_0(x - x')\Psi(x - x')dx' = Q\langle\Psi\rangle , \tag{7.9.36}$$

where the tensor Q is represented in the form of a converging series

$$Q = -D + \sum_m{}' \oint S_0(x)\left[\frac{1}{2l} J(x-m) L'_m(x) - 1\right] dx. \qquad (7.9.37)$$

Here D has the form of (7.9.16) and the mean value of $\Psi(x)$ is

$$\langle \Psi \rangle = \frac{\pi l^2}{2\omega_0} P(n). \qquad (7.9.38)$$

Expressions for the tensors $\bar{\sigma}^1$ and B_* have in this case the form which is analogous to (7.8.29, 30).

Let us consider a lattice of cracks of one orientation, which has two orthogonal symmetry axes. The expression for the vector $\bar{\sigma}^1 n$ (which, here, determines the state of all cracks) has the form

$$\bar{\sigma}^{1\lambda\mu} n_\mu = \frac{1}{(1-\beta)}\left(\delta^\lambda_\mu + \frac{\alpha}{1-\alpha-\beta} n^\lambda n_\mu\right) \sigma_0^{\mu\lambda} n_\nu. \qquad (7.9.39)$$

The expression for the tensor of effective elastic compliance B_* in terms of the basis tensors E^i defined in Appendix A1, is

$$B_* = B_0 + \frac{q}{\mu_0(1+\nu_0)} \cdot \frac{1}{(1-\beta)}\left[E^5(n) + \frac{\alpha}{1-\alpha-\beta} E^6(n)\right]. \qquad (7.9.40)$$

The dimensionless coefficients α and β have the form of converging double series, and their explicit expressions were presented in [7.17].

The stress intensity factors K_{I} and K_{II} on cracks (the first is under tension along the direction of the normal, the second is under pure shear) are determined by

$$K_{\mathrm{I}} = \sqrt{2\pi l}\, \sigma_0 k_{\mathrm{I}}, \quad K_{\mathrm{II}} = \sqrt{2\pi l}\, \tau_0 k_{\mathrm{II}}, \qquad (7.9.41)$$

where σ_0 and τ_0 are the values of external tensile stress and shear stress, respectively,

$$k_{\mathrm{I}} = \frac{1}{1-\alpha-\beta}, \quad k_{\mathrm{II}} = \frac{1}{1-\alpha}. \qquad (7.9.42)$$

Let us consider a few particular cases.

a) *Regular triangular lattice of cracks.* Fig. 7.8, illustrates E_*/E_0 (curve 1) and ν_*/ν_0 (curve 2) as functions of the parameter $\lambda = 2l/\rho$ of the lattice, where ρ is the distance between the centers of cracks, these functions being obtained by the method of an effective field. The curves for the intensity factors k_{I} and k_{II} are shown in Fig. 7.9. The continuous curves in this figure correspond to the exact solutions obtained in [7.18].

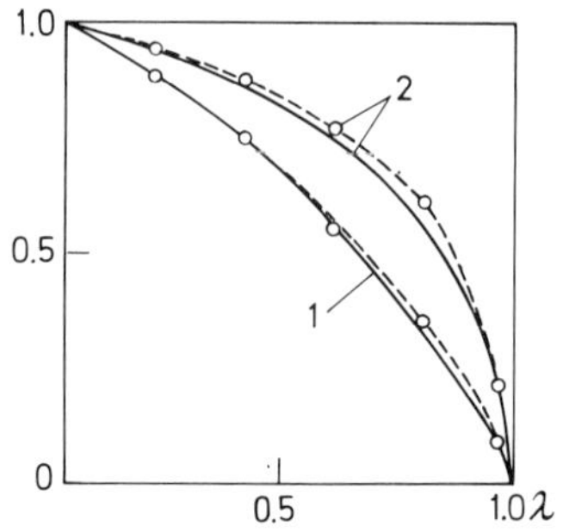

Fig. 7.8. Elastic moduli for triangle lattice of cracks

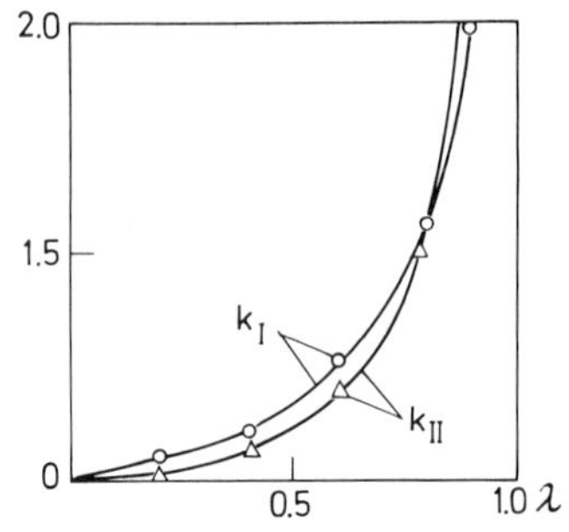

Fig. 7.9. Stress intensity factors for triangle lattice of cracks

b) *A system of collinear cracks on one straight line.* The values of the coefficients α and β in (7.9.42), which correspond to this case, can be obtained from the solution for a rectangular lattice when one of the characteristic dimensions of the latter tends to infinity. Then α and β are expressed as simple series [7.17]. The value of the coefficient k_I is in agreement with its exact value, which has the form [7.19]

$$k_I^0 = \frac{2 \sin \frac{\pi}{2} \lambda}{\sqrt{\lambda\pi \sin \pi\lambda}} . \tag{7.9.43}$$

(c) *A row of parallel cracks.* In Fig. 7.10, the values of the intensity factors k_{I} and k_{II}, obtained by the method of an effective field (dotted line with small circles), are compared with results of numerical solution of the corresponding integral equations [7.20] (solid curves).

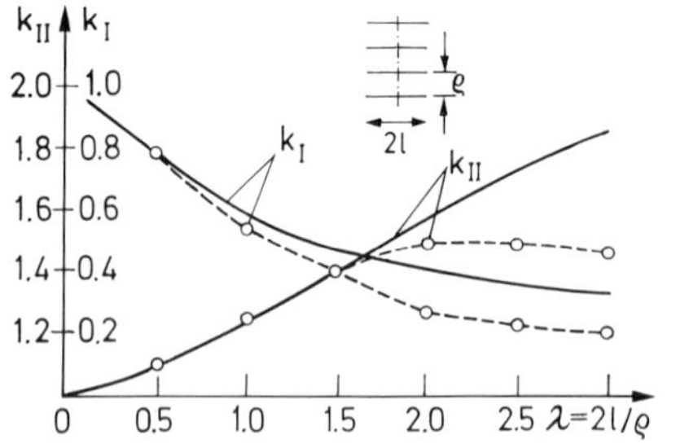

Fig. 7.10. Stress intensity factors for a row of parallel cracks

As it can be seen from the above examples, in the two-dimensional case, the first approximation of the method of an effective field gives good agreement with the results of more precise calculations of elastic constants and of intensity factors, when the lengths of the cracks are smaller than the distances between their centers. This statement appears to remain valid for other types of regular systems of cracks as well.

When the centers of cracks in a lattice become closer (see case c), the external field $\bar{\sigma}(x)$, in which each crack is contained, begins to differ more and more from a constant one. Therefore, the error which is given here by the method of the effective field, is connected with violating the hypothesis H1 (Sect. 7.3). However, a way of making the computational part of the method more precise can be suggested.

Let the effective field $\bar{\sigma}(x)$ be not constant and have the form of a polynomial. The solution for an isolated inhomogeneity can be found, using results of Sect. 4.9, 12. This will enable one to express the corresponding induced density $m(x)$ of dislocation moments on defects in terms of the coefficients of the polynomial $\bar{\sigma}(x)$.

Equations for these coefficients can be obtained, substituting $\bar{\sigma}(x)$ and explicit expressions for $m(x)$ into (7.3.3) for the effective field and averaging a necessary number of times the expression obtained, over the region occupied by defects. As weights for the averaging, a system of linearly independent functions (polynomials, for example) is to be chosen. An analogous modification of the method can be used also in the case of stochastic fields of ellipsoidal inclusions and cracks.

7.10 Random Field of Point Defects

In the previous sections the effective field method was applied only to the determination of the first moments of stress-strain fields in composite materials. The calculation of the second moments is technically a more complicated problem. Finding the averages in (7.5.9, 13), determined by the second moments of $\bar{\sigma}(x)$, and the solution of the system of these equations is only possible by using numerical methods, even for simple cases. However, many technical difficulties disappear if we pass to the model of point inhomogeneities, where inclusions of finite sizes are replaced by isolated point defects.

In this and the following section the general scheme of the effective field method is applied to the description of the elastic medium containing a homogeneous random field of point defects. Note that in this case the application of the method has a better foundation than for inhomogeneities of finite sizes. As a matter of fact the main hypothesis H1 (Sect. 7.3) about the constancy of the effective field $\bar{\sigma}(x)$ in the region occupied by an arbitrary inclusion, turns into a trivial fact in case of point defects.

Let us begin with the problem of replacing an isolated finite inclusion by a point defect. The basis for such a substitution is the following convention.

The perturbed fields of the isolated inclusion and of the corresponding point defect should have the same asymptotics at infinity. Thus in order to pass to the point-defects model from the model of the composite with inclusions of finite sizes, we should substitute the functions $P_k V_k(x)$ and $P_k(x)\Omega_k(x)$ in the previous relations by their first terms in the multipole series expansions

$$P_k V_k(x) = P_k^0 \delta(x - \xi_k) + \cdots , \tag{7.10.1}$$

where ξ_k is the center of the region V_k. Note that $P_k V_k(x)$ characterizes the density of dislocation moments induced in the region V_k by applying an external field. The tensor P_k^0 in (7.10.1) has the form

$$P_k^0 = \int P_k V_k(x) dx = v_k B_0 C_{1k}(I + A_k C_{1k})^{-1} B_0 . \tag{7.10.2}$$

where v_k is the volume of the region V_k; (7.3.6) being used here.

In case of cracks we deal with the function $P_k(x)\Omega_k(x)$ which is also represented in the form of (7.10.1), where

$$P_k^0 = \int P_k(x)\Omega_k(x) dx = \frac{2\pi}{3} a_k^3 P_k . \tag{7.10.3}$$

Here $P_k(x)$ and P_k are defined by (7.3.9, 11), a_k is the larger half-axis of the ellips Ω_k.

Let X be a set of points ξ_k with the point defects specified by the set X_{x0} and defined by

$$X_{x_0} = \begin{cases} X = \bigcup_i \xi_i & \text{by } x_0 \,\bar{\in}\, X \\ \bigcup_{i \neq k} \xi_i & \text{by } x_0 = \xi_k \in X , \end{cases} \tag{7.10.4}$$

by the analogy with V_{x_0}, see (7.3.14). We consider the generalized functions

$$X(x) = \sum_{\xi_i \in X} \delta(x - \xi_i) , \tag{7.10.5}$$

$$X(x; x') = \sum_{\xi_i \in X_x} \delta(x' - \xi_i) , \tag{7.10.6}$$

concentrated on the sets X and X_x, respectively.

By the transition to the point defects the governing equation (7.3.17 or 19) for the effective field $\bar{\sigma}(x)$ is reduced to

$$\bar{\sigma}(x) = \sigma_0 + \int S_0(x - x')\, P^0(x')\bar{\sigma}(x')\, X(x; x')\, dx', \; x \in X , \tag{7.10.7}$$

where $P_0(x)$ is a continuous function coinciding with P_k^0, see (7.10.2, 3), at the point $x = \xi_k$.

For the determination of the effective field's first moment $\bar{\sigma}^1$ both parts of (7.10.7) should be averaged with the convention $x \in X$. Then the hypothesis H2 (Sect. 7.3) about the statistical independence of $\bar{\sigma}(x)$ from $P^0(x')X(x; x')$ at the point x together with an assumption similar to (7.5.5), should be used. As a result the equation for $\bar{\sigma}^1$ is presented by

$$\bar{\sigma}^1 = \sigma_0 + \int S_0(x - x') \langle P^0(x')X(x; x')|x\rangle dx' \bar{\sigma}^1 , \tag{7.10.8}$$

where the operation of the average with the conventions $x \in X$ and $x_1, \ldots, x_n \in X_x$, is denoted by $\langle 1x; x_1, \ldots, x_n\rangle$, see also (Sect. 7.4).

In case of the point-defects model, (7.5.17) for the effective compliance tensor B_* is reduced to

$$B_* = B_0 - \langle P^0(x)X(x)\rangle \Lambda , \tag{7.10.9}$$

where the tensor Λ, defined by (7.5.19) in case of inclusions of finite size, is represented in the form

$$\Lambda = \left[I - \int S_0(x - x') \langle P^0(x')X(x; x')|x\rangle dx' \right]^{-1} , \tag{7.10.10}$$

which is a consequence of (7.10.8).

As above, we assume that random sizes and elastic moduli of inclusions do not depend on their position in space. Thus the functions $P^0(x)$ and $X(x)$ are statistically independent and consequently

$$\langle P^0(x)X(x)\rangle = P\langle X(x)\rangle , \tag{7.10.11}$$

$$\langle P^0(x')X(x; x')|x\rangle = P\langle X(x; x')|x\rangle , \tag{7.10.12}$$

where $P = \langle P^0\rangle$ is the average over the ensemble of random variables (7.10.2 or 3).

Let us consider two stochastic point-detects models in space (a Poisson field and a random point lattice) and determine the tensors $\bar{\sigma}^1$ and B_* in these cases.

1) *The Poisson point field.* An example is the Poisson field of the set of the centers of the cracks described in Sect. 7.8.

In case of the Poisson field the averages on the right-hand sides of (7.10.11, 12) have the following forms, see (A4.6, 17),

$$\langle X(x)\rangle = \langle X(x; x')|x\rangle = \frac{1}{v_0} = \text{const} . \tag{7.10.13}$$

As a consequence of this relation the integral in (7.10.10) vanishes and when using (7.10.9) we obtain

$$B_* = B_0 - \frac{1}{v_0} P . \tag{7.10.14}$$

If the point-defect model describes a random field of cracks and therefore P has the form of (7.10.3), the last relation coincides with (7.8.19) which was presented in the case of the finite crack field in space.

As it was emphasized in Sect. 4.8 and 4.13, caution is necessary in employing the point-defect model within the framework of the local theory of elasticity. Strictly speaking, if inclusions are replaced by the point defects, the characteristic length l is introduced in the continuous medium. The length l is approximately equal to the mean size of the defects and if the distances between the neighboring point defects become smaller than l, phenomena show up, which have no analogy in the case of finite inhomogeneities (Sect. 7.11). It turns out that the point-detects model cannot describe the interaction between finite inclusions for such distances even qualitatively.

On the basis of such considerations let us complete the model of the Poisson point field. It will be assumed hereafter that there is a neighborhood (spherical, for example) with the diameter $\sim l$ surrounding every point, such that the other points of the field cannot enter it.

For this new model the properties of the average $\Psi(x - x') = \langle X(x; x')|x\rangle$ do not differ from the properties of $\langle V(x; x')|x\rangle$ where $V(x)$ is the characteristic function of the random field of finite regions in space. In this case $\Psi(x)$ is a continuous function which is equal to zero at $x = 0$. The asymptotic value of this function at infinity is v^{-1}.

Because this model is isotropic, $\Psi(x)$ has a spherical symmetry if the density of defects is not too large. As a consequence of these properties of $\Psi(x)$, the integral in (7.10.10) is represented in the form (Sect. 7.6)

$$\int S_0(x - x')P\langle X(x; x')|x\rangle \, dx' = -\frac{1}{v_0} DP , \tag{7.10.15}$$

where the tensor D is defined by (7.6.5) and $\alpha^{\alpha\beta} = \delta^{\alpha\beta}$. This relation together with (7.10.9) yields

$$B_* = B_0 - \frac{1}{v_0} P\left(I + \frac{1}{v_0} DP\right)^{-1} . \tag{7.10.16}$$

It coincides with (7.6.11) in the case of inclusions and with (7.8.19) in the case of cracks.

Thus, the point-defects model with the necessary restrictions yields the same expressions for the effective constants tensors as the model for finite sizes of inhomogeneities.

2) *Random lattice of point defects.* Let point defects constitute a random lattice in space. The set of the lattice nodes are described by the random vector ξ_m

$$\xi_m = m + \rho_m + r, \tag{7.10.17}$$

where m is the vector of the fixed deterministic lattice, ρ_m are independent random vectors with the same density functions $f(x)$ and zero expectations, r is a random vector the same for all m and uniformly distributed in the whole space. Obviously, if $f(x) = \delta(x)$, then (7.10.17) defines a deterministic lattice to within random translations in space. If all ρ_m are distributed uniformly in space we get a Poisson point field.

The averages (7.10.11, 12) for random point fields are considered in Appendix A4. The relations (7.7.8, 29) yield

$$\langle X(x)\rangle = \frac{1}{v_0}, \tag{7.10.18}$$

$$\Psi(x - x') = \langle X(x; x')|x\rangle - \sum_m{}' g(x - x' - m), \tag{7.10.19}$$

where the primed summation sign means the omission of the term $m = 0$. The functions $g(x)$ and $f(x)$ are connected with each other by

$$g(x) = \int f(x - x')f(-x')\,dx'. \tag{7.10.20}$$

The integral entering (7.10.10) for B_* has the form

$$\begin{aligned} &\int S_0(x - x')\,\langle P^0(x')\,X(x; x')|x\rangle\,dx' \\ &= \frac{1}{v_0}\int S_0(x - x')\,[v_0\Psi(x) - 1]\,dx\,P. \end{aligned} \tag{7.10.21}$$

Note that $\langle \Psi\rangle = 1/v$ and this equality may be proved as (7.6.5). Thus (7.10.21) is a consequence of (7.10.19) and (7.2.6).

In order to calculate the integral (7.10.21) let us pass to the k-representation of the functions in the integrand. Using Parseval's formula we obtain

$$Q = \int S_0(x)[v_0\Psi(x) - 1]\,dx = \frac{1}{(2\pi)^3}\int S_0(k)[v_0\Psi(k) - (2\pi)^3\delta(k)]\,dk. \tag{7.10.22}$$

Let $f(x)$ be the Gaussian density with the dispersion σ

$$f(x) = \frac{1}{\sqrt{2\pi}\,\sigma}\exp(-x^2/2\sigma^2). \tag{7.10.23}$$

It may be shown for this case that

$$v_0\Psi(k) - (2\pi)^3\delta(k) = \exp(-\frac{1}{2}\sigma^2k^2)[(2\pi)^3 \sum_{m'} \delta(k - 2\pi m') - 1], \quad (7.10.24)$$

where m' is the vector of the reciprocal lattice.

If the basic medium is isotropic the tensor $S_0(k)$ is defined by (4.6.28, 16). After substituting the k-representation of $S_0(k)$ the integral (7.10.22) is expressed in terms of the following three integrals:

$$I_0 = (2\pi)^{-3} \int [v_0\Psi(k) - (2\pi)^3 \delta(k) \, dk ,$$

$$I_2^{\alpha\beta} = (2\pi)^{-3} \int \frac{k^\alpha k^\beta}{k^2} [v_0\Psi(k) - (2\pi)^3\delta(k)] \, dk ,$$

$$I_4^{\alpha\beta\lambda\mu} = (2\pi)^{-3} \int \frac{k^\alpha k^\beta k^\lambda k^\mu}{k^4} [v_0\Psi(k) - (2\pi)^3\delta(k)] \, dk .$$

If the expectation of the random point field is the cubical lattice these integrals become

$$I_0 = \alpha_0, \; I_2^{\alpha\beta} = \frac{1}{3}\alpha_0\delta^{\alpha\beta}, \; I_4 = \alpha_1(E^2 + 2E^1) + \alpha_2\Gamma , \quad (7.10.25)$$

where the components of the tensor Γ are defined in the system of the basic vectors of the cubical lattice by

$$\Gamma^{\alpha\beta\lambda\mu} = \sum_{i=1}^{3} \delta_i^\alpha\delta_i^\beta\delta_i^\lambda\delta_i^\mu . \quad (7.10.26)$$

The coefficients $\alpha_0, \alpha_1, \alpha_2$ have the form of the convergent series (1 is the lattice parameter)

$$\alpha_1 = \sum_{i,j,k=-\infty}^{\infty} \frac{(i^2 + j^2)(j^2 + k^2)(k^2 + i^2)}{(i^2 + j^2 + k^2)^2} \exp\left[-2\left(\frac{\pi\sigma}{l}\right)(i^2 + j^2 + k^2)\right]$$
$$- 1 - \frac{l^3}{10\sqrt{2\pi}\,\sigma^3}, \quad \alpha_2 = 3\alpha_0 - 5\alpha_1 . \quad (7.10.27)$$

Taking into consideration the previous relations, (7.10.22) for Q may be represented by

$$\frac{1}{2\mu_0} Q = -\left(\frac{v_0\alpha_0}{3(1 - 2\nu_0)} + \frac{\alpha_1}{1 - v_0}\right) E^2 - \left(\alpha_0 + \frac{2\alpha_1}{1 - v_0}\right)E^1 - \frac{\alpha_2}{1 - v_0}\Gamma . \quad (7.10.28)$$

It should be noted that in the case of regular lattices of point defects ($\bar{\sigma} \to 0$), it is more convenient to calculate the integral (7.10.22) using the x-representations of the integrand functions.

The effective moduli tensor for the random cubical lattice of defects in the isotropic medium is given by

$$B_* = B_0 - \frac{1}{v_0} P\left(I - \frac{1}{v_0} QP\right)^{-1} \tag{7.10.29}$$

where Q is defined by (10.28).

Note that in the case of regular lattices of identical point defects, the expression for B_* obtained by the effective field method is the exact solution of the problem. In fact, in the case of simple lattices all hypotheses of the method become exact statements. If we consider a complex lattice consisting of several simple lattices, the exact solutions may be found in much the same way as in the case of regular lattices of finite inclusions (see the end of Sect. 7.8).

It is interesting to estimate the error due to the substitution of finite inclusions for the point defects by the calculation of effective constants. Let us consider a cubical lattice of isotropic spheres embedded in an isotropic matrix. It may be shown that the point-defects model corresponding to this case gives the expression for the effective moduli tensor which is similar to (7.7.11) but the coefficient $\alpha(p)$ should be replaced by its value at zero, i.e. $\alpha(0)$. The analogous fact is encountered in the plane case for the square lattice of circular inclusions.

In Fig. 7.11 the curves for the relative shear modulus μ_*/μ_0 in the case of square lattices of circular holes (curve 1) and circular rigid inclusions (curve 2) are represented. The continuous curves correspond to the exact solution of the problem [7.13], the broken curves are obtained from the corresponding point-defects model and the broken curves with ringlets represent the effective-field approximation in the case of finite inclusions (Sect. 7.7).

Analyzing the curves in the Fig. 7.11 we arrive at the following conclusion. The substitution of inclusions for the point defects gives an error less than 10% if the distances between the centers of inclusions are larger by $1.5 \div 2$ times than the diameters of the inclusions.

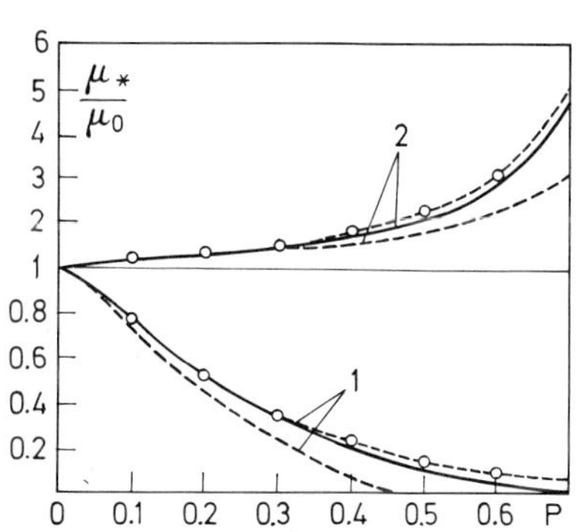

Fig. 7.11. Shear modulus for square lattices of holes (1) and rigid inclusions (2)

In the case of a regular triangular lattice the transition to the point-defects model gives the same values for effective constants as by an application of the effective field method to the lattice of finite inclusions.

Thus the point-defects model yields good results by the calculation of the effective constants of composites and simplifies simultaneously the process of constructing the solution.

7.11 Correlation Functions in the Approximation by Point Defects

Let us proceed to constructing correlation functions of fields of stress and strain in a medium with point inhomogeneities. Since these correlation functions are expressed in terms of the second moment $\bar{\sigma}^2$ of the effective field, see (7.5.20), the problem is reduced to constructing this very quantity.

When approximating inclusions by point inhomogeneities (7.5.10) for $\bar{\sigma}^2(x)$ becomes

$$\bar{\sigma}^2(x_1 - x_2) = \bar{\sigma}_0 \varphi(x_2 - x_1) + \int S_0(x_1 - x') \langle P^0(x')X(x_1; x')\bar{\sigma}(x')\bar{\sigma}(x_2)|x_1, x_2\rangle \, dx' \, , \tag{7.11.1}$$

where the mean

$$\varphi(x_2 - x_1) = \langle \sigma(x_2)|x_1, x_2\rangle \tag{7.11.2}$$

by virtue of (7.10.7) is represented in the form

$$\varphi(x_2 - x_1) = \sigma_0 + \int S_0(x_2 - x')\langle P^0(x')X(x_2; x')\bar{\sigma}(x')|x_1, x_2\rangle \, dx' \, . \tag{7.11.3}$$

The function

$$\bar{\sigma}^2(x_1 - x_2) = \langle \bar{\sigma}(x_1)\bar{\sigma}(x_2)|x_1, x_2\rangle \tag{7.11.4}$$

is the mean under the condition that $x_1, x_2 \in X$, where X is the set of points, at which isolated defects are located. Obviously, if $x_1 \neq x_2$, then in x_1 and x_2 only two different defects can occur. Therefore, $\bar{\sigma}^2(x_1 - x_2)$ pertains to the pair interaction in a random field of point defects.

Along with the two-point moment of the effective field (7.11.4), let us introduce the mean of $\bar{\sigma}^2$ at a point x_1 under the condition that there is a defect at x_2

$$d(x_1 - x_2) = \langle \bar{\sigma}(x_1)\bar{\sigma}(x_1)|x_1; x_2\rangle \, . \tag{7.11.5}$$

The limit of $d(x)$, as $|x| \to \infty$, will be denoted by d_∞. Since at large distances the dependence on x_2 in (7.11.5) disappears, we have

$$d_\infty = \langle \bar{\sigma}(x)\bar{\sigma}(x)|x\rangle . \tag{7.11.6}$$

The expression for $d(x)$ can be obtained analogously to (7.11.1) and has the form

$$d(x_1 - x_2) = \sigma_0\varphi(x_1 - x_2) + \int S_0(x_1 - x')\langle P^0(x')X(x_1; x')\bar{\sigma}(x')\bar{\sigma}(x_1)|x_1; x_2\rangle\, dx' . \tag{7.11.7}$$

Our next aim is to construct closed equations for the functions $\bar{\sigma}^2(x)$, $d(x)$ and $\varphi(x)$. Here we shall make use of (7.11.1, 2 and 7), uncoupling the means in the integrands with the help of statistical hypotheses of the type of H2 (Sect. 7.3) and some additional assumptions concerning the structure of conditional means.

Let $X_{x_1x_2}$ be X with points ξ_i, which appeared at x_1 or x_2. If

$$X(x_1, x_2; x') = \sum_{\xi_i \in X_{x_1x_2}} \delta(x' - \xi_i) , \tag{7.11.8}$$

then, by the virtue of the definition (7.10.6) of the function $X(x_1, x')$, we have the equality

$$X(x_1; x') = X(x_1, x_2; x') + \delta(x' - x_2) \quad \text{with } x_2 \in X. \tag{7.11.9}$$

We consider the conditional mean in the integrand in (7.11.1). Taking into account (7.11.9), we have

$$\langle P^0(x')X(x_1; x')\bar{\sigma}(x')\bar{\sigma}(x_2)|x_1; x_2\rangle = \langle P^0(x')X(x_1, x_2; x')\bar{\sigma}(x')\bar{\sigma}(x_2)|x_1; x_2\rangle + \delta(x' - x_2)\langle P^0(x_2)\bar{\sigma}(x_2)\bar{\sigma}(x_2)|x_1; x_2\rangle . \tag{7.11.10}$$

Using the hypothesis H2 (Sect. 7.3) and the assumption (7.5.12)

$$\langle \bar{\sigma}(x')\bar{\sigma}(x_2)|x', x_1, x_2\rangle = \langle \bar{\sigma}(x')\bar{\sigma}(x_2)|x', x_2\rangle , \tag{7.11.11}$$

the expression for each term on the right-hand side of (7.11.10) can be represented in the form

$$\langle P^0(x')X(x_1, x_2; x')\bar{\sigma}(x')\bar{\sigma}(x_2)|x_1; x_2\rangle = P\langle X(x_1, x_2; x')|x_1; x_2\rangle\, \bar{\sigma}^2(x' - x_2) , \tag{7.11.12}$$

$$\langle P^0(x_2)\bar{\sigma}(x_2)\bar{\sigma}(x_2)|x_1; x_2\rangle = Pd(x_2 - x_1) , \tag{7.11.13}$$

where P is the mean with respect to the ensemble of the stochastic variables (7.10.2, 3).

Substituting these expressions into (7.11.10) and then the result into (7.11.1), we find

$$\bar{\sigma}^2(x_1 - x_2) = \sigma_0\varphi(x_2 - x_1) + S_0(x_1 - x_2)Pd(x_2 - x_1)$$
$$+ \int S_0(x_1 - x')\, P\bar{\sigma}^2(x' - x_2)\, F(x', x_1, x_2)\, dx' , \tag{7.11.14}$$

where

$$F(x', x_1, x_2) = \langle X(x_1, x_2; x')|x_1; x_2\rangle . \tag{7.11.15}$$

The equation for the function $\varphi(x)$ follows from (7.11.3), if one represents the mean in the integrand in the form

$$\langle P^0(x')\, X(x_1; x')\bar{\sigma}(x')|x_1; x_2\rangle = P\langle X(x_1, x_2; x')|x_1; x_2\rangle\langle\bar{\sigma}(x')|x', x_1, x_2\rangle$$
$$+ \delta(x' - x_2)\, P\langle\bar{\sigma}(x_2)|x_1; x_2\rangle . \tag{7.11.16}$$

Here the hypothesis H2 was used again.

Substituting this result into (7.11.2) with the assumption (7.5.2) taken into account, we obtain

$$\varphi(x_1 - x_2) = \sigma_0 + S_0(x_1 - x_2)\, P\varphi(x_2 - x_1)$$
$$+ \int S_0(x_1 - x')\, P\varphi(x' - x_2)\, F(x', x_1, x_2)\, dx' . \tag{7.11.17}$$

Analogously, transforming (7.11.7) for $d(x_1 - x_2)$, we have

$$d(x_1 - x_2) = \sigma_0\varphi(x_1 - x_2) + S_0(x_1 - x_2)\, P\bar{\sigma}^2(x_1 - x_2)$$
$$+ \int S_0(x_1 - x')\, P\bar{\sigma}^2(x' - x_1)F(x', x_1, x_2)\, dx' . \tag{7.11.18}$$

The equations (7.11.14, 17 and 18) form a closed system with respect to the required three functions. A specific structure of the random field of defects is contained in these equations through the function $F(x', x_1, x_2)$, which is defied by (7.11.15). Let us consider this function in more detail.

From (7.11.15) and the definition of the conditional mean (Sect. 7.4), we obtain

$$F(x', x_1, x_2) = \frac{\langle X(x_1, x_2; x')\, X(x_2; x')\, X(x')\rangle}{\langle X(x_2; x_1)\, X(x_1)\rangle} . \tag{7.11.19}$$

This expression for the function F is the point of departure for the construction of the explicit expressions for random point fields of different types. Specific examples of such constructions are carried out in Appendix A4.

In the case of a Poisson point field, the function F turns out to be equal to the constant

$$F(x', x_1, x_2) = \frac{1}{v_0}\,. \qquad (7.11.20)$$

If the set X is a random spatial lattice (Sect. 7.10), then F is determined from (A4.30) and has the form

$$F(x', x_1, x_2) = \frac{\sum_m' g(x' - x_1 - m) \sum_{n \neq m}' g(x_2 - x_1 - n)}{\sum_m' g(x_1 - x_2 - m)}\,. \qquad (7.11.21)$$

Here m, n are vectors of a fixed regular lattice (the expectation of the random one under consideration), the primed summation signs mean exclusion of the terms with $m = 0$ and $n = 0$, respectively, the function $g(x)$ is defined by (7.10.20).

The system of equations (7.11.14, 17 and 18) will be solved for the example of the two-dimensional linear chain of defects. Let a system of parallel rectilinear cuts of a stochastic length $2l$ located on one straight line be modeled by point defects. The coordinates of the centers of cuts form a onedimensional homogeneous random point field. Let us assume that the external stress σ_0 is a tension along the normal to the line of cuts, i.e.

$$\sigma_0^{\alpha\beta} = \sigma_0 n^\alpha n^\beta\,, \qquad (7.11.22)$$

where σ_0 is a scalar.

Note that in the given problem, it is not the field $\sigma(x)$ itself, that is of interest, but $n\bar{\sigma}(x)$, the latter determining completely the state of each crack. From the symmetry properties it follows that

$$\bar{\sigma}^{\alpha\beta}(x) n_\beta = \bar{\sigma}(x) n^\alpha\,, \qquad (7.11.23)$$

where $\bar{\sigma}(x)$ is a scalar.

In the case of an isotropic medium, the tensors P and $nS_0(x)n$ take the forms (x is the coordinate along the straight line of the defects)

$$P^{\alpha\beta}_{\cdot\cdot\lambda\mu} = \frac{1 - \nu_0}{\mu_0} \langle \pi l^2 \rangle\, n^\alpha \delta^\beta_\lambda n_\mu\,, \qquad (7.11.24)$$

$$n_\alpha S_0^{\alpha\beta\lambda\mu}(x) n_\mu = \frac{\mu_0}{2\pi(1 - \nu_0)} \frac{1}{x^2} \delta^{\beta\lambda}\,, \qquad (7.11.25)$$

where $1/x^2$ is a generalized function, the inverse Fourier transform of which is $-J_1|k|$.

Multiplying (7.11.14, 17 and 18) by the normal n and taking into account (7.11.24, 25), we arrive at the system of equations ($b^2 = \langle l^2 \rangle / 2$)

$$\begin{cases} \bar{\sigma}^2(x) = \sigma_0 \varphi(x) + b^2 \dfrac{d(x)}{x^2} + b^2 \int\limits_{-\infty}^{\infty} \dfrac{1}{(x-x')^2} F(x', x, 0) \bar{\sigma}^2(x')\, dx' , \\ d(x) = \sigma_0 \varphi(x) + b^2 \dfrac{\bar{\sigma}^2(x)}{x^2} + b^2 \int\limits_{-\infty}^{\infty} \dfrac{1}{(x-x')^2} F(x', x, 0) \bar{\sigma}^2(x')\, dx' , \\ \varphi(x) = \sigma_0 + b^2 \dfrac{\varphi(x)}{x^2} + b^2 \int\limits_{-\infty}^{\infty} \dfrac{1}{(x-x')^2} F(x', x, 0) \varphi(x')\, dx' \end{cases} \tag{7.11.26}$$

with respect to the three scalar functions

$$\begin{aligned} &\bar{\sigma}^2(x) = \langle \bar{\sigma}(x) \bar{\sigma}(0) | x, 0 \rangle, \; d(x) = \langle \bar{\sigma}(x) \bar{\sigma}(x) | x, 0 \rangle , \\ &\varphi(x) = \langle \bar{\sigma}(x) | x, 0 \rangle . \end{aligned} \tag{7.11.27}$$

If the point-defects density tends to zero, then the integral terms in these equations vanish, and the equations describe interaction between two isolate defects on the straight line. At the same time, the solution of the system (7.11.26) has the form

$$\bar{\sigma}^2(x) = d(x) = \varphi^2(x) , \quad \varphi(x) = \sigma_0 \frac{x^2}{x^2 - b^2} . \tag{7.11.28}$$

Here the expression for $\varphi(x)$ is the normal component of the stress field, in which there are located two point defects, the distance between them being x. It is obvious, that $\varphi(x)$ is the asymptotics, as $x > b$, of the solution for the interaction of two rectilinear cracks, which lie on the same straight line. If $x \leq b$, the solution (7.11.28) has no physical sense.

As pointed out in Sect. 7.10, in the case of finite defects, there is no analog to the point field, in which defects can occur at arbitrary small distances. For example, in the case under consideration, the centers of cracks must be not closer to each other, than the sum of their half lengths. This circumstance can be partially taken into account, if one makes use of the model of a one-dimensional point field, which is analogous to a random lattice of defects in the space.

Let x_k be a coordinate of the k-th defect and the differences $x_{k+1} - x_k$ be independent stochastic variables (for different k's), which have the same normal distribution

$$f(x) = \frac{1}{\sqrt{2\pi}\sigma} \exp\left[-\frac{(x-l_0)^2}{2\sigma^2} \right]. \tag{7.11.29}$$

Here l_0 is the average distance between defects, σ^2 is the dispersion. If $\sigma \to 0$, then we have a regular chain of defects, and if $\sigma \to \infty$, then this is a Poisson point field on the straight line. To restrict the probability of defects coming too close to each other, let us assume that the dispersion of stochastic variables $x_{k+1} - x_k$ is sufficiently small. Since the quantity $(x_{k+1} - x_k)$, distributed according to the law (7.11.29), lies in the interval $(l_0 - 3\sigma, l_0 + 3\sigma)$ with a probability practically equal to 1, let us assume $\varkappa \equiv l_0/\sigma \geq 3$.

The function $F(x', x_1, x_2)$ in the integrands of (7.11.26) has, in the case under consideration, the form

$$F(x', x_1, x_2) = \frac{\sum\limits_{k=-\infty}^{\infty}{}' f_k(x' - x_1) \sum\limits_{k=-\infty}^{\infty}{}'' f_n(x_1 - x_2)}{\sum\limits_{k=-\infty}^{\infty}{}' f_k(x_1 - x_2)}, \tag{7.11.30}$$

where the primed summation sign means exclusion of the term with $k = 0$, and the double prime means the same for $n = 0$ and $n = k$,

$$f_k(x) = \frac{3}{\sqrt{2\pi |k|}\sigma} \exp\left[- \frac{(x - \varkappa l_0)^2}{2\sigma^2} \right]. \tag{7.11.31}$$

Let us substitute (7.11.30) into (7.11.26) and search for the solution of the latter in the form

$$\bar{\sigma}^2(x) = \bar{\sigma}_1^2(x) \left(\frac{x^2}{x^2 - b^2} \right)^2, \quad d(x) = d_1(x) \left(\frac{x^2}{x^2 - b^2} \right)^2,$$

$$\varphi(x) = \varphi_1(x) \frac{x^2}{x^2 - b^2}. \tag{7.11.32}$$

This choice of the structure of the solution is due to the following reasons. The functions φ, $\bar{\sigma}^2$ and d characterize an interaction between two defects in a random field of point defects. In the first approximation one can assume that their form coincides with that of the corresponding functions for two isolated defects (7.11.28), and the presence of other defects can be taken into account by correcting the external field. Thus we immediately arrive at (7.11.32).

Numerical computations show that the functions $\bar{\sigma}_1^2(x)$, $d_1(x)$ and $\varphi_1(x)$ are well approximated by constants, the values of the latter depending on the parameters $p = 2b/l_0$ and $\varkappa = l/\sigma$. Let us denote these constants by $\bar{\sigma}_\infty^2(p, \varkappa)$, $d_\infty(p, \varkappa)$ and $\varphi_\infty(p, \varkappa)$, respectively; one can prove that

$$\bar{\sigma}_\infty^2(p, \varkappa) = \varphi_\infty^2(p, \varkappa). \tag{7.11.33}$$

The dependence of $\varphi_\infty(p, \varkappa)$ and $d_\infty(p, \varkappa) - \varphi_\infty^2(p, \varkappa)$ on the parameters is shown in Figs 7.12 and 13.

Since $\varphi_\infty = \bar{\sigma}^1$ the mean value of the effective field, as it is seen from Fig. 7.12, attains its maximum for large relative dispersion of distance between

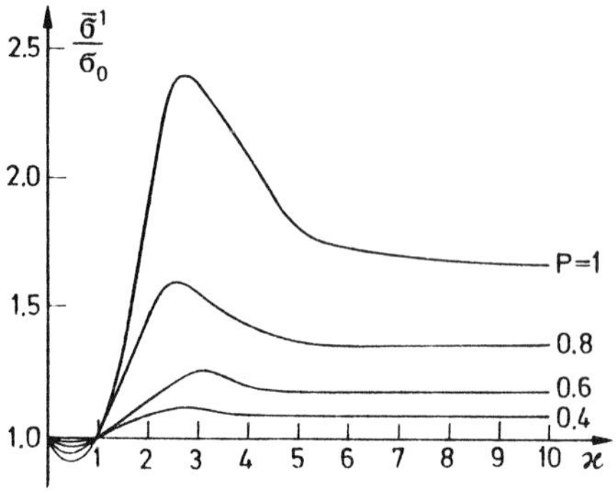

Fig. 7.12. Dependence $\bar{\sigma}_1/\sigma_0$ on κ and p

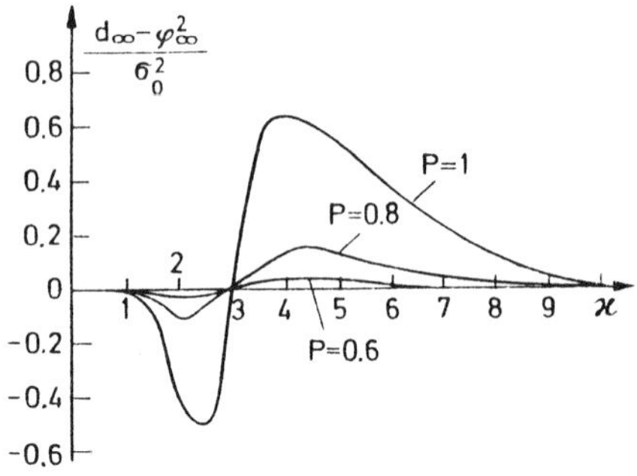

Fig. 7.13. Dependence $(d_\infty - y_\infty^2)/\sigma_0^2$ on κ and p

defects ($\kappa = 2.5 \div 3.5$). With $\kappa = 5 \div 6$, the quantity $\bar{\sigma}^1$ no more coincides with the value, which corresponds to a regular chain of defects ($\kappa = \infty$).

The difference $d_\infty - \varphi_\infty^2$ is equal to the dispersion of the effective field $\bar{\sigma}$. From Fig. 7.13, it is seen that the dispersion has its maximum for $\kappa = 4 \div 5$, and for $\kappa = 9 \div 10$, is practically zero, which corresponds to the transition to a regular structure. For $\kappa < 3$, the quantity $d_\infty - \varphi_\infty^2$ becomes negative. This loss of physical meaning is connected with an increase of the probability of the defects approaching each other at distances less than b for small κ.

As it is seen from (7.11.32) for $\bar{\sigma}^2(x)$, the correlation radius of the effective field has the order of the average value of the dimension b of a defect and weakly depends on the relative dispersion of the distance κ^{-2} between defects.

Let us now proceed to calculating the second moment of the stress field $\sigma(x)$ in a medium with point defects. The relation (7.5.20) becomes

$$\langle \sigma(x_1)\sigma(x_2)\rangle = \sigma_0^2 + \int S_0(x_1 - x')dx' \int S_0(x_2 - x'')\langle P^0(x')P^0(x'')X(x')X(x'')\bar{\sigma}(x')\bar{\sigma}(x'')\rangle \, dx'. \tag{7.11.34}$$

Taking into account the equality analogous to (7.11.8), namely,

$$X(x') = X(x''; x') + \delta(x' - x'') \text{ with } x'' \in X \tag{7.11.35}$$

and the hypothesis H2 about the statistical independence (Sect. 7.3), we represent the mean in the integrand in (7.11.34) by

$$\langle P^0(x')P^0(x'')X(x')X(x'')\bar{\sigma}(x')\bar{\sigma}(x'')\rangle = \langle X(x'', x')X(x'')\rangle P\bar{\sigma}^2(x' - x'')P + \frac{1}{v_0}\delta(x' - x'')\langle P^2\rangle d_\infty\,, \tag{7.11.36}$$

where d_∞ has the form of (7.11.6).[4]

Substituting (7.11.36) into (7.11.34) we obtain

$$\langle\sigma(x)\sigma(0)\rangle = \sigma_0^2 + \frac{1}{v_0}\Pi(x)d_\infty + v_0\int \Pi(x-x')\psi(x')\bar{\sigma}^2(x')\,dx' , \quad (7.11.37)$$

where

$$\Pi(x) = \int S_0(x-x')PS_0(x')\,Pdx' , \quad (7.11.38)$$

$$\psi(x-x') = \langle X(x;x')|x\rangle . \quad (7.11.39)$$

Let us return to the one-dimensional field of point defects and calculate the second moment of the normal component $\rho(x)$ of the stress tensor, when x is a point on the line of the defects

$$\rho(x) = \langle\sigma_{nn}(x)\sigma_{nn}(0)\rangle , \quad (7.11.40)$$

where

$$\sigma_{nn}(x) = n_\alpha\sigma^{\alpha\beta}(x)n_\beta . \quad (7.11.41)$$

From (7.11.37), for the given case, we obtain

$$\rho(x) = \sigma_0^2 + \frac{d_\infty}{l_0}\pi(x) + l_0\int_{-\infty}^{\infty}\pi(x-x')\psi(x')\bar{\sigma}^2(x')\,dx' , \quad (7.11.42)$$

where $d_\infty \equiv d(\infty)$ and $\bar{\sigma}^2(x)$ have the form of (7.11.32), the function $\psi(x)$ is represented by

$$\psi(x) = \sum_{k=-\infty}^{\infty}{}' f_k(x) . \quad (7.11.43)$$

The primed summation sign means exclusion of the term with $k = 0$, and $f_k(x)$ has the form of (7.11.31). The function $\pi(x)$ is an analog of $\Pi(x)$ from (7.11.37) and is defined by the relation

$$\pi(x) = b^4\int_{-\infty}^{\infty}\frac{1}{(x-x')^2}\cdot\frac{1}{x'^2}\,dx' = -b^4\delta''(x) . \quad (7.11.44)$$

From here and from (7.11.39), we finally obtain

---

[4]It is taken into account here that at one and the same point there cannot be two different defects; therefore, we have $\langle\bar{\sigma}(x')\bar{\sigma}(x'')|x,x''\rangle = \langle\bar{\sigma}(x')\bar{\sigma}(x')|x'\rangle = d_\infty$ if $x' = x''$.

$$\rho(x) = \sigma_0^2 - b^4 \frac{d^2}{dx^2}\left[\frac{d_\infty}{l_0}\,\delta(x) + \psi(x)\bar{\sigma}^2(x)\right]. \tag{7.11.45}$$

The graph of the continuous part of the function $\rho(x) - \sigma_2^0$ is presented in the Fig. 7.14. The presence of a singular part and of a singularity at $x = b$ in

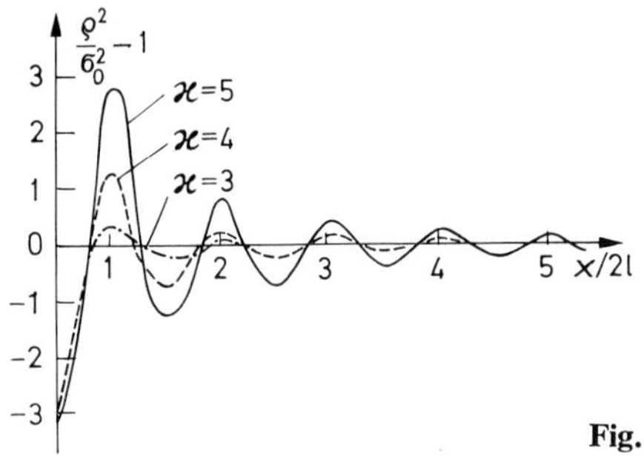

Fig. 7.14. Dependence of correlation radius ρ on $\kappa/2l$

the correlation function $\sigma_{nn}(x)$ of the random field on the line of defects is connected with replacing real defects by point inhomogeneities. For a random field of defects with finite dimensions, the correlation function is smooth, bound and has the minimum correlation radius of the order of the average dimension of a defect. When the random field of defects tends to a regular structure, the correlation radius of the stress field increases; this is easily seen from Fig. 7.14.

7.12 Conclusions

Let us now estimate the accuracy of the method of an effective field and discuss a possibility of using the first moments of the solution of the stochastic elastic problem, for strength calculations of structures made of composites.

It was pointed out above that in the general case it is impossible to obtain an explicit solution of the elastic problem for stochastically inhomogeneous medium. Each statistical moment of the solution is expressed in terms of a set of moments of higher orders, and to truncate the chain of equations, one has to introduce additional assumptions, usually without rigourous proofs of their validity.

There exists a number of approaches to solving such problems; surveys can be found, for example, in [7.2, 21–24]. The method of an effective field considered in this chapter, stands somewhere between self-consistent schemes and methods of the type of smoothing approximation or stochastic phases approximation [7.21, 27].

Introducing the local external field $\bar{\sigma}$ and assuming $\bar{\sigma}$ to be constant in a neighborhood of each particle (hypothesis H1 of Sect. 7.3)[5] are in agreement with the method of the self-consistent field (MSF). The distinction from the classical scheme of MSF is that we consider the field $\bar{\sigma}$ as stochastic and, when constructing the closed equations for statistical moments $\bar{\sigma}$, the procedure of decomposing complex means is used, this being typical for the method of smoothing approximation.

Note that the first approximation of the method of an effective field (Sect. 7.6–9) yields the same results as MSF does, when applied to computations of effective parameters of inhomogeneous media [7.12, 25, 26]. However, this does not exhaust all possibilities of the method of an effective field. The method enables one to obtain expressions, generally speaking, for any statistical moment of the solution, while it appears to be possible to assign physical sense to hypotheses and corrections of the employed calculating scheme. This is the advantage of the method of an effective field in comparison with, for example, an approximate summation of the formal series of perturbation theory [7.2, 24].

An estimate of the accuracy of the method of an effective field is important. It is known [7.27] that when describing a system of point particles in quantum and stochastic physics in the region where interaction is essential, the slower the potential of an individual particle decays at infinity and the larger the density of particles, the better the classical MSF works. For particles with a potential of the Coulomb type, the self-consistent solution gives a good coincidence with experimental data, though rigorous estimates cannot be obtained. Such estimates are usually deduced on the basis of physical reasonings.

Particles of composite media considered in this chapter have the following peculiarities. First, their dimensions are finite and, secondly, the state of each particle is described by its own field which depends on the local external field that contains the particle. The method of an effective field is based on hypotheses which concern the approximation of a state of each particle in a deformable composite. The range of the variation of the parameters of inhomogeneities, in which these hypotheses are precise statements, is small: this covers low concentrations or absence of interaction. In fact, in this case each particle is contained in a constant external field (hypothesis H1 of Sect. 7.3), which statistically does not depend on stochastic dimensions and elastic constants of the particle (hypothesis H2 of Sect. 7.3). Hence, all the results of the preceeding sections become precise if the concentration of defects is low enough.

With increasing the concentration, the validity of the hypotheses H1 and H2 is, generally speaking, broken. The hypothesis H1 becomes invalid, if the effective field essentially differs from a constant one in regions occupied by

[5]The latter is, generally speaking, not obligatory. One can assume only an identical behavior of the field $\bar{\sigma}$ in a neighborhood of each particle; see the end of Sect. 7.9.

the particles. This can occur when particles are situated close enough to each other, i.e. when the concentration is high. On the other hand, the more the defects affect each other, the more reasonable the hypothesis H2 is.

However, invalidity of the hypotheses H1 and H2 can weakly influence the first and the second moments of the solution, i.e. the quantities of most interest for applications. These moments are extremely rough statistical characteristics of a solution, and a part of the information about detailed behavior of the particles is here lost. Therefore, a detailed consideration of the state of each particle is not reasonable. Of course, this argument can be confirmed only by comparing the derived results with exact solutions or experiments.

Since exact solutions of stochastic problems are unknown, comparisons with experiments play an important role. From the results of Sect. 7.7 and 9, it follows that the method of an effective field enables one to describe sufficiently the results of experimental measurements of effective elastic constants in the two-dimensional case, and, for regular composites, yields a fair agreement with exact solutions.

In Sect. 7.7. a dimensionless scalar parameter χ was introduced, with the help of which the range of applicability of the method can be described. The size of the region of values χ, where the method gives an error more than 10%, is relatively small. This region contains, for example, χ for the cases of circular holes or of absolutely rigid inclusions at concentrations of $p < 0.4$. The estimate seems to be valid not only in the case of a regular triangular lattice of inclusions, but also for other types of structures, in which distributions of inclusions differ weakly from the isotropic one. Analogous estimates for stochastic fields of cracks follow from the results of Sect. 7.9.

In the spatial case, the method of an effective field has qualitatively the same region of applicability. However, in the three-dimensional medium, the potential of a particle decays more rapidly than in the planar case ($\sim r^{-1}$ and $\sim r^{-2}$, respectively). Another peculiarity of the three-dimensional problem is, that here more particles can interact directly. All this can exhibit quantitative boundaries for the applicability of the method.

Unfortunately, the absence of sufficient data in the literature, concerning the determination of correlation functions of elastic fields in an inhomogeneous medium, does not allow us to estimate the error of the method when calculating the second moments of the solution.

In conclusion, let us dwell briefly upon the scheme of using the obtained results for strength computations of structures made of composites. Such a computation can be divided into two steps: computing the macro-stress and estimating the level of micro-stress.

When computing the macro-stress, the composite is considered as a homogeneous medium, the elastic properties of which are determined by effective constants. Here one can use the classical elasticity and consider the macro-stress obtained as an external one when estimating the level of micro-stress connected with an inhomogeneity of the material.

Even the first approximation of the method of an effective field enables one to calculate effective constants and to estimate the stress concentration on defects under the assumption that each of them is contained in a constant external field $\bar{\sigma}^1$ (Sect. 7.6–9).

More precise results follow by taking into account fluctuations of the effective field (Sect. 7.5, 11). Knowledge of the values of dispersion and of correlation radii of the effective field $\bar{\sigma}(x)$ and knowledge of the real micro-stress field $\sigma(x)$ enables one to calculate the probability of fracture of the structure and the value of critical loadings. When using a criterion of fracture, for example, through the value of maximum stress, one has, first of all, to determine the validity of the inequality $\sigma(x) > \sigma_*$, where σ_* is a critical level of stress, and to estimate the set of points, where this condition is valid. If one assumes that $\sigma(x)$ is a Gauss random field, then the last problem is solved by standard methods of probability theory.

When the condition of fracture is fulfilled at a point, a whole neighborhood of the point will become fractured, its dimension being of the order of the correlation radius of the random field of micro-stress. In such a way dimensions and concentration of fractured regions are determined, the level of macro-stress being given. Replacing the material of the fractured regions by defects with appropriate properties, one can repeat the computation and find the probability of fracture of the whole structure.

An analogous scheme can be used for estimating the probability of transforming the composite into the plastic state.

7.13 Notes

The method of an effective medium as a tool for calculating macro-constants of heterogeneous media has a long history. With the help of this method, effective coefficients of electroconductivity, and the dielectric and magnetic conductivities of inhomogeneous materials of various structures were found. A listing of corresponding works can be found, for example, in the surveys [B7.5, 17] where other approaches to these problems are also discussed. The work of Hori [B7.11] in which practically all the existing methods of constructing effective constants are considered, is of particular interest.

A number of basic concepts, concerning the calculation of elastic macro-constants of polycrystal metals were suggested by Kröner [B7.19, 20]. The method of an effective medium was treated by Kröner in [B7,19] (see also the work of Hershey [B7.9]). In the paper of Hill [B7.10], effective constants of composites with ellipsoidal inclusions were found with the help of this method, textured polycrystals were considered by Kneer [B7.18] and Morris [B7.27], media with cracks were studied by Budjansky and O'Connel [B7.3]. A subsequent development of static and dynamic problems is due to Willis [B7.33,34].

Application of the method of self-consistent fields to solving multi-particle problems of quantum mechanics (the Hartree-Fock method) and to a description of phase transitions (Weiss method) can be found in standard courses of quantum mechanics and statistical physics. Some estimates of the applicability of the method were considered by Kac [B7.12].

The notion of a constant self-consistent field, in which each particle is situated, was used in a number of works for solving the problem of a wave propagating in an inhomogeneous medium. The calculation of effective characteristics of propagation and attenuation of sound in a medium with bubbles was described in the monograph of Morse and Feshbach [B8.22], an analogous problem for liquid suspensions was solved by Chaban by the method of a self-consistent field [B7.4].

The method of a self-consistent field was also used implicitly in works of Walpole [B7.32] and Levin [B7.24] for the calculation of effective elastic constants of composites, which contain an isotropic random field of ellipsoidal inclusions. A generalization to arbitrary fields of random ellipsoidal inhomogeneities (in particular, on regular structures) was obtained by Kanaun [B7.14, 15].

An application of the method to the description of stochastic and regular systems of cracks in the space and in the plane was considered by Kanaun [B7.13], and Kanaun and Iablokova [B7.16] (see also [B7.24]).

Appendices

A1 Fourth-Order Tensors of Special Structure

When constructing Green's operators for isotropic media and in a number of other cases, one has to deal with tensors of a special type, which depend on the Kronecker delta and on a unit vector. Here, the formulae are presented, which enable one to simplify the operation with these tensors, and in particular, to obtain explicit expressions for inverse tensors.

Let us consider tensors of the forth order made up of Kronecker deltas and the unit vector n, which are symmetric in the first and the second pairs of indices, but which are, generally speaking, not symmetric with respect to permutation of the pairs. It is easy to show that all tensors having such a structure can be represented as linear combinations of the following six linearly independent tensors (these thus play the role of a basis):

$$\begin{aligned} E^1_{\alpha\beta\lambda\mu} &= \frac{1}{2}(\delta_{\alpha\lambda}\delta_{\beta\mu} + \delta_{\alpha\mu}\delta_{\beta\lambda}), \; E^2_{\alpha\beta\lambda\mu} = \delta_{\alpha\beta}\delta_{\lambda\mu}, \\ E^3_{\alpha\beta\lambda\mu} &= \delta_{\alpha\beta}n_\lambda n_\mu, \qquad\qquad E^4_{\alpha\beta\lambda\mu} = n_\alpha n_\beta \delta_{\lambda\mu}, \\ E^5_{\alpha\beta\lambda\mu} &= \frac{1}{4}(n_\alpha n_\lambda \delta_{\beta\mu} + n_\alpha n_\mu \delta_{\beta\lambda} + n_\beta n_\lambda \delta_{\alpha\mu} + n_\beta n_\mu \delta_{\alpha\lambda}), \\ E^6_{\alpha\beta\lambda\mu} &= n_\alpha n_\beta n_\lambda n_\mu . \end{aligned} \tag{A1.1}$$

Let us define a multiplication-convolution of two tensors

$$E^{ik} = E^i \cdot E^k , \tag{A1.2}$$

where

$$E^{ik}_{\alpha\beta\lambda\mu} = E^{i\cdot\cdot\nu\rho}_{\alpha\beta} E^k_{\nu\rho\lambda\mu} . \tag{A1.3}$$

It is easy to verify that the tensors E^{ik} can be expressed as linear combinations of the basis elements E^i.

The linear tensor space, spanned by the basis E^i is closed with respect to the product introduced, and hence forms an algebra which we denote $A(n)$. We set

$$E^i \cdot E^k = \alpha^{ik}_{\cdot\cdot j} E^j ,$$

where $\alpha^{ik}_{\cdot\cdot j}$ is a matrix defined by the table

	E^1	E^2	E^3	E^4	E^5	E^6
E^1	E^1	E^2	E^3	E^4	E^5	E^6
E^2	E^2	$3E^2$	$3E^3$	E^2	E^3	E^3
E^3	E^3	E^2	E^3	E^2	E^3	E^3
E^4	E^4	$3E^4$	$3E^6$	E^4	E^6	E^6
E^5	E^5	E^4	E^6	E^4	$\frac{1}{2}(E^5+E^6)$	E^6
E^6	E^6	E^4	E^6	E^4	E^6	E^6

(A1.4)

However, for practical calculations, it is more convenient to change the basis to the set J^i which is connected with the former by the relation

$$J^i = A^i_k E^k\,, \tag{A1.5}$$

where A^i_k is the following matrix (i is the number of a row):

$$A^i_k = \begin{pmatrix} 0 & \frac{1}{2} & -\frac{1}{2} & -\frac{1}{2} & 0 & \frac{3}{2} \\ 0 & -\frac{1}{2} & \frac{1}{2} & \frac{1}{2} & 0 & \frac{1}{2} \\ 0 & 0 & -1 & -\frac{1}{2} & 0 & \frac{3}{2} \\ 0 & 0 & -1 & \frac{1}{2} & 0 & \frac{1}{2} \\ 1 & -\frac{1}{2} & \frac{1}{2} & \frac{1}{2} & -2 & \frac{1}{2} \end{pmatrix} \tag{A1.6}$$

The inverse matrix a^i_k has the form

$$a^i_k = \begin{pmatrix} 1 & 0 & 0 & 0 & 1 & 1 \\ \frac{3}{2} & -\frac{1}{2} & -\frac{3}{2} & \frac{1}{2} & 0 & 0 \\ \frac{1}{2} & \frac{1}{2} & -\frac{1}{2} & -\frac{1}{2} & 0 & 0 \\ \frac{1}{2} & \frac{1}{2} & -1 & 1 & 0 & 0 \\ \frac{1}{2} & \frac{1}{2} & 0 & 0 & \frac{1}{2} & 0 \\ \frac{1}{2} & \frac{1}{2} & 0 & 0 & 0 & 0 \end{pmatrix} \tag{A1.7}$$

The multiplication table for the basis tensors J^i has a simpler form

	J^1	J^2	J^3	J^4	J^5	J^6
J^1	J^1	J^2	J^3	J^4	0	0
J^2	J^2	J^1	$-J^4$	$-J^3$	0	0
J^3	J^3	J^4	J^1	J^2	0	0
J^4	J^4	J^3	$-J^2$	$-J^1$	0	0
J^5	0	0	0	0	J^5	0
J^6	0	0	0	0	0	J^6

(A1.8)

It is seen from the table that the algebra $A(n)$ is split into the direct sum of two subalgebras with the basis J^5, J^6 and $J^1, \ldots, J^4$, the second subalgebra being isomorphic to a real form of the complex quaternion algebra (the usual real quaternion algebra will be obtained if one substitutes $J^2 \to J^2_* = \mathrm{i}J^2$, $J^3 \to J^3_* = \mathrm{i}J^3$).

A tensors $Q_{\alpha\beta\lambda\mu}$ is inverse to a tensor $P_{\alpha\beta\lambda\mu}$ if these tensors satisfy the relation

$$P_{\alpha\beta}^{\cdot\cdot\nu\rho} Q_{\nu\rho\lambda\mu} = \frac{1}{2}\left(\delta_{\alpha\lambda}\,\delta_{\beta\mu} + \delta_{\alpha\mu}\,\delta_{\beta\lambda}\right) = E^1_{\alpha\beta\lambda\mu}\,. \tag{A1.9}$$

Let a tensor $P = p_k J^k$, and a tensor $Q = q_k J^k$ be inverse to P. Then, for the components of the tensor Q, we have

$$(q_1, \ldots, q_6) = \left(\frac{p_1}{\Delta}, -\frac{p_2}{\Delta}, -\frac{p_3}{\Delta}, -\frac{p_4}{\Delta}, \frac{1}{p_5}, \frac{1}{p^6}\right), \tag{A1.10}$$

where

$$\Delta = p_1^2 - p_2^2 - p_3^2 + p_4^2\,. \tag{A1.11}$$

It is easily seen that $q_1, \ldots, q_4$ are connected with $p_1, \ldots, p_4$, as in the quaternion case, and only difference being that "modulus squared" is not of constant sign. The conditions of existence of the inverse tensor are: $\Delta \neq 0$, $p_5 \neq 0$, $p_6 = 0$.

A special case, when the tensor $P_{\alpha\beta\lambda\mu}$ is symmetric with respect to permutation of the pairs $\alpha\beta$ and $\lambda\mu$ is of interest, since the tensor of elastic constants has this symmetry. One can show that in this case, the inverse tensor $Q_{\alpha\beta\lambda\mu}$ possesses the same symmetry.

As usual, one can define a scalar product of two elements of the algebra $A(n)$ as the complete contraction of the two tensors. Then, the scalar product of the unit tensor and an aribtrary one coincides with the trace of the latter. In particular, for the basis elements J^k, we have

	J^1	J^2	J^3	J^4	J^5	J^6
Tr J^k	2	0	0	0	2	2

(A1.12)

The algebra $A(n)$ and its applications to elasticity is considered in more detail in [B6.25].

A2 Green's Operators of Elasticity

Here, the Green's operators of static elasticity will be considered in the k-representation. The elastic medium is assumed to be unbound and homogeneous. The tensors of the elastic moduli C and of the elastic compliance $B = C^{-1}$ are assumed to be given. In a local theory, C and B are constants, while in a nonlocal theory they are functions of the vector k. For the sake of simplicity the dependence of C and B on k is not pointed out explicitly.

As a preliminary, let us consider the operator Rot in the k-representation. According to (5.3.11), the Fourier transform $R(k)$ of the operator Rot can be represented in the form

$$R(k) = -k^2 \mathscr{R}(n)\,, \tag{A2.1}$$

where $n_\alpha = k_\alpha / \mid k \mid$ and

$$\mathscr{R}^{\alpha\beta\lambda\mu}(n) = (\varepsilon^{\alpha\nu\lambda}\varepsilon^{\beta\rho\mu} n_\nu n_\rho)_{(\lambda\mu)}\,. \tag{A2.2}$$

It is easily seen that $\mathscr{R}(n)$ belongs to the class of tensors considered in Appendix A1. In fact, one can directly verify the identity

$$\begin{aligned} \varepsilon^{\alpha\nu\lambda}\varepsilon^{\beta\rho\mu} = {} & \delta^{\alpha\beta}\delta^{\nu\rho}\delta^{\lambda\mu} + \delta^{\nu\beta}\delta^{\lambda\rho}\delta^{\alpha\mu} + \delta^{\lambda\beta}\delta^{\alpha\rho}\delta^{\nu\mu} - \delta^{\alpha\beta}\delta^{\lambda\rho}\delta^{\nu\mu} \\ & - \delta^{\lambda\beta}\delta^{\nu\rho}\delta^{\alpha\mu} - \delta^{\nu\beta}\delta^{\alpha\rho}\delta^{\lambda\mu} \end{aligned} \tag{A2.3}$$

and taking into account the formulae (A1.1) for the basis E^i, we find

$$\mathscr{R}(n) = -E^1 + E^2 - E^3 - E^4 + 2E^5\,. \tag{A2.4}$$

With the help of the matrix a^i_k given by the table (A1.7), we can rewrite (A2.4) in the basis J^i of the algebra $A(n)$

$$\mathscr{R}(n) = \frac{1}{2}(J^1 - J^2) - J^6\,. \tag{A2.5}$$

Note the following important identities for $\mathscr{R}(n)$, being easily verified with the help of the Table (A1.8):

$$\mathscr{R}^3(n) = \mathscr{R}(n), \quad \mathscr{R}^4(n) = \mathscr{R}^2(n) . \tag{A2.6}$$

From this, it follows, in particular, that $\mathscr{R}^2(n)$ is a projection operator. Let us denote this projector by Θ_0 and introduce the complementary projector $\Pi_0 = I - \Theta_0$ ($I \equiv E_1$). It is easily seen that the projections Π_0 and Θ_0 are self-adjoint and their representation in the basis J^i has the form

$$\Pi_0 = \frac{1}{2}(J^1 + J^2) + J^5, \quad \Theta_0 = \frac{1}{2}(J^1 - J^2) + J^6 . \tag{A2.7}$$

Let $\xi(x)$ be a second-order tensor field, which vanishes at infinity, and let $\xi(k)$ be its Fourier transform. Then, as is easily verified, the following equivalence relations

$$\begin{aligned} &\operatorname{div} \xi = 0 \Leftrightarrow \Pi_0 \xi = 0 , \\ &\operatorname{Rot} \xi = 0 \Leftrightarrow \Theta_0 \xi = 0 \end{aligned} \tag{A2.8}$$

hold.

Let us proceed to the Green's operators of elasticity. The Green's operator for displacements $G(k)$ is defined by (4.6.4, 5). Its explicit expression in the isotropic case is given by (2.8.6), where λ and μ are to be taken as constants.

From the results of Sect. 4.6, it follows that the Green's operator for the strain $K(k)$ is connected with $G(k)$ by the relation

$$K_{\alpha\beta\lambda\mu}(k) = [k_\alpha k_\mu G_{\beta\lambda}(k)]_{(\alpha\beta)\,(\lambda\mu)} ; \tag{A2.9}$$

it satisfies the equations

$$\mathscr{R}K = 0, \quad k \cdot CKC = k \cdot C \tag{A2.10}$$

and the identity

$$KCK = K . \tag{A2.11}$$

In the isotropic case, its explicit expression is given by (4.6.16).

The Green's operator for internal stress $S(k)$ is connected with $K(k)$ by

$$S = C - CKC, \quad K = B - BSB , \tag{A2.12}$$

satisfies

$$k \cdot S = 0, \; RBS = R \tag{A2.13}$$

and the identity

$$SBS = S\,. \tag{A2.14}$$

In the isotropic case, its expression is given by (5.3.13).

The relations given above express K and S in terms of G. Conversely, one can accept K and S as the main Green's operators and express G in terms of them. A complete system of equations, which determines the operators K and S, has the form

$$\text{Rot}\ K = 0,\ \text{div}\ S = 0\,, \tag{A2.15}$$

$$S = C - CKC \text{ or } K = B - BSB\,. \tag{A2.16}$$

Taking into account (A2.8), (A2.15) can be rewritten in an equivalent form

$$\Theta_0 K = 0,\ \Pi_0 S = 0\,. \tag{A2.17}$$

In order to show the existence and uniqueness of the solution of the algebraic (in the k-representation) system of equations (A2.16, 17), let us introduce some useful notations. Let L be an arbitrary linear operator, L_1 and L_2 its restrictions to the subspaces of the projections Π_0 and Θ_0, respectively, and

$$L_{1i} = \Pi_0 L_i,\ L_{2i} = \Theta_0 L_i \quad (i = 1, 2)\,. \tag{A2.18}$$

Then the operator L can be represented in the form of a two-by-two matrix with operator components L_{ij} $(i, j = 1, 2)$.

Let us write down all the operators in the matrix form and substitute them into (A2.16, 17). Elementary computations show that the only components of K and S which differ from zero, are K_{11} and S_{22}, respectively, and

$$K_{11} = C_{11}^{-1},\ S_{22} = B_{22}^{-1}\,, \tag{A2.19}$$

where C_{11} and B_{22} are the inverses in the corresponding subspaces. Finally, the existence of the inverse operators follows from the positive-definiteness of the operators C and B. The Green's tensor $G(k)$ for displacement is, in turn, expressed in terms of K or S (or Π and Θ) with the help of simple algebraic operations. At the same time, obtaining explicit formulae in the x-representation is possible only in the simplest cases, in particular for the isotropic or hexagonal symmetries.

The Green's tensors and projection operators for the isotropic medium were obtained in [B6.25]. In particular, it is shown that the projections Π_0 and Θ_0 admit an additional invariant decomposition

$$\Pi_0 = \Pi_0' + \Pi_0'',\ \Theta_0 = \Theta_0' + \Theta_0''\,, \tag{A2.20}$$

where Π_0'' and Θ_0'' are projections onto subspaces of traceless tensors.

A3 Green's Operators K and S in x-Representation [1]

For a homogeneous medium, the action of the Green's operators K and S in the x-representation is reduced to convolution with the corresponding kernels $K(x)$ and $S(x)$. Therefore we shall begin with a consideration of these kernels.

Generalized functions $K(x)$ and $S(x)$ can be defined as the inverse Fourier transforms of the functions $K(k)$ and $S(k)$, or, according to (4.6.13) and (4.6.29), as the second derivatives of $G(x)$. Let us base our argument on the second approach and consider $K(x)$.

We shall use a known form of regularization, which is applicable to generalized functions of the type of the second derivatives of the Green's functions [B8]. In the case given, for $K(x)$, we have in shorthand notation

$$K(x) = -\nabla G(x)\nabla = \tilde{K}(x) + A\delta(x)\,, \tag{A3.1}$$

where $\tilde{K}(x)$ is a regular functional and A is a constant. This representation of $K(x)$ in the form of a sum of regular and singular components depends on the choice of a special form of regularization, although $K(x)$ itself does not depend on the latter.

Let V be a region with a smooth boundary Ω and $x = 0$ be an internal point of V. Then $\tilde{K}(x)$ acts on the test functions $\varphi(x)$ according to the rule

$$(\tilde{K}, \varphi) = \int_V K(x)[\varphi(x) - \varphi(0)]\, dx + \int_{\bar{V}} K(x)\varphi(x)\, dx\,, \tag{A3.2}$$

where $K(x)$ is, according to (4.6.13) a formal second derivative of $G(x)$ and $\bar{V}$ is the complement of V in three-space. The constant A is given by the expression

$$A = -\int_\Omega \nabla G(x) n(x)\, d\Omega\,, \tag{A3.3}$$

where $n(x)$ is the normal to Ω at a point x.

Our first task is to obtain a more convenient expression for A directly in terms of $K(k)$. For this purpose, not that, making use of Gauss' theorem, one can rewrite (A3.3) in the form

$$A = (\nabla G(x), \nabla V(x))\,, \tag{A3.4}$$

where $V(x)$ is the characteristic function of the region V. Further, taking into account Parceval's equality (2.1.6), we find

$$A = \frac{1}{(2\pi)^3} \int kG(k)kV(k)\, dk\,, \tag{A3.5}$$

[1] Appendix A3 was written in collaboration with S. K. Kanaun.

or, taking into account (9.6.13),

$$A = \frac{1}{(2\pi)^3} \int K(k) V(k)\, dk\,. \tag{A3.6}$$

Let us assume now that Ω is an ellipsoid given by (4.9.13) and perform the substitution $x = ay$. Then $V'(y) = V(ay)$ is the characteristic function of the unit sphere and $V(k) = \det\{a\} V'(ak)$. Substituting into (A3.6) and replacing k by $a^{-1}k$, we obtain

$$A = \frac{1}{(2\pi)^3} \int K(a^{-1}k) V'(k) dk\,. \tag{A3.7}$$

Let $k = |k|\,\omega$. Then $K(a^{-1}k) = K(a^{-1}\omega)$, since $K(k)$ is a homogeneous function of the zeroth degree. On the other hand, it is obvious that $V'(k) = V'(|k|)$. This enables us to carry out the integration in (A3.7) with respect to $|k|$ and the unit vector ω separately. Using formulae from Appendix A1 for the Fourier transform of spherically symmetric functions, we find

$$\int_0^\infty V'(k)\; |k|^2 d\,|k| = 2\pi^2 V'(y) \Big|_{y=0} = 2\pi^2 \tag{A3.8}$$

and, hence,

$$A = \frac{1}{4\pi} \int K(a^{-1}\omega)\, d\omega\,, \tag{A3.9}$$

where the integration is carried out over the unit sphere $\omega^2 = 1$.

Note that the constant A does not depend on the absolute dimensions of the ellipsoid Ω, since $K(k)$ is a homogeneous function of the zeroth degree. Therefore the functional defined by (A3.2) does not depend on absolute dimensions of the ellipsoid. This enables one to pass to the limit by contracting the ellipsoid to a point; for definiteness, let this point be $x = 0$). Then the integral over V vanishes, since $K(x) \sim |x|^3$, and the second integral is expressed as an integral in the sense of principal value, which exists since $(\tilde{K}, \varphi)$ is well defined.

Thus, finally,

$$K(x) = \tilde{K}(x) + A\delta(x)\,, \tag{A3.10}$$

where A is given by (A3.9) and the regular generalized function is defined by the principal value of the integral

$$\tilde{K}, \varphi = \det\{a\} \fint K(ax)\varphi(ax)\, dx\,. \tag{A3.11}$$

Here and in (A3.9), the tensor a can be considered as an arbitrary nondegenerate linear transformation of x-space.

Quite analogously, for the generalized function $S(x)$ we have

$$S(x) = \tilde{S}(x) + D\delta(x)\,, \tag{A3.12}$$

where

$$(\tilde{S}, \varphi) = \det\{a\} \int S(ax)\, \varphi\,(ax)\, dx\,, \tag{A3.13}$$

$$D = \frac{1}{4\pi} \int S(a^{-1}\omega)\, d\omega\,, \tag{A3.14}$$

while, according to (4.6.9),

$$D = C - CAC\,. \tag{A3.15}$$

The regularization of $K(x)$ and $S(x)$ can be carried over to the plane case in an obvious way, while

$$A = \frac{1}{2\pi} \int K(a^{-1}\omega)\, d\omega, \quad D = \frac{1}{2\pi} \int S(a^{-1}\omega)\, d\omega\,, \tag{A3.16}$$

where the integration is performed over the unit circle $\omega^2 = 1$.

Let us now consider a generalized function $F(x) = \nabla\nabla\Phi(x)$, where $\Phi(x)$ is a homogeneous function of degree $-m+1$, m being the dimension of the space. It is obvious that $F(x)$ is a homogeneous function of degree $-m-1$ and that its Fourier transform is a homogeneous function of degree 1. The regularization formula for $F(x)$ can be obtained analogously to (A3.1, 2) and has the form

$$\begin{aligned}(F, \varphi) = &- \int_V \nabla\Phi\,(x)\, [\nabla\varphi(x) - \nabla\varphi(0)]\, dx + \int_\Omega \nabla\Phi(x)[\varphi(x) - \varphi(0)]\, n(x)\, d\Omega \\ &- \int_\Omega \Phi(x) n(x) d\Omega \nabla\varphi(0) + \int_{\bar{V}} F(x)\, [\varphi(x) - \varphi(0)]\, dx\,.\end{aligned} \tag{A3.17}$$

Let V be a sphere of radius ε. If $\varepsilon \to 0$, then the first integral vanishes, since $\nabla\Phi \sim |x|^{-m}$, the integrals over Ω are reduced to integrals of the function $F(k)k$ over the unit sphere in k-space, and the last integral tends to an integral in the sense of a principal value and exists since (F, φ) is well defined.

If $F(-x) = F(x)$, then the integrals over Ω in (A3.17) vanish also, and the regularization becomes

$$(F, \varphi) = \int F(x)\, [\varphi(x) - \varphi(0)]\, dx\,. \tag{A3.18}$$

Let x^1, x^2, x^3 be cartesian coordinates in the three-dimensional x-space, and let n be a normal to the plane $x^3 = 0$. We consider a generalized function T over the plane x^1, x^2 which is generated by the function $S(x)$,

$$T(x^1, x^2) = nS(x^1, x^2, x^3)n \mid_{x^3=0} . \tag{A3.19}$$

It is easily verified that the k-representation of the function T has the form

$$T(k_1, k_2) = \frac{1}{2\pi} \int_{-\infty}^{\infty} nS(k_1, k_2, k_3)n \, dk_3 . \tag{A3.20}$$

Using the explicit expression for the k-representation of the function S (Appendix A2), one can show that this integral converges absolutely and determines an even, homogeneous function of degree 1. Since, in the x-representation, regularization of such functions has the form of (A3.6, 18), the functional T acts on a (two-space) test function φ, according to the rule

$$(T, \varphi) = ⨍ T(x)[\varphi(x) - \varphi(0)] \, dx , \tag{A3.21}$$

where the integral over the plane is understood in the sense of principal value and exists.

It is not difficult to verify that in the Fourier representation,

$$(T, \varphi) = \frac{2}{(2\pi)^2} \int T(k)\varphi(k) \, dk , \tag{A3.22}$$

corresponds to the regularization (A3.21). The integral converges absolutely.

Let Ω be a smooth, closed surface with the normal $n(x)$. We consider the integral

$$I(x) = \int n(x)S(x - x')n(x')b(x')\delta(\Omega) \, dx' , \tag{A3.23}$$

where $b(x)\delta(\Omega)$ is a δ-function with a smooth weight $b(x)$ (Sect. 6.1). The function $I(x)$ is defined for all x, other than $x \in \Omega$, since, in the latter case, the integral (A3.23) formally diverges. Let us find a regular representation of the integral $I(x)$ for $x \in \Omega$. For this purpose, we rewrite $I(x)$ in the form

$$I(x) = \int n(x)S(x - x') \, n(x')[b(x') - b(x)]\delta(\Omega) \, dx'$$
$$+ \int n(x)S(x - x')n(x')\delta(\Omega)dx'b(x) . \tag{A3.24}$$

From Gauss' formula and from the property $\operatorname{div} S = 0$ of the function $S(x)$, follows the equality

$$\int S(x - x') \, n(x')\delta(\Omega)dx' = \int_V \operatorname{div} S(x - x') \, dx' = 0 , \tag{A3.25}$$

where V is the region bounded by Ω. Thus, the second integral in (A3.24) is equal to zero.

Let us consider a sequence of finite functions φ_i, which converges to $\delta(\Omega)$, as $i \to \infty$. Replacing $\delta(\Omega)$ in (A3.12) for $S(x)$ and then take the limit as $i \to \infty$. The final expression for $I(x)$ is

$$I(x) = \fint n(x)S(x - x')n(x')[b(x') - b(x)]\delta(\Omega)\, dx' . \tag{A3.26}$$

This representation is valid for all x, while, for $x \in \Omega$ the existence of the integral in the sense of a principal value follows from (A3.21).

Now let Ω in (A3.23) be a smooth, not closed, simply-connected surface bound by a smooth contour Γ. This case is reduced to the previous one, if we consider Ω to be a part of the closed surface Ω_1, and $b(x)$ to be zero for all $x \in \bar{\Omega} = \Omega_1 \backslash \Omega$. In this case, the integral (A3.26) is represented by

$$I(x) = \fint_{\Omega} n(x)S(x - x')n(x')[b(x') - b(x)]\, d\Omega'$$
$$- \fint_{\bar{\Omega}} n(x)S(x - x')n(x')d\Omega' b(x) . \tag{A3.27}$$

With the help of Stokes' formula, the second integral is transformed into an integral along the boundary Γ. Here one has to take into account that, due to (5.3.3)[1]

$$S(x) = \mathrm{Rot}\, Z^{+}(x) , \tag{A3.28}$$

where Z is the Green's tensor for internal stress (Sect. 5.3).

Finally, the regularization of the integral (A3.23) becomes

$$\int n(x)S(x - x')n(x')b(x')\delta(\Omega)\, dx'$$
$$= \fint_{\Omega} n(x)S(x - x')n(x')[b(x') - b(x)]d\Omega' - \int_{\Gamma} n(x)\mathrm{rot}\, Z^{+}(x - x')d\Gamma b(x), \tag{A3.29}$$

where $d\Gamma$ is a vector element of length on Γ. For the integrals on the right-hand side to exist, it is sufficient that $b(x)$ be continuously differentiable on Ω.

A4 Calculation of Certain Conditional Means

We consider a composite medium which is a homogeneous matrix contaning a homogeneously distributed set of ellipsoidal inhomogeneities. Let $V(x)$ be the characteristic function of the whole region occupied by the inclusions. When using the method of an effective field in the case of composites, the problem of

[1]Recall that, in Chaps. 5 and 6, the Green's function S is denoted by G.

constructing different conditional means of stochastic functions, which are analogous to $V(x)$, arises. The one-dimensional analog of the problem is considered in describing stochastic pulse flows in radio engineering [B7.17]; a number of general results was obtained in papers on geometrical probabilities [B8.19, 26].

In a number of cases, it is convenient to approximate inclusions in the composite by point defects (Sect. 7.10). Therefore, we start with the the simplest case of a homogeneous point field in three-space. Let X be a random point field of ξ_i, $X_{x_1 \ldots x_n}$ reduces to X after excluding from the latter those ξ_i which coincided with one of the points x_i $(i = 1, \ldots, n)$ in the space; $X(x)$ and $X(x_1, \ldots, X_n; x)$ are the following generalized functions, concentrated on X $X_{x_1 \ldots x_n}$, respectively:

$$X(x) = \sum_{\xi_i \in X} \delta(x - \xi_i), \tag{A4.1}$$

$$X(x_1, x_2, \ldots, x_n, x) = \sum_{\xi_i \in X_{x_1 x_2 \ldots x_n}} \delta(k - \xi_i). \tag{A4.2}$$

Let us consider the following averages of the functions $X(x)$ and $X(x_1; x)$ with respect to the ensemble of realizations of the point field X:

$$\langle X(x) \rangle = \langle \sum_{\xi_i \in X} \delta(x - \xi_i \rangle, \tag{A4.3}$$

$$\langle X(x_1; x) \mid x_1; x \rangle = \frac{\langle X(x_1; x) X(x_1) X(x_1; x_2) \rangle}{\langle X(x_1) X(x_1, x_2) \rangle}, \tag{A4.4}$$

$$\langle X(x_1; x) \mid x_1 \rangle = \frac{\langle X(x_1; x) X(x_1) \rangle}{\langle X(x_1) \rangle} \tag{A4.5}$$

(with regard to the interpretation of these quantities as conditional means, Sect. 7.4).

In what follows, all random fields are assumed to be ergodic. The standard method of constructing the means of the type (A4.3–5), which we shall use, consists of using the ergodic property with the subsequent averaging with respect to the ensemble, if necessary. For example,

$$\langle X(x) \rangle = \lim_{v \to \infty} \frac{1}{v} \int_G \sum_{i=1}^{N} \delta(x - \xi_i)\, dx = \lim_{v \to \infty} \frac{N}{v} = \frac{1}{v_0}. \tag{A4.6}$$

Here G is the region of averaging, its limit coinciding with the whole space, v is the volume of G, N is a number of points of a fixed realization of the point field, which are contained in G, and v_0 is an average volume per point.

Analogously,

$$\langle X(x_1, x) \rangle = \lim_{v \to \infty} \frac{1}{v} \int_G \sum_{\xi_i = x_1} \delta(x - \xi_i) dx = \begin{cases} \dfrac{1}{v_0} \text{ if } x \neq x_1, \\ 0 \text{ if } x = x_1, \end{cases} \tag{A4.7}$$

where it is taken into account that, by definition $X(x_1; x_1) \equiv 0$.
Hence

$$\langle X(x)\rangle = \langle X(x_1; x)\rangle , \tag{A4.8}$$

to within a null-set. Note that all the equalities will be understood in this sense.

Let us proceed to the more complicated problem of constructing the two-point moment of the function $X(x)$. For ergodic fields

$$\begin{aligned}\langle X(x)X(x + x_1)\rangle &= \lim_{v\to\infty} \frac{1}{v} \int_G \sum_{i,j} \delta(x - \xi_i)\delta(x + x_1 - \xi_j)\, dx \\ &= \lim_{v\to\infty} \sum_{\xi_i, \xi_j \in G} \delta(x - \xi_j + \xi_i)\,.\end{aligned} \tag{A4.9}$$

We introduce a stochastic vector $\xi_{ij} = \xi_i - \xi_j$ and let its distribution density be $g_{ij}(x)$. We then average (A4.9) once more with respect to the ensemble of realizations. The means of spearate terms in (A4.9) are

$$\langle \delta(x_1 - \xi_{ij})\rangle = \begin{cases} \int \delta(x_1 - x) g_{ji}(x)dx = g_{ji}(x_1) \text{ if } d \neq i \\ \delta(x) \quad \text{if } j = i \end{cases} \tag{A4.10}$$

where it is taken into account that $\xi_{ii} = 0$.

Separating on the right-hand side of (A4.9) the terms with $i = j$, we have

$$\langle X(x)X(x + x_1)\rangle = \frac{1}{v_0}\,\delta(x_1) + \lim_{v\to\infty} \frac{1}{v} \sum_{\substack{i,j \\ (i\neq j)}} g_{ij}(x_1)\,. \tag{A4.11}$$

From here we find the expression for the numerator in the conditional mean (A4.4). By virtue of the definition (A4.2) of the function $X(x_1; x)$, the equality

$$X(x) = \delta(x - x_1) + X(x_1; x) \text{ with } x_1 \in X \tag{A4.12}$$

is valid, from which it follows that

$$\langle X(x)X(x + x_1)\rangle = \frac{1}{v_0}\,\delta(x_1) + \langle X(x; x + x_1)X(x)\rangle\,. \tag{A4.13}$$

Comparing this result with (A4.11), we obtain

$$\langle X(x_1; x)\rangle = \lim_{v\to\infty} \frac{1}{v} \sum_{\substack{i,j \\ i\neq j}} g_{ij}(x - x_1)\,. \tag{A4.14}$$

Let us consider two examples of random point fields in space.

1) *Poisson's field* (Sect. 7.8, 10). To Poisson's point field in space, there corresponds in (A4.14) the density $g_{ij} = 1/v$ with any finite v. It is easy to verify that

then

$$\langle X(x_1;\, x) X(x_1) \rangle = \frac{1}{v^2}\,, \tag{A4.15}$$

and the conditional mean (A4.4) becomes

$$\langle X(x_1;\, x) \mid x \rangle = \frac{1}{v_0}\,. \tag{A4.16}$$

In order to construct the conditional mean (A4.5), note that, due to the absence of correlation between positions of the points (Poisson's field!), the equalities

$$\langle X(x_1;\, x) X(x_1) X(x_1;\, x_2) \rangle = \langle X(x_1) \rangle \, \langle X(x_1;\, x) X(x_1;\, x_2) \rangle, \tag{A4.17}$$

$$\langle X(x_1;\, x) X(x_1) \rangle = \langle X(x_1) \rangle \, \langle X(x_1;\, x) \rangle \tag{A4.18}$$

hold.

Calculating the right means analogously to the preceding and substituting the result into (A4.5), we obtain

$$\langle X(x_1;\, x) \mid x_1;\, x_2 \rangle = \delta(x - x_2) + \frac{1}{v_0}\,. \tag{A4.19}$$

2) *Random spatial lattice.* Let us consider a random point field, to each element ξ_i of which there corresponds a stochastic vector

$$\xi_m = m + \rho_m + r\,, \tag{A4.20}$$

where m is a vector of a fixed regular lattice, ρ_m are independent stochastic vectors with zero mathematical expectation, r is a stochastic vector with a homogeneous distribution.

The characteristic function $g_{mn}(k)$ of the stochastic vector

$$\xi_{mn} = \xi_m - \xi_n = m - n + \rho_m - \rho_n \tag{A4.21}$$

has the form

$$g_{mn}(k) = f(k) f(-k) \mathrm{e}^{-\mathrm{i}k(m-n)}\,, \tag{A4.22}$$

where $f(k)$ is the characteristic function of the stochastic vector ρ_m.

If one introduces

$$g(x) = (2\pi)^{-3} \int f(k) f(-k) \mathrm{e}^{\mathrm{i}kx}\, dk\,, \tag{A4.23}$$

then the function $g_{mn}(x)$ is represented in the form

$$g_{mn}(x) = g(x - m + n)\,. \tag{A4.24}$$

Let us consider the double sum in (A4.14) and proceed from two-index quantities g_{mn} to one-index ones, and rewrite this double sum as a single one. The vectors $m - n$, with $m, n \in G$, take on N^2 values, some of which may be identical. Each of these vectors coincides with one of the vectors of the fixed lattice. The number of vectors, which are equal to a fixed one m, depends on the relations between the length $|m|$ of m and dimensions of G. If the dimensions of G are much larger than $|m|$, then this number is $\sim N$. Taking this into account, (A4.14) can be represented by

$$\langle X(x_1; x)X(x_1)\rangle = \lim_{v\to\infty}\frac{1}{v}\sum_{\substack{m,n\in G\\(m\neq n)}} g(x - x_1 - m + n) = \frac{1}{v_0}\sum_m{}' g(x - m)\,, \tag{A4.25}$$

where the primed summation sign means exclusion of the term with $m = 0$.

From here and from (A4.6) follows the desired expression for the mean (A4.4)

$$\langle X(x_1; x)\mid x_1\rangle = \sum_m{}' g(x - x_1 - m)\,. \tag{A4.26}$$

Let us proceed to constructing the mean (A4.5) for the random lattice. The mean in the numerator of (A4.5) has the form

$$\begin{aligned}
&\langle X(x_1; x_1 + x_2)X(x_1; x_1 + x_3)X(x_1)\rangle\\
&=\lim_{v\to\infty}\frac{1}{v}\int_G \sum_{\substack{j,j,k\\(j,k\neq i)}} \delta(x_1 - \xi_1)\delta(x_1 + x_2 - \xi_j)\delta(x_1 + x_3 - \xi_k)dx_1\\
&=\lim_{v\to\infty} \sum_{\substack{\xi_i,\xi_j,\xi_k\in G\\(j,k\neq i)}} \delta(x_2 - \xi_{ji})\delta(x_3 - \xi_{ki})\,,
\end{aligned} \tag{A.4.27}$$

where ξ_{ij} is given by (A4.21).

Let us average both sides of (A4.27) with respect to the ensemble of realizations of the point field (A4.20). Since ξ_{ji} and ξ_{ki} with $j \neq k$ are independant stochastic vectors, then their joint distribution function is $g_{ji}(x)g_{ki}(x)$. Averaging with respect to the ensemble of the separate terms in (A4.27) yields

$$\begin{aligned}
\langle\delta(x_2 - \xi_{ji})\delta(x_3 - \xi_{ki})\rangle &= \int \delta(x_2 - x')g_{ji}(x')dx' \int \delta(x_3 - x'')g_{ki}(x'')dx''\\
&= g_{ji}(x_2)g_{ki}(x_3) \text{ if } j \neq k\,,
\end{aligned} \tag{A4.28}$$

$$\langle\delta(x_2-\xi_{ji})\delta(x_3-\xi_{ji})\rangle = \int \delta(x_2-x)\delta(x_3-x)g_{ji}(x)dx$$
$$= \delta(x_2-x_3)g_{ji}(x_3) \quad \text{if } j=k\,. \tag{A4.29}$$

We substitute this result into (A4.27) and transform the double sums to single ones. Using (A4.25), we find the expression for the mean (A4.5) in the form

$$\langle X(x_1;x)\,|\,x_1;x_2\rangle = \frac{\sum_m' g(x-x_1-m)\sum_{n\neq m}' g(x_2-x_1-n)}{\sum_m' g(x_1-x_2-m)} + \delta(x-x_2)\,, \tag{A4.30}$$

where the primed summation sign means exclusion of the terms with $m=0$ or $n=0$.

We complete the consideration of random point fields and proceed to the construction of analogous means for homogeneous random fields of nonintersecting ellipsoids. Let us confine ourselves to the problem of constructing the means of the characteristic functions $V(x)$ and $V(x_1;x)$ of the form

$$\langle V(x)\rangle = \langle\sum_i V_i(x)\rangle\,, \tag{A4.31}$$

$$\langle V(x_1;x)\rangle = \langle\sum_{V_i\subset V x_1} V_i(x)\rangle\,, \tag{A4.32}$$

$$\langle V(x_1;x)\,|\,x_1\rangle = \frac{\langle V(x_1;x)V(x_1)\rangle}{\langle V(x_1)\rangle}\,, \tag{A4.33}$$

where the quantities appearing here are defined in Sects. 7.3, 4.

For ergodic fields, the means (A 4.31, 32) are most simply calculated according to

$$\langle V(x)\rangle = \langle V(x_1;x)\rangle = \lim_{v\to\infty}\frac{1}{v}\int_G \sum_k V_k(x)\,dx = p\,, \tag{A4.34}$$

where p is the concentration of inclusions.

Let us consider the second moment of the function $V(x)$. By definition,

$$\langle V(x)V(x+x_1\rangle = \lim_{v\to\infty}\frac{1}{v}\int_G \sum_i V_i(x)V_i(x+x_1) + \sum_{\substack{i,j\\(i\neq j)}} V_i(x)V_j(x+x_1]\,dx. \tag{A4.35}$$

Note that $\int V_i(x)V_i(x+x_1)\,dx$ is the volume of intersection of two identical ellipsoids, the centers of which differ by the vector x_1. From here, we have

$$\int V_i(x)V_i(x+x_1)dx = v_iJ_i(x_1)\,, \tag{A4.36}$$

where

$$J_i(x) = \begin{cases} \left(1 - \dfrac{|a_i^{-1}x|}{2}\right)^2 \left(1 + \dfrac{|a_i^{-1}x|}{4}\right) & \text{if } |a_i^{-1}x| \leq 2 \\ 0 & \text{if } |a_i^{-1}x| > 2 \end{cases} \tag{A4.37}$$

The tensor a_i^{-1} determines a linear transformation which transforms the ellipsoid V_i to the unit sphere.

For homogeneous fields of inclusions, the stochastic variables $v_i J_i(x)$ with any fixed x have the same distributions for all i. Therefore, the first term in (A4.35) is equal to the mean value of this stochastic variable, divided by the volume v_0 per inclusion.

The mean $\langle V(x_1, x) V(x_1) \rangle$ coincides with the mean (A4.35), if one does not consider those realizations, for which the points x and x_1 occur inside the same inclusion. It is obvious that this mean is equal to the second item in (A4.35). From here and from (A4.33, 34) we find

$$\Psi(x) = \langle V(x_1, x_1 + x \mid x_1) \rangle - \frac{1}{p} \lim_{v \to \infty} \frac{1}{v} \int_G \sum_{\substack{i=j \\ (i \neq j)}} V_i(x_1) V_j(x_1 + x)\, dx_1 . \tag{A4.38}$$

Starting with this expression, one can show that $\Psi(x)$ is a continuous function with the mean value p. Since the different V_i do not intersect $\Psi(0) = 0$. For each realization it follows from the definition of the function $V(x; x_1)$ that $V(x; x) = 0$.

The form of the function $\Psi(x)$ depends on specific structure of the random field of inclusions. In particular, for regular lattices of identical inclusions, the mean $\langle V(x) V(x_1) \rangle$ is a periodic function of the difference $x - x_1$. In a neighborhood of the origin, the form of this function is determined by the first term in (A4.35). From here, it follows that $\Psi(x)$ can be represented in the form

$$\Psi(x) = p \sum_m{}' J(x - m) , \tag{A4.39}$$

where m is a vector of the regular lattice, formed by the centers of inclusions, the primed summation sign means that the term with $m = 0$ is omitted, and the function $J(x)$ is defined by (A4.37).

In conclusion, we consider the means which appear in the problem of a random field of cracks in an elastic medium (Sect. 7.8).

Let $\Omega(x)$ be the sum of δ-functions concentrated on the set of flat elliptic surfaces, the latter forming a random field in space (Sect. 7.4). For ergodic fields, we have

$$\langle \Omega(x) \rangle = \lim_{v \to \infty} \frac{1}{v} \int_G \sum_i \Omega_i(x) dx = \lim_{v \to \infty} \frac{1}{v} \sum_{i=1}^{N} \pi a_i b_i , \tag{A4.40}$$

where the a_i, b_i are semiaxes of the ellipse Ω_i. Since, for a homogeneous field, all stochastic variables a_i, b_i have identical distributions with the same mean value of $\langle a\, b\rangle$, then

$$\langle \Omega \rangle = \frac{\pi \langle ab \rangle}{v_0}. \tag{A4.41}$$

Further, let us consider the mean

$$\langle h(x)\Omega(x)\Omega(x_1)\rangle = \langle \sum_{ij} h_i(x)\Omega_i(x)\Omega_j(x_1)\rangle, \tag{A4.42}$$

where the functions $h_i(x)$ are given on Ω_i by the relations

$$h_i(x^1, x^2) = \sqrt{1 - \left(\frac{x^1}{a_i}\right)^2 - \left(\frac{x^2}{b_i}\right)^2}\ . \tag{A4.43}$$

The axes x^1, x^2 coincide with the principal axes of the ellipse Ω_i.

Analogously to (A4.35), let us write (A4.42) in the form

$$\langle h(x)\Omega(x)\Omega(x + x_1)\rangle$$
$$= \lim_{v\to\infty} \frac{1}{v} \sum_i \int_G h_i(x)\Omega_i(x)\Omega_i(x + x_1)dx + \frac{1}{\langle \Omega \rangle} \Psi(x_1), \tag{A4.44}$$

where

$$\Psi(x_1) = \langle \Omega \rangle \lim_{v\to\infty} \frac{1}{v} \sum_{\substack{i,j\\(i\neq j)}} \int_G h_i(x)\Omega_i(x)\Omega_j(x + x_1)\, dx. \tag{A4.45}$$

In order to calculate each of the integrals in the first term in (A 4.44), let us choose a cartesian coordinate system on Ω_i, by taking the x^3-axis along the normal to Ω_i and the x^2 and x^3-axes along the principal axes of the ellipse. Integrating first with respect to x^3 and then with respect to x^1 and x^2, we obtain

$$\int h_i(x')\Omega_i(x')\Omega_i(x' + x)dx' = \pi a_i b_i J_i(x^1, x^2)\delta(x^3), \tag{A4.46}$$

where

$$J_i(x^1, x^2) = \begin{cases} \displaystyle\int_{\frac{1}{2}\eta_i}^{1} \frac{(\zeta - \eta_i)\sqrt{\eta_i(2\zeta - \eta_i)} + \zeta^2\left[\sin^{-1}\left(1 - \frac{\eta_i}{\zeta}\right) + \frac{\pi}{2}\right]}{\sqrt{1 - \zeta^2}}\, \zeta\, d\zeta \ \text{if}\ \eta_i \le 2 \\ 0 \ \text{if}\ \eta_i > 2 \qquad \left(\eta_i = \sqrt{\left(\frac{x^1}{a_i}\right)^2 + \left(\frac{x^2}{b_i}\right)^2}\right). \end{cases} \tag{A4.47}$$

Substituting this result into (A4.44), we have

$$\langle h(x)\Omega(x)\Omega(x+x_1)\rangle = \pi\langle abJ(x_1)\Omega'(x_1)\rangle + \frac{1}{\langle\Omega\rangle}\Psi(x_1)\,, \qquad \text{(A4.48)}$$

where $\langle abJ(x)\Omega(x)\rangle$ is the mean value of the stochastic variables $a_ib_iJ_i(x)\ \Omega'(x, a_i, b_i)$, $\Omega'(x, a_i, b_i)$ is the δ-function, concentrated on the flat elliptic surface, whose normal coincides with that of Ω_i, the values of the semiaxes are $2a_i$, $2b_i$ and the center is located at the origin.

The function $\Psi(x)$ depends on a specific model of a random field of cracks. For a regular lattice of identically oriented surfaces, $\Psi(x)$ has the form

$$\Psi(x) = \frac{\pi ab}{v_0}\sum_m{}' J(x-m)\Omega'(x-m)\,, \qquad \text{(A4.49)}$$

where m is a vector of the lattice, formed by the centers of the ellipses Ω_m, the primed summation sign denoting omission of the term with $m = 0$.

In Chap. 7, the simplified notation

$$\Omega'(x-m) = \Omega'_m(x) \qquad \text{(A4.50)}$$

was introduced.

References

Chapter 1

1.1 E. Cosserat, F. Cosserat: *Theorie des Corps Deformables* (Hermann, Paris 1909)

1.2 C. Truesdell, R.A. Toupin: "The Classical Field Theories", in: *Principles of Classical Mechanics and Field Theory,* Encyclopedia of Physics, Vol. III/1 (Springer, Berlin, Göttingen, Heidelberg 1960) pp. 226–793

1.3 Cz. Woźniak: Dynamic models of certain bodies with discrete-continuous structure. Arch. Mech. Stosow. *21*, 707–724 (1969)

1.4 G.N. Savin, Iu. N. Nemish: Investigations on stress concentration in couple-stress elasticity (survey). Prikl. Mekh. *6*, 1–17 (1968) [in Russian]

1.5 A.A. Il'jushin, V.A. Lomakin: "Couple-Stress Theories in the Mechanics of Solid Deformable Bodies", in *Prochnost i plastichnost* (Strength and Plasticity), ed. by A. Il'jushin (Nauka, Moscow 1971) pp. 54–61 [in Russian]

1.6 M. Misicu: *Mecanica Mediilor Deformabile. Fundamentele Eleaticitátii Structurale* (Continuum Mechanics. Foundations of Elasticity) (Acad. R.S.R., Bucharest 1967) [in Rumanian]

1.7 W. Nowacki: *Theory of Micropolar Elasticity* (Wydawn. uczeln. politechn., Poznań 1970) [in Polish]
W. Nowacki: *Theory of Nonsymmetric Elasticity* (Zaklad narod. im Ossolinskich PAN, Wroclaw 1970) [in Polish]

1.8 E. Kröner (ed.): *Mechanics of Generalized Continua.* Proc IUTAM Symposium on the Generalized Cosserat Continuum and the Continuum Theory of Dislocations with Applications, Freudenstadt and Stuttgart 1967 (Springer, Berlin, Heidelberg, New York 1968)

1.9 I.A. Kunin, V.G. Kosilova (eds): *Teoria Sred s Mikrostrukturoi* (Theory of Media with Microstructure, Bibliography) (Sib. Otdelenie AN SSSR, Institut teplofiziki, Novosibirsk 1976)

1.10 E. Kröner: On the physical reality of torque stresses in continuum mechanics. Int. J. Eng. Sci. *1*, 261–278 (1963)

1.11 I.A. Kunin: *Teoria Uprugikh Sred c Mikrostrukturoi* (Theory of Elastic Media with Microstructure) (Nauka, Moscow 1975)

1.12 I.A. Kunin: *Elastic Media with Microstructure I. One-Dimensional Models,* Springer Series in Solid-State Sciences, Vol. 26 (Springer, Berlin, Heidelberg, New York 1982)

1.13 D.G.B. Edelen: Irreversible thermodynamics of nonlocal systems. Int. J. Eng. Sci. *12*, 607–631 (1974)

Chapter 2

2.1 I.M. Gel'fand, G.E. Shilov: *Generalized Functions* (Academic, New York 1964)
2.2 M.Born, K. Huang: *Dynamical Theory of Crystal Lattices* (Clarendon Press, Oxford 1954)
2.3 E.L. Aero, E.V. Kuvshinskii: Fundamental equations of the theory of elastic media with rotationally interacting particles. Sov. Phys. Solid State *2*, 1272–1281 (1960)
2.4 G. Grioli: Elasticita asimmetrica. Annali di Matematica Pura e Applicata *4*, 389–418 (1960)
2.5 R.D. Mindlin, H.E. Tiersten: Effects of couple-stresses in linear elasticity. Arch. Rat. Mech. Anal. *11*, 415–448 (1962)
2.6 G. Leibfried: "Gittertheorie der mechanischen und thermischen Eigenschaften der Kristalle", in *Encyclopedia of Physics*, Vol.7, Part 1 (Springer, Berlin 1955)
2.7 J.M. Ziman: *Electrons and Phonons* (University Press, Oxford 1960)

Chapter 3

3.1 V.S. Oskotskii, A.L. Efros: On the theory of crystal lattices with noncentral interatomic interaction. Sov. Phys.-Solid State *3*, 448–457 (1961)
3.2 V.E.Vdovin, I.A. Kunin: A theory of elasticity with spatial dispersion. Three-dimensional complex structure. J. Appl. Math. Mech. *30*, 1272–1281 (1966)
3.3 I.M. Gel'fand, S.V. Fomin: *Calculus of Variations* (Prentice-Hall, Englewood Cliffs, NJ 1963)
3.4 V.E. Vdovin: On energy relations in the theory of media with microstructure. Dinamika Sploshnoi Sredy 7, 94–104 (1971)
3.5 M.Born, K. Huang: *Dynamical Theory of Crystal Lattices* (Clarendon Press, Oxford 1954)
3.6 K. Huang: On the atomic theory of elasticity. Proc. R. Soc. *A203*, No. 1072, 178–194 (1950)
3.7 A.O. Gel'fond: *Calcul des Differences Finies* (Dunod, Paris 1963)
3.8 V.A. Pal'mov: Fundamental equations of the theory of asymmetric elasticity. Prikl. Math. and Mech. *28*, 401–408 (1964)

Chapter 4

4.1 I.M. Lifshits, L.N. Rosentsveig: Green's tensor for anisotropic unbounded elastic medium. Zh. Exsp. Teor. Fiz. *17*, 783–791 (1947)
4.2 I.M. Gel'fand, G.E. Shilov: *Generalized Functions* (Academic Press, New York 1964)
4.3 G.I. Eskin: *Boundary-Value Problems for Elliptic Pseudo-Differential Equations* (Math. Soc. of USA, N.Y. 1981)

4.4 J.R. Willis: The stress field around an elliptical crack in an anisotropic elastic medium. Int. J. Eng. Sci. *6*, 253–263 (1968)
4.5 J.D. Eshelby: "Elastic inclusions and inhomogeneities", in *Prog. Solid Mech.*, Vol. 2, ed. by I.N. Sneddon, R. Hill (1961) pp. 88–140
4.6 J.D. Eshelby: The elastic field outside an ellipsoidal inclusion. Proc. R. Soc. *A252*, 561–569 (1959)
4.7 J.D. Eshelby: The determination of the elastic field of an ellipsoidal inclusion, and related problems. Proc. R. Soc. *A241*, 376–396 (1957)
4.8 I.M. Lifshifts, L.V. Tanatarov: On the elastic interaction of impurity atoms in crystal. J. Phys. Metals Metallography *12*, 331–338 (1961)

Chapter 5

5.1 E. Kröner: *Kontinuumstheorie der Versetzungen und Eigenspannungen*, (Springer, Berlin, Göttingen, Heidelberg 1958)
5.2 B.A. Bilby: "Geometry and continuum mechanics", in [Ref. B. 2.1, pp. 180–199]
5.3 E. Kröner: Allgemeine Kontinuumstheorie der Versetzungen und Eigenspannungen, Arch. Rat. Mech. Anal. *4* , 273–334 (1960)
5.4 I.A. Kunin: "Methods of tensor analysis in the theory of dislocations". A supplement to the Russian translation of J.A. Schouten: *Tensor Analysis for Physicists* (Nauka, Moscow 1965), pp. 374–443. (The supplement is available in English from the U.S. Department of Commerce, Clearing house for Federal Sci. and Techn. Information, Springfield, VA 22151)
5.5 *RAAG Memoirs of the Unifying Study of Basic Problems in Engineering and Physical Sciences by Means of Geometry*, ed. by K. Kondo (Gakujutsu Bunken Fukyu-Kai, Tokyo, V.I. 1955, V. II 1958; V. III 1962, V. IV 1968)
5.6 C.C. Wang: "On the geometric structure of simple bodies, a mathematical foundation for the theories of continuous distributions of dislocations", in [Ref. B2.1, pp. 247–250]
5.7 R. de Wit: "Differential geometry of a nonlinear continuum theory of dislocations", in [Ref. B2.1, pp. 251–261]
5.8 I.A. Kunin: Internal stresses in media with microstructure. J. Appl. Math. Mech. *31*, 898–906 (1967)
5.9 I.A. Kunin: An algebra of tensor operators and its applications to elasticity. Int. J. Eng. Sci., *19*, 1551–1561 (1981)
5.10 A.I. Lur'e: *Three dimensional Problems of the Theory of Elasticity* (Interscience, New York 1964)
5.11 I.M. Lifshits, L.V. Tanatarov: On the elastic interaction of impurity atoms in crystals. J. Phys. of Metals and Metallography *12*, 331–338 (1961)

Chapter 6

6.1 J. Friedel: *Les Dislocations* (Gauthier-Villars, Paris 1956)

6.2 E. Kröner: *Kontinuumstheorie der Versetzungen und Eigenspannungen,* (Springer, Berlin, Göttingen, Heidelberg 1958)

6.3 I.A. Kunin: Fields due to arbitrary distributions of dislocations in an anisotropic elastic medium. J. Appl. Mech. Tech. Phys. *6*, 76 (1955)

6.4 I.A. Kunin: "Methods of tensor analysis in the theory of dislocations". A supplement to the Russian translation of J.A. Schouten. *Tensor Analysis for Physicists* (Nauka, Moscow 1965), pp. 374–443. (The supplement is available in English from the U.S. Department of Commerce, Clearing house for Federal Sci. and Techn. Information, Springfield, VA 22151)

6.5 I.M. Gel'fand, G.E. Shilov: *Generalized Functions* (Academic Press, New York 1964)

6.6 A.D. Brailsford: Stress field of a dislocation. Phys. Rev. *142*, 383–392 (1966)

Chapter 7

7.1 I.I. Gikhman, A.V. Skorokhod: *Theory of Random Processes*, Vol. 1, (Nauka, Moscow 1971) [in Russian]

7.2 T.D. Shermergor: *Theory of Elastic Inhomogeneous Media* (Nauka, Moskow 1977) [in Russian]

7.3 I.A. Chaban: Self-consistent field approach to calculation of effective parameters of microinhomogeneous media. Sov. Phys.-Acoust. *10*, 298–304 (1964)

7.4 E. Kröner: Bounds for effective elastic moduli of disordered materials. J. Mech. and Phys. Solids, *25*, 137–155 (1977)

7.5 I.I. Gikhman, A.V. Skorokhod: *Theory of Random Processes,* Vol.2 (Nauka, Moscow 1971) [in Russian]

7.6 V.M. Levin: On the stress concentration in inclusions in composite materials. J. Appl. Math. and Mech. *41*, 753–761 (1977)

7.7 A.V. Hershey: The elasticity of an isotropic aggregate of anisotropic cubic crystals. J. Appl. Mech. *21*, 236 (1954)

7.8 G. Kneer: Über die Berechnung der Elastizitätmoduls vielkristaller Aggregate mit Textur. Phys. Stat. Sol. *9*, 825–838 (1965)

7.9 E. Kröner: Berechnung der elastischen Konstanten des Vielkristalls aus den Konstanten des Einkristalls, Z. Phys. *151*, 504 (1958)

7.10 P.R. Morris: Elastic constants of polycrystals. Int. J. Eng. Sci. *8*, 49 (1970)

7.11 Z. Hashin, S. Shtrikman: On some variational principles in anisotropic and nonhomogeneous elasticity; J. Mech. Phys. Solids *10*, 343 (1962)

7.12 S.K. Kanaun: Self-consistent field approximation for an elastic composite medium. J. Appl. Mech. Tech. Phys. *18*, 274–282 (1977)

7.13 E.I. Grigoliuk, L.A. Fil'shtinskii: *Perforated Plates and Shells* (Nauka, Moscow 1970) [in Russian]

7.14 A.E. Andreikiv, V.V. Panasiuk, M.M. Stadnik: Fracture of prismatic bars with cracks. Problemy Prochnosti, No. 10. 10–16 (1972)
7.15 B. Budjanski, R.J. O'Connel: Elastic moduli of a cracked solid. Int. J. Solids Struct. *12*, (1976)
7.16 A.S. Vavakin, R.L. Salganik: On effective characteristics of inhomogenous media with local inhomogeneities. Izv. Akad. Nauk SSSR, MTT No.3 (1975)
7.17 S.K. Kanaun, G.I. Iablokova: "Self-consistent field approximation in plane problem for systems of interacting cracks", in *Mekhanika Sterzhnevykh Sistem i Sploshnykh Sred,* No. 9, (Leningr. Inshenernostroitel'nyi Institut, Leningrad 1976) pp. 194–203
7.18 L.A. Fil'shtinskii: Interaction of a doubly-periodic system of rectilinear cracks in an isotropic medium. J. Appl. Math. and Mech. *38* , 853–861 (1974)
7.19 P.C. Daris, G.C.M. Sih: "Stress analysis of cracks" in *Fracture Toughness Testing and its Applications* (Am. Soc. for Testing and Materials, 1965) pp. 30–83
7.20 V.V.Panasiuk, M. Savruk, A.P. Datsishin: *Stress Distribution Around Cracks in Plates and Shells* (Naukova Dumka, Kiev 1976)
7.21 G. Adomian: Linear random operator equations in mathematical physics. J. Math. Phys. *12*, 1944 (1971)
7.22 R.J. Elliott, J.A. Krumhansl, P.L. Leath: The theory and properties of randomly disordered crystals. Rev. Mod. Phys. *46*, 465 (1974)
7.23 M. Hori: Statistical theory of effective electrical, thermal and magnetic properties of random heterogeneous materials. J. Math. Phys. *14*, 514–523; 1942–1948 (1973); *15*, 2177–2185 (1974); *16*, 352–364; 1772–1775 (1975); *18*, 487–501 (1977)
7.24 A.H. Nayfeh: *Pertubation Methods* (Wiley, New York 1970)
7.25 V.M. Levin: On calculation of effective moduli of composites. Dokl. Akad. Nauk SSSR *220*, No. 5 (1975)
7.26 L.T. Walpole: On bounds for the overall elastic moduli of inhomogeneous systems – I, II. J. Mech. Phys. Solids *14*, 151 and 289 (1966)
7.27 M. Kac: "Mathematical mechanisms of phase transitions" in *Statistical Physics. Phase Transitions and Superfluidity,* Vol. 1 ed. by M. Chretien, (Gordon & Breach, New York 1968) pp. 241–305

Bibliography

B1 General Topics in Continuum Mechanics

B1.1 M.Y. Beran: *Statistical Continuum Theories* (Interscience, New York, 1968)

B1.2 I.M. Lifshits, L.N. Rosentsveig: Green's tensor for anisotropic unbounded elastic medium. Zh. Exsp. Teor. Fiz. *17*, 783–791 (1947)

B1.3 A.I. Lur'e: Three-dimensional Problems of the Theory of Elasticity (Interscience , New York 1964)

B1.4 N.I. Muskhelishvili: *Singular Integral Equations: Boundary Problems of Functions Theory and their Applications to Mathematical Physics*, (Wolters-Noordhoff, Groningen 1972)

B1.5 S. Timoshenko, D.H. Young, W. Weaver: *Vibration Problems in Engineering*, (Wiley, New York 1974)

B1.6 C. Truesdell, R.A. Toupin: "The Classical Field Theories", in *Handbuch der Physik*, Vol. III/1, ed. by S. Flügge (Springer, Berlin, Göttingen, Heidelberg 1960)

B2 Couple-Stress Theories

B2.1 E. Kröner (ed.): *Mechanics of Generalized Continua.* Proc. IUTAM Symposium on the Generalized Cosserat Continuum and the Continuum Theory of Dislocations with Applications, Freudenstadt and Stuttgart 1967 (Springer, Berlin, Heidelberg, New York 1968)

B2.2 I.A. Kunin, V.G. Kosilova (eds.): *Theory of Media with Microstructure.* Bibliography (Sib. Otdelenie AN SSSR, Institut Teplofiziki, Novosibirsk 1976)

B2.3 E.L. Aero, E.V. Kuvshinskii: Continuum theory of asymmetric elasticity. The problem of internal rotation. Sov. Phys.-Solid State *5*, 1892–1897 (1963)

B2.4 E.L. Aero, E.V. Kuvshinskii: Continuum theory of asymmetric elasticity. Equilibrium of an isotropic body. Sov. Phys.-Solid State *6*, 2141–2148 (1964)

B2.5 E.L. Aero, E.V. Kuvshinskii: Fundamental equations of the theory of elastic media with rotationally interacting particles. Sov. Phys.-Solid State *2*, 1272–1281 (1960)

B2.6 A. Askar: Molecular crystals and the polar theories of the continua. Experimental values of material coefficients for KNO_3. Int. J. Eng. Sci. *10*, 233–300 (1972)

B2.7 E. Cosserat, F. Cosserat: *Theorie des Corps Deformables* (Hermann, Paris 1909)

B2.8 A.C. Eringen: Balance laws of micromorphic mechanics. Int. J. Eng. Sci. *8*, 819–828 (1970)

B2.9 A.C. Eringen: Linear theory of micropolar elasticity. J. Math. and Mech. *15*, 909–924 (1966)

B2.10 A.C. Eringen: "Mechanics of micromorphic continua", in [Ref. B2.1, pp. 18–35]

B2.11 A.C. Eringen, E.S. Suhubi: Nonlinear theory of simple micro-elastic solids. Int. J. Eng. Sci. *2*, 189–203 and 389–404 (1964)

B2.12 G. Herrmann, J.D. Achenbach: Applications of theories of generalized Cosserat continua to the dynamics of composite materials, in [Ref. B2.1, pp. 69–79]

B2.13 A.E. Green: Micro-materials and multipolar continuum mechanics. Int. J. Eng. Sci. *3*, 533–537 (1965)

B2.14 A.E. Green, P.M. Naghdi: "The Cosserat surface", in [Ref. B2.1, pp. 36–48]

B2.15 A.E. Green, R.S. Rivlin: Multipolar continuum mechanics. Arch. Rat. Mech. Anal. *17*, 113–147 (1964)

B2.16 G. Grioli: Elasticita asimmetrica. Annali di matematica pura e applicata *4*, 389–418 (1960)

B2.17 W. Günter: Zur Statik und Kinematik des Cosseratschen Kontinuum. Abh. Braunschw. Wiss. Ges. *10*, 195–213 (1958)

B2.18 A.A. Il'jushin, V.A. Lomakin: "Couple-stress theories in the mechanics of solid deformable bodies", in "Prochnost' i plastichnost'" (Nauka, Moscow 1971) pp. 54–61

B2.19 W.T. Koiter: Couple-stresses in the theory of elasticity. Proc. K. Ned. Acad. Wet. Ser. *B67*, 17 (1964)

B2.20 M.R. Korotkina: Note about couple-stresses in discrete media. Vestn. Mosk. Univ.-Math. Mech. No. 5, 103–109 (1969)

B2.21 R.D. Mindlin: Micro-structure in linear elasticity, Arch. Rat. Mech. Anal. *16*, 51–78 (1964)

B2.22 R.D. Mindlin, H.E. Tiersten: Effects of couple-stresses in linear elasticity. Arch. Rat. Mech. Anal. *11*, 415–448 (1962)

B2.23 M. Misicu: *Mecanica Medulor Deformabile Fundamentele Elasticitatii Structurale* (Acad. R. S. R., Bucuresti 1967) [in Roumanian]

B2.24 W. Nowacki: *Theory of Micropolar Elasticity* (Wydawn. uczeln. politechn. Poznan 1970) [in Polish]

B2.25 W. Nowacki: *Theory of Nonsymmetric Elasticity* (Zaklad narod. im Ossolinskich PAN, Wroclaw 1970) [in Polish]

B2.26 V.A. Pal'mov: Fundamental equations of the theory of asymmetric elasticity. Prikl. Math. and Mech. *28*, 401–408 (1964)

B2.27 V.A. Pal'mov: On a model of a medium with complex structure. J. Appl. Math. and Mechanics *33*, 747–753 (1969)

B2.28 R.S. Rivlin: "Generalized Mechanics of continuous media", in [Ref. B2.1, pp. 1–17]

B2.29 D. Rogula: Moment stresses and the symmetry of the stress tensor in bodies with non-local structure. Bull Acad. Pol. Sci. Ser. Sci. Techn. *18*, 159–164 (1970)

B2.30 G.N. Savin, Iu. N. Nemish: Investigations on stress concentration in couple-stress elasticity (survey). Prikl. Mekhanika *6*, 1–17 (1968)

B2.31 H. Schaefer: The Cosserat continuum. Z. Angew. Math. und Mech. *47*, 485–498 (1967)

B2.32 Ia. M. Shiriaev: Investigation of the influence of a scale factor on stress concentrations around holes. Mekhanika Polimerov, No. 3, 565 (Akad. Nauk Latv. SSR, 1970)

B2.33 E. Sternberg: "Couple stresses and singular stress concentrations in elastic solids", in [Ref. B2.1, pp. 95–108]

B2.34 R. Stojanovic: "Dislocations in the generalized elastic Cosserat continuum", in [Ref. B2.1, pp. 152–155]

B2.35 R.A. Toupin: Theories of elasticity with couple-stresses. Arch. Rat. Mech. Anal. *17*, 85–112 (1964)

B2.36 R.A. Toupin: "Dislocated and oriented media", in [Ref. B2.1, pp. 126–140]

B2.37 K. Wilmanski, Cz. Wozniak: On geometry of continuous medium with microstructure. Arch. Mech. Stosowanej *19*, 715–723 (1967)

B2.38 Cz. Wozniak: Dynamic models of certain bodies with discrete-continuous structure. Arch. Mech. Stosowanej *21*, 707–724 (1969)

B3 Nonlocal Theories

B3.1 D.G.B. Edelen, A.E. Green, N. Laws: Nonlocal continuum mechanics. Arch. Rat. Mech. Anal. *43*, 36–44 (1971)

B3.2 A.C. Eringen, D.G.B. Edelen: On non-local elasticity. Int. J. Engng. Sci. *10*, 233–248 (1972)

B3.3 A.C. Eringen: Nonlocal polar elastic continua. Int. J. Eng. Sci *10*, 1–6 (1972)

B3.4 A.C. Eringen: Linear theory of nonlocal microelasticity and dispersion of plane waves. Lett. Appl. Sci. *1*, 129–146 (1973)

B3.5 A.C. Eringen, C.G. Speziale, B.S. Kim: Crack-tip problem in nonlocal elasticity. J. Mech. Phys. Solid. *25*, 339–355 (1977)

B3.6 A.C. Eringen: "Nonlocal continuum mechanics and some applications", in *Nonlinear Equations in Physics and Mathematics*, ed. by A.O. Barut (Reidel, Dordrecht, Holland 1978) pp. 271–318

B3.7 F. Hehl, E. Kröner: Zum Materialgesetz eines elastichen Mediums mit Momentenspannungen. Z. Naturforsch. *20*, 336–350 (1965)

B3.8 S. Kaliski, C. Rymarz: Surface waves in continua with nonlocal interaction. Biul. WAT J. Dabrowskiego *20*, 17–39 and 25–37 (1971)

B3.9 D. Kessel, E. Kröner: Nichtlokale Elastizitätstheorie der Ionenkristalle vom Typ A^+B^-. Z. Naturforsch. *25*, 1046–1053 (1970)

B3.10 V.G. Kosilova, I.A. Kunin: Dynamics of generalized Cosserat models. Dinamika Sploshnoi Sredy No. 4, 73–82 (1970)

B3.11 V.G. Kosilova, I.A. Kunin: On a scattering problem for media with dispersion. Chislennye Metody Mekhaniki Sploshnoi Sredy, *5,* No. 2, 51–72 (1974)

B3.12 V.G. Kosilova, I.A. Kunin: Wave packets propagation in one-dimensional medium with space dispersion. Dinamika Sploshnoi Sredy, No. 6, 127–142 (1970)

B3.13 E. Kröner: On the physical reality of torque stresses in continuum mechanics. Int. J. Eng. Sci. *1,* 261–278 (1963)

B3.14 E. Kröner: Elasticity theory of materials with long range cohesive forces. Int. J. Structures, *3,* 731–742 (1967)

B3.15 E. Kröner: "Interrelations between various branches of continuum mechanics", in [Ref. B2.1, pp. 330–340]

B3.16 E. Kröner: " Über die Symmetrie-Eigenschaften elastischer Materialien mit Kohäsionskräften endlicher Reichweite," in *"Problemy Gidrodinamiki i Mekhaniki Sploshnoi Sredy"* (Nauka, Moskva 1969) pp. 293–300

B3.17 E. Kröner, B.K. Datta: Nichtlokale Elastostatik: Ableitung aus der Gittertheorie. Z. Physik *196,* 203–211 (1966)

B3.18 E. Kröner, B.K. Datta: "Non-local theory of elasticity for a finite inhomogeneous medium - a derivation from lattice theory", in *Fundamental Aspects of Dislocation Theory,* Vol. 2, ed. by J. Simmons and R. de Wit (National Bureau of Standards, Washington 1970) pp. 737–746

B3.19 J.A. Krumhansl: "Some considerations of the relation between solid state physics and generalized continuum mechanics", in [Ref. B2.1, pp. 298–311]

B3.20 I.A. Kunin: Inhomogeneous elastic medium with nonlocal interaction. J. Appl. Mech. Tech. Phys. *8,* 41–44 (1967)

B3.21 I.A. Kunin: Internal stresses in media with microstructure. J. Appl. Math. Mech. *31,* 898–906 (1967)

B3.22 I.A. Kunin: Model of an elastic medium of simple structure with three-dimensional dispersion. J. Appl. Math. Mech. *30,* 642–652 (1966)

B3.23 I.A. Kunin: Theory of elasticity with spatial dispersion. One-dimensional complex structure. J. Appl. Math. Mech. *30,* 1025–1034 (1966)

B3.24 I.A. Kunin: The Green's tensor for an elastic isotropic medium with spatial dispersion. J. Appl. Mech. Tech. Phys. *8,* 103 (1967)

B3.25 I.A. Kunin: Theories of elastic media with microstructure. Proc. Vibr. Probl. *9,* 323–336 (1968)

B3.26 I.A. Kunin: "The theories of elastic media with microstructure and the theory of dislocations", in [Ref. B2.1, pp. 321–329]

B3.27 I.A. Kunin: "The theory of elastic media with microstructure", in *Prochnost'i Plastichnost'* (Nauka, Moskva 1971) pp. 65–70

B3.28 I.A. Kunin, A.M. Vaisman: "On problems of the non-local theory of elasticity", in: *Fundamental Aspects of Dislocation Theory,* Vol. 2, ed. by J. Simmons and R. de Wit (National Bureau of Standards, Washington 1970) pp. 747–759

B3.29 R.D. Mindlin: "Theories of elastic continua and crystal lattice theories", in [Ref. B2.1, pp. 312–320]

B3.30 E. Soos: One model of pseudo-continuum for ionic crystal lattices. Proc. Vibrat. Probl. Pol. Acad. Sci. *11,* 251–265 (1970)

B3.31 A.M. Vaisman, I.A. Kunin: Boundary value problems in nonlocal theory of elasticity. J. Appl. Math. Mech. *33,* 765–777 (1969)

B3.32 A.M. Vaisman, A.P. Putintseva: Certain effects of a nonlocal nature in a chain involving long-range interaction, J. Appl. Mech. Tech. Phys. *11,* 621–626 (1970)

B3.33 V.E. Vdovin: On energy relations in the theory of media with microstructure. Dinamika Sploshnoi Sredy No. 7, 94–104 (1971)

B3.34 V.E. Vdovin, I.A. Kunin: A theory of elasticity with spatial dispersion. Three-dimensional complex structure. J. Appl. Math. Mech. *30,* 1272–1281 (1966)

B4 Nonlinear Waves in Media with Dispersion

B4.1 *A Discussion on Nonlinear Theory of Wave Propagation in Dispersive Systems.* Proc. R. Soc. *A 299,* No. 1456 (1967)

B4.2 M.J. Ablowitz, D.S. Kaup, A.C. Newel, H. Segur: Method for solving the Sine-Gordon equation. Phys. Rev. Lett. *30,* 1262 (1973)

B4.3 M.J. Ablowitz, D.S. Kaup, A.S. Newell, H. Segur: The inverse scattering transform – Fourier analysis for nonlinear problems. Studies in Appl. Math. *53,* 249 (1974)

B4.4 *Bäcklund Tranformations, the Inverse Scattering Method, Solitons and their Applications,* ed. by R.M. Miura, Lecture Notes Math., Vol. 515 (Springer, Berlin, Heidelberg, New York 1976)

B4.5 A.A. Belavin, V.E. Zakharov: Multi-dimensional method of the inverse scattering problem and duality equations for the Yang-Mills fields. JETP Lett. *25,* 567–570 (1977)

B4.6 V.A. Belinskii, V.E. Zakharov: Integration of the Einstein equations by means of the inverse scattering problem technique and construction of exact soliton solutions. Sov. Phys. JETP *48,* 985–994 (1978)

B4.7 A.S. Budagov: "A completely integrable model of a classical field theory with non-trivial particles interaction in two-dimensional space-time", in *Voprosy Kvantovoi Teorii Polia i Statisticheskoi Fiziki* (Nauka, Leningrad 1978)

B4.8 F. Calogero, A. Degasperis: Nonlinear evaluation equations solvable by the inverse spectral transform. Nuovo Cimento *32B,* 201 (1976); *39B,* 1 (1977)

B4.9 J. Corones: An illustration of the Lie group framework for soliton equations: generalization of the Lund-Regge model. J. Math. Phys. *19,* 2431–2436 (1978)

B4.10 J. Corones: Solitons and simple pseudopotentials. J. Math. Phys. *17,* 756–759 (1976)

B4.11 J. Corones: Solitons, pseudopotentials and certain Lie algebras. J. Math. Phys. *18,* 163–164 (1977)

B4.12 J. Corones, B.L. Markovski, V.A. Risov: A Lie group framework for soliton equations. I. Path independent case. J. Math. Phys. *18,* 2207–2213 (1977)

B4.13 J. Corones, B.L. Markovski, V.A. Risov: Bilocal Lie groups and solitons. Phys. Lett. *61A,* 439–440 (1977)

B4.14 M. Crampin: Solitons and SL (2, R). Phys. Lett. *66A,* 170–172 (1978)

B4.15 M. Crampin, F.A.E. Pirani, D.C. Robinson: The soliton connection. Lett. Math. Phys. *2,* 15–19 (1977)

B4.16 B.A. Dubrovin: Inverse problem for periodic finite-zones potentials in the theory of scattering. Funct. Anal. Appl. *9,* 61–62 (1975)

B4.17 B.A. Dubrovin, V.B. Matveev, S.P. Novikov: Non-linear equations of Korteweg-de Vries type, finite-zone linear operations, and Abelian varieties, Russian Math. Surveys *31,* 59–146 (1976)

B4.18 B.A. Dubrovin, S.P. Novikov: A periodicity problem for the Korteweg-de Vries and Sturm-Liouville equations. Their connection with algebraic geometry. Soviet Math. Dokl. *15,* 1597–1601 (1974)

B4.19 G. Eilenberger: *Solitons,* Springer Ser. Solid-State Sci., Vol. 19 (Springer, Berlin, Heidelberg, New York 1981)

B4.20 *Exact Treatment of Nonlinear Waves,* Prog. Theor. Phys. (Suppl.) *59,* 1–161 (1976)

B4.21 C.S. Gardner, J.M. Green, M. Kruskal, R.M. Miura: Method for solving the Korteweg-de Vries equation. Phys. Rev. Lett. *19,* 1095–1097 (1967)

B4.22 I.M. Gel'fand, L.A. Dikii: A Lie Algebra structure in a formal variational calculus. Funct. Anal. Appl. *10,* 16–22 (1976)

B4.23 I.M. Gel'fand, L.A. Dikii: Asymptotic behavior of the resolvent of Sturm-Lioville equations and the algebra of the Korteweg-de Vries equations. Russian Math. Surveys *30,* No. 5, 77–113 (1975)

B4.24 I.M. Gel'fand, L.A. Dikii: The calculus of jets and nonlinear Hamiltonian systems. Funct. Anal. Appl. *12,* 81–94 (1978)

B4.25 I.M. Gel'fand, L.A. Dikii: The resolvent and Hamiltonian systems. Funct. Anal. Appl. *11,* 93–105 (1977)

B4.26 R. Hermann: Pseudopotentials of Estabrook and Wahlquist, the geometry of solitons and the theory of connections. Phys. Rev. Lett. *37,* 835–836 (1976)

B4.27 R. Hirota: Exact N-soliton solutions of the wave equation of long waves in shallow-water and in nonlinear lattices. J. Math. Phys. *14,* 810–814 (1973)

B4.28 B.B. Kadomtsev, V.I. Karpman: Nonlinear waves. Sov. Phys.-Uspekhi *14,* 40–60 (1971)

B4.29 V.I. Karpman: *Non-linear Waves in Dispersive Media* (Pergamon, Oxford, New York 1975)

B4.30 I. Kay, U.E. Moses: Reflectionless transmission through dielectrics and scattering potentials. J. Appl. Phys. *27,* 1503–1508 (1956)

B4.31 B.G. Konopelchenko: Completely integrable equations: dynamical groups and their nonlinear realizations. J. Phys. *A12*, 1937–1949 (1979)

B4.32 I.M. Krichever: Algebro-geometric construction of the Zakharov-Shabat equations and their periodic solution. Sov. Math. Dokl. *17*, 394–397 (1976)

B4.33 I.M. Krichever: Methods of algebraic geometry in the theory of non-linear equations. Russian Math. Surveys *32*, No. 6, 185–213 (1977)

B4.34 E.A. Kuznetsov, A.V. Mikhailov: On the complete integrability of the two-dimensional classical Thirring model. Theor. Math. Phys. *30*, 193–200 (1977)

B4.35 M. Lakshmanan: Geometric interpretation of solitons. Phys. Lett. *64A*, 354–356 (1978)

B4.36 P.D. Lax: Integrals of nonlinear equations of evolution and solitary waves. Commun. Pure Appl. Math. *21*, 467–490 (1968)

B4.37 P. Lax: Periodic solutions of the Korteweg-de Vries equations. Commun Pure Appl. Math. *28*, 141–188 (1975)

B4.38 P. Lax: Periodic solutions of the Korteweg-de Vries equations. Lect. Appl. Math. *15*, 85–95 (1974)

B4.39 S.C. Lowell: Wave propagation in monoatomic lattices with anharmonic potential. Proc. R. Soc. *A318*, No. 1532, 93–106 (1970)

B4.40 F. Lund: Example of a relativistic completely integrable Hamiltonian system. Phys. Rev. Lett. *38*, 1178 (1977)

B4.41 F. Lund: Classically solvable field theory model. Ann. Phys. *115*, 251–268 (1978)

B4.42 Iu. I. Manin: *Algebraic Aspects of Nonlinear Differential Equations* (VINITI, Moskva 1978)

B4.43 S.V. Manakov: Complete integrability and stochastization of discrete dynamical systems. Sov. Phys.-JETP *40*, 269–274 (1974)

B4.44 S.V. Manakov: Nonlinear Fraunhofer diffraction. Sov. Phys.-JETP *38*, 693–696 (1974)

B4.45 S.V. Manakov: On the theory of two-dimensional stationary self-focusing of electromagnetic waves. Sov. Phys.-JETP *38*, 248–253 (1974)

B4.46 V.A. Marchenko: *Spectral Theory of Sturm-Liouville Operators* (Naukova Dumka, Kiev 1972)

B4.47 V.A. Marchenko: The periodic Korteweg-de Vries problem. Math. of the USSR – Sbornik *24*, 319–344 (1974)

B4.48 M.S. Marinov: Nonlinear wave equations and differential geometry. Sov. J. Nucl. Phys. *28*, 125–132 (1978)

B4.49 R.M. Miura: The Korteweg-de Vries equation: a survey of the results. SIAM Rev. *18*, 412–459 (1976)

B4.50 R.M. Miura, C.S. Gardner, M.D. Kruskal: Korteweg-de Vries equation and generalization. II. Existence of conservation laws and constants of motion. J. Math. Phys. *9*, 1204–1209 (1968)

B4.51 H.C. Morris: A prolongation structure for the AKNS system and its generalization. J. Math. Phys. *18,* 533–536 (1977)
B4.52 H.C. Morris: Prolongation structures and a generalized inverse scattering problem. J. Math. Phys. *17,* 1867–1869 (1976)
B4.53 S.P. Novikov: The periodic problem for the Korteweg-de Vries equation. Funct. Anal. Appl. *8,* 236–246 (1974)
B4.54 N. Ooyama, N. Saito: On the stability of lattice solitons. Supplement of the Progress of Theoretical Physics No. 45, 201–208 (1970)
B4.55 A.C. Scott, F.Y. Chu, D.W. Laughlin: The soliton: a new concept in applied science. Proc. IEEE *61,* 1443–1483 (1973)
B4.56 A.B. Shabat: On non-scattering potentials in the Sturm-Liouville equation. Dinamika Sploshnoi Sredy, No. 5, 130 (1970)
B4.57 A.B. Shabat: On the Korteweg-de Vries equation. Sov. Math. Dokl. *14,* 1266–1270 (1973)
B4.58 *Solitons,* ed. by R. Bullough, P. Caudrey, Topics Current Phys. Vol. 17 (Springer, Berlin, Heidelberg, New York 1980)
B4.59 L.A. Takhtadzhyan: Exact theory of propagation of ultrashort optical pulses in two-level media. Sov. Phys.-JETP *39,* 228–233 (1974)
B4.60 L.A. Takhtadzhyan, L.D. Faddeev: Essentially nonlinear one-dimensional model of classical field theory. Theor. Math. Phys. *21,* 1046–1057 (1974)
B4.61 S. Tanaka: On the N-tuple wave solutions of the Korteweg-de Vries equation. Publ. Res. Inst. Math. Sci. Kyoto University *8,* 419–427 (1972/1973)
B4.62 *Theory of Solitons,* ed. by S.P. Novikov (Nauka, Moscow 1980) [in Russian]
B4.63 M. Toda: Waves in nonlinear lattice. Supplement of the Progress of Theoretical Physics No. 45, 174–200 (1970)
B4.64 M. Toda: *Theory of Nonlinear Lattices,* Springer Ser. Solid-State Sci., Vol. 20 (Springer, Berlin, Heidelberg, New York 1981)
B4.65 M. Wadafi, H. Sanuki, K. Kondo: Relationships among inverse methods, Bäcklung transformation and an infinite number of conservation laws. Progr. Theor. Phys. *53,* 419–436 (1975)
B4.66 M. Wadati, M. Toda: The exact N-solution of the Korteweg-de Vries equation. J. Phys. Soc. Jpn. *32* (1972)
B4.67 H.D. Wahlquist, F.B. Estabrook: Prolongation structures of nonlinear evolution equations. J. Math. Phys. *16,* 1–7 (1975); *17,* 1293–1297 (1976)
B4.68 Y. Watanabe: Note on gauge invariance and conservation laws for a class of nonlinear partial differential equations. J. Math. Phys. *15,* 453–457 (1974)
B4.69 N.J. Zabusky: Solitons and bound states of the time-independent Schrödinger equation. Phys. Rev. *168,* 124–128 (1968)
B4.70 V.E. Zakharov: Kinetic equation for solitons. Sov. Phys.-JETP *33,* 538–541 (1971)

B4.71 V.E. Zakharov: On stochastization of one-dimensional chains of nonlinear oscillators. Sov. Phys.-JETP *38*, 108–110 (1974)

B4.72 V.E. Zakharov: *The Method of Inverse Scattering Problem*, Lecture Notes in Mathematics (Springer, Berlin, Heidelberg, New York 1978)

B4.73 V.E. Zakharov, L.D. Faddeev: Korteweg-de Vries equation: a completely integrable Hamiltonian system. Funct. Anal. Appl. *5*, 280–287 (1971)

B4.74 V.E. Zakharov, S.V. Manakov: Resonant interaction of wave packets in nonlinear media. JETP Lett. *18*, 243–245 (1973)

B4.75 V.E. Zakharov, S.V. Manakov: The theory of resonance interaction of wave packets in nonlinear media. Sov. Phys. JETP *42*, 842–851 (1975)

B4.76 V.E. Zakharov, A.B. Shabat: A scheme for integrating the nonlinear equations of mathematical physics by the method of the inverse scattering problem I. Funct. Anal. Appl. *8*, 226–235 (1975)

B4.77 V.E. Zakharov, A.B. Shabat: Exact theory of two-dimensional self-focusing and one-dimensional self-modulation of waves in nonlinear media. Sov. Phys.-JETP *34*, 62–69 (1972)

B4.78 V.E. Zakharov, L.A. Takhtadzhyan: Equivalence of the nonlinear Schrödinger equation and the equation of a Heisenberg ferromagnetics. Theor. Math. Phys. *38*, 17–23 (1979)

B4.79 V.E. Zakharov, L.A. Takhtadzhyan, L.D. Faddeev: Complete description of solutions of Sine-Gordon equation. Dokl. Akad. Nauk SSSR *219*, 1334–1337 (1974)

B4.80 A. Seeger: *Solitons in crystals, in Continuum Models of Discrete Systems*, ed. by E. Kröner, K.-H. Anthony and H.H.E. Leipholz, (University of Waterloo Press, Waterloo 1980) pp. 253–327

B5 Crystal Lattice

B5.1 V.M. Agranovich, V.L. Ginzburg: *Crystal-Optics with Spatial Dispersion and the Theory of Excitons*. Springer Ser. Solid-State Sci., Vol. 42 (Springer, Berlin, Heidelberg, New York 1983)

B5.2 M. Born, K. Huang: *Dynamical Theory of Crystal Lattices* (Clarendon Press, Oxford 1954)

B5.3 L. Brillouin, M. Parodi: *Propagation des Ondes dans les Milieux Periodiques* (Masson, Paris 1956)

B5.4 K. Huang: On the atomic theory of elasticity. Proc. R. Soc. *A203*, No. 1072, 178–194 (1950)

B5.5 P.N. Keating: On the sufficiency of the Born-Huang relations. Phys. Lett. *A25*, 496–497 (1967)

B5.6 A.M. Kosevich: *Fundamentals of Crystal Lattice Mechanics* (Nauka, Moscow 1972)

B5.7 G. Leibfried: „Gittertheorie der mechanischen und thermischen Eigenschaften der Kristalle", in *Encyclopedia of Physics*, Vol. 7, Part 1 (Springer, Berlin 1955)

B5.8 A.N. Luzin: Some problems in the dynamics of crystal lattices and of the theory of elasticity. Sov. Phys.-JETP *24*, 615–620 (1966)

B5.9 H.J. Maris, C. Elbaum: Validity of the Born-Huang relations in lattice dynamics. Phys. Lett. *A25*, 96–97 (1967)

B5.10 A.A. Maradudin, E.M. Montroll, G.H. Weiss, I.P. Ipatova: *Theory of Lattice Dynamics in the Harmonic Approximation*, Solid State Physics, Suppl. 3, 2nd ed. (Academic, New York 1971)

B5.11 V.S. Oskotskii, A.L. Efros: On the theory of crystal lattices with noncentral interatomic interaction. Sov. Phys.-Solid State *3*, 448–457 (1961)

B5.12 J.M. Ziman: *Electrons and Phonons* (University Press, Oxford 1960)

B6 Dislocations and Local Defects

B6.1 B.A. Bilby: "Geometry and continuum mechanics", in [Ref. B2.1, pp. 180–199]

B6.2 A.D. Brailsford: Interaction between localized defects in an isotropic elastic medium. J. Appl. Phys. *40*, 3087–3088 (1969)

B6.3 A.D. Brailsford: Stress field of a dislocation. Phys. Rev. *142*, 383–392 (1966)

B6.4 A.H. Cottrell: *Theory of Crystal Dislocations* (Gordon and Breach, New York 1964)

B6.5 J.D. Eshelby: "Elastic inclusions and inhomogeneities", in *Prog. Solid Mech.*, Vol. 2, ed. by I.N. Sneddon, R. Hill (1961) pp. 88–140

B6.6 J.D. Eshelby: The elastic field outside an ellipsoidal inclusion. Proc. R. Soc. *A252*, 561–569 (1959)

B6.7 J.D. Eshelby: The determination of the elastic field of an ellipsoidal inclusion, and related problems. Proc. R. Soc. *A241*, 376–396 (1957)

B6.8 *Fracture: an Advanced Treatise*, Vol. 1–7, ed. by H. Liebowitz (Academic, New York 1968)

B6.9 J. Friedel: *Les Dislocations* (Gauthier-Villars, Paris 1956)

B6.10 V.L. Indenbom, A.N. Orlov: "Physical foundations of dislocation theory", in [Ref. B2.1, pp. 166–179]

B6.11 V.L. Indenbom, A.N. Orlov: Physical theory of plasticity and strength. Sov. Phys.-Uspekhi *5*, 272–291 (1962)

B6.12 N. Kinoshita, T. Mura: Elastic fields of inclusions in anisotropic media. Phys. Stat. Sol. (A) *5*, 759–768 (1971)

B6.13 A.M. Kosevich, V.D. Natsik: Dislocation damping in a medium having dispersion of the elastic moduli. Sov. Phys.-Solid State *8*, 993–999 (1966)

B6.14 V.S. Kosilova, I.A. Kunin, E.G. Sosnina: Interaction of point defects taking into account spatial dispersion. Sov. Phys.-Solid State *10*, 291–296 (1968)

B6.15 E. Kröner: *Kontinuumstheorie der Versetzungen und Eigenspannungen* (Springer, Berlin, Göttingen, Heidelberg, 1958)

B6.16 E. Kröner: Allgemeine Kontinuumstheorie der Versetzungen und Eigenspannungen, Arch. Rat. Mech. Anal. *4*, 273–334 (1960)

B6.17 I.A. Kunin: Green's tensor for anisotropic elastic medium with internal stress sources. Dokl. Akad. Nauk SSSR *157*, 1319–1320 (1964)

B6.18 I.A. Kunin: Internal stress in an anisotropic elastic medium. J. Appl. Math. Mech. *28*, 612 (1964)
B6.19 I.A. Kunin: Fields due to arbitrary distributions of dislocations in an anisotropic elastic medium. J. Appl. Mech. Tech. Phys. *6*. 76 (1955)
B6.20 I.A. Kunin: "Methods of tensor analysis in the theory of dislocations". A supplement to the Russian translation of J.A. Schouten: *Tensor Analysis for Physicists* (Nauka, Moscow 1965) pp. 374–443. (The supplement is available in English from the U.S. Department of Commerce, Clearing house for Federal Sci. and Techn. Information, Springfield, VA 22151)
B6.21 I.A. Kunin, G.N. Mirenkova, E.G. Sosnina: An ellipsoidal crack and needle in an anisotropic elastic medium. J. Appl. Math. Mech. *37*, 501–508 (1973)
B6.22 I.A. Kunin, E.G. Sosnina: An ellipsoidal inhomogeneity in an elastic medium. Dokl. Akad Nauk SSSR *199*, 571–574 (1971)
B6.23 I.A. Kunin, E.G. Sosnina: Local inhomogeneities in an elastic medium. J. Appl. Math. Mech. *34*, 399–405 (1970)
B6.24 I.A. Kunin, E.G. Sosnina: Stress concentration on an ellipsoidal inhomogeneity in an anisotropic elastic medium. J. Appl. Math. Mech. *37*, 287–305 (1973)
B6.25 I.A. Kunin: An algebra of tensor operators and its applications to elasticity. Int. J. Eng. Sci., *19*, 1551–1561 (1981)
B6.26 I.A. Kunin: "Projection operator method in continuum mechanics", in *Continuum Models of Discrete Systems 4* (North-Holland, Amsterdam 1981) pp. 179–187
B6.27 G. Leibfried, N. Breuer: *Point Defects in Metals I,* Springer Tracts Mod. Phys., Vol. 81 (Springer, Berlin, Heidelberg, New York 1978)
B6.28 I.M. Lifshifts, L.V. Tanatarov: On the elastic interaction of impurity atoms in crystal. J. Phys. Metals Metallography *12*, 331–338 (1961)
B6.29 *Mechanics of Fracture,* Vols. 1–4, ed. by G.C. Sih (Nordhoff, Leyden 1973)
B6.30 T. Mura: Continuum theory of plasticity and dislocations. Int. J. Eng. Sci. *5*, 341–351 (1967)
B6.31 T. Mura, P.C. Cheng: The elastic field outside an ellipsoidal inclusion. J. Appl. Mech. *44*, 591–594 (1977)
B6.32 T. Mura, T. Mori: Elastic fields produced by dislocations in anisotropic media. Phil. Mag. *33*, 1021–1027 (1976)
B6.33 F.R.V. Nabarro: The mathematical theory of stationary dislocations. Adv. Phys. *1*, 271–395 (1952)
B6.34 W. Noll: "Inhomogeneities in materially uniform simple bodies", in [Ref. B2.1, pp. 239–246]
B6.35 *RAAG Memoirs of the Unifying Study of Basic Problems in Engineering and Physical Sciences by Means of Geometry*, ed. by K. Kondo (Gakujutsu Bunken Fukyu-Kai, Tokyo, V.I. 1955, V.II 1958, V.III 1962, V.IV 1968)

B6.36 D. Rogula: Influence of spatial acoustic dispersion on dynamical properties of dislocation. Bull. Akad. Pol. Sci., Ser. Sci. Techn. *13*, 337–585 (1965)

B6.37 M.A. Sadowsky, E. Sternberg: Stress concentration around a triaxial ellipsoidal cavity. J. Appl. Mech. *16*, 149–157 (1949)

B6.38 A.M. Vaisman, I.A. Kunin: On boundary-value problems of the continuum theory of dislocations. Dokl. Akad. Nauk SSSR *173*, 1024–1027 (1967)

B6.39 H.G. van Bueren: *Imperfections in Crystals* (North-Holland, Amsterdam 1960)

B6.40 V.E. Vdovin, I.A. Kunin: Dislocation interactions taking into account spatial dispersion. Sov. Phys.-Solid State *10*, 297–303 (1968)

B6.41 C.C. Wang: "On the geometric structure of simple bodies, a mathematical foundation for the theories of continuous distributions of dislocations", in [Ref. B2.1, pp. 247–250]

B6.42 L.T. Wheeler, I.A. Kunin: On voids of minimum stress concentration. Int. J. Sol. Struct. *18*, 85–89 (1982)

B6.43 J.R. Willis: The stress field around an elliptical crack in an anisotropic elastic medium. Int. J. Eng. Sci. *6*, 253–263 (1968)

B6.44 R. de Wit: "Differential geometry of a nonlinear continuum theory of dislocations" in [Ref. B2.1, pp. 251–261]

B6.45 H. Zorski: Statistical theory of dislocations. Int. J. Solids Struct. *4*, 959–974 (1968)

B7 Random Fields of Inhomogeneities

B7.1 G. Adomian: Linear random operator equations in mathematical physics. J. Math. Phys. *12*, 1944 (1971)

B7.2 A.E. Andreikiv, V.V. Panasiuk, M.M. Stadnik: Fracture of prismatic bars with cracks. Problemy Prochnosti *10*, 10–16 (1972)

B7.3 B. Budjansky, R.J. O'Connel: Elastic moduli of a cracked solid. Int. J. Solids Struct. *12*, No. 2 (1976)

B7.4 I.A. Chaban: Self-consistent field approach to calculation of effective parameters of microinhomogeneous media. Sov. Phys.-Acoust. *10*, 298–304 (1964)

B7.5 R.J. Elliott, J.A. Krumhansl, P.L. Leath: The theory and properties of randomly disordered crystals. Rev. Mod. Phys. *46*, 465 (1974)

B7.6 L.A. Fil'shtinskii: Interaction of a doubly-periodic system of rectilinear cracks in an isotropic medium. J. Appl. Math. and Mech. *38*, 853–861 (1974)

B7.7 E.I. Grigoliuk, L.A. Fil'shtinskii: *Perforated Plates and Shells* (Nauka, Moskow 1970)

B7.8 Z. Hashin, S. Shtrikman: On some variational principles in anisotropic and nonhomogeneous elasticity. J. Mech. Phys. Solids *10*, 343 (1962)

B7.9 A.V. Hershey: The elasticity of an isotropic aggregate of anisotropic cubic crystals. J. Appl. Mech. *21*, 236 (1954)

B7.10 R. Hill: A self-consistent mechanics of composite materials. J. Mech; Phys. Solids *13*, 213 (1965)
B7.11 M. Hori: Statistical theory of effective electrical, thermal and magnetic properties of random heterogeneous materials. J. Math. Phys. *14*, 514–523; 1942–1948 (1973); *15*, 2177–2185 (1974); *16*, 352–364; 1772–1775 (1975); *18*, 487–501 (1977)
B7.12 M. Kac: "Mathematical mechanisms of phase transitions", in *Statistical Physics. Phase Transitions and Superfluidity*, Vol. 1, ed. by M. Chretien, (Gordon & Breach, New York 1968) pp. 241–305
B7.13 S.K. Kanaun: "Random field of cracks in elastic solid medium", in: "Issledovaniia po uprugosti i plastichnosti", No. 10 (Izd. Leningr. Univ. Leningrad 1974)
B7.14 S.K. Kanaun: Self-consistent field approximation for an elastic composite medium. J. Appl. Mech. Tech. Phys. *18*, 274–282 (1977)
B7.15 S.K. Kanaun: Self-consistent field method in the problem of effective properties of an elastic composite. J. Appl. Mech. Tech. Phys. *16*, 649–657 (1975)
B7.16 S.K. Kanaun, G.I. Iablokova: "Self-consistent field approximation in plane problem for systems of interacting cracks", in: Mekhanika sterzhnevykh sistem u sploshnykh sred No. 9, (Leningr. Inshenerno-stroitel'nyi Institut, Leningrad 1976) pp. 194–203
B7.17 S. Kirkpatrick: Percolation and conductivity. Rev. Mod. Phys. *45*, 574 (1973)
B7.18 G. Kneer: Über die Berechnung des Elastizitätsmodules vielkristaller Aggregate mit Textur. Phys. Stat. Sol. *9*, 825–838 (1965)
B7.19 E. Kröner: Berechnung der elastischen Konstanten des Vielkristalls aus den Konstanten des Einkristalls, Z. Physik *151*, 504 (1958)
D7.20 E. Kröner: Elastic moduli of perfectly disordered composite materials. J. Mech. Phys. Solids *15*, 319 (1967)
B7.21 V.M. Levin: On calculation of effective moduli of composites. Dokl. Akad. Nauk SSSR *220*, No. 5 (1975)
B7.22 E. Kröner: Bounds for effective elastic moduli of disordered materials. J. Mech. Phys. Solids *25*, 137–155 (1977)
B7.23 E. Kröner: Graded and perfect disorder in random media elasticity. ASCE J. Eng. Mech. Div. *106*, 889–914 (1980)
B7.24 V.M. Levin: "On calculation of microstress in polycrystals", in: Issledovaniia po uprugosti i plastichnosti, No. 10 (Izd. Leningr. Univ., Leningrad 1974)
B7.25 V.M. Levin: On the stress concentration in inclusions in composite materials. J. Appl. Math. Mech. *41*, 753–761 (1977)
B7.26 B.R. Levin: *Theoretical Foundations of Statistical Radioengineering*, (Sovetskoe Radio, Moskow 1969)
B7.27 P.R. Morris: Elastic constants of polycrystals. Int. J. Eng. Sci. *8*, 49 (1970)

B7.28 V.V. Panasiuk, M. Savruk, A.P. Datsishin: *Stress Distribution Around Cracks in Plates and Shells*, (Naukova Dumka, Kiev 1976)

B7.29 P.C. Daris, G.C.M. Sih: "Stress analysis of cracks", in: *Fracture Toughness Testing and its Applications* (Am. Soc. for Testing and Materials, New York 1965) pp. 30–83

B7.30 T.D. Shermergor: *Theory of Elastic Inhomogeneous Media* (Nauka, Moskow 1977)

B7.31 A.S. Vavakin, R.L. Salganik: On effective characteristics of inhomogeneous media with local inhomogeneities. Izv. Akad. Nauk SSSR, MTT No. 3 (1975)

B7.32 L.T. Walpole: On bounds for the overall elastic moduli of inhomogeneous systems I, II. J. Mech. Phys. Solids *14*, 151 and 289 (1966)

B7.33 J.R. Willis: Bounds and self- consistent estimates for the overall properties of anisotropic composites. J. Mech. Phys. Solids *25*, 185–202 (1977)

B7.34 J.R. Willis: A polarization approach to the scattering of elastic waves I, II. J. Mech. Phys. Solids *28*, 287–305 and 307–327 (1980)

B8 References of General Nature

B8.1 N.I. Achieser, I.M. Glasmann: *Theorie der linearen Operatoren im Hilbert Raum*, (Adademie, Berlin 1968)

B8.2 D. Bohm: *Quantum Theory* (Prentice-Hall, New-York 1952)

B8.3 H. Bremermann: *Distributions, Complex Variables, and Fourier Transforms* (Addison-Wesley, Reading, MA 1965)

B8.4 E.T. Copson: *Asymptotic Expansions* (University Press, Cambridge 1965)

B8.5 A. Erdelyi: *Asymptotic Expansions* (Dover Publications, New York 1956)

B8.6 G.I. Eskin: *Boundary-Value Problems for Elliptic Pseudo-Differential Equations,* (Math. Soc. of USA, N.Y. 1981)

B8.7 I.M. Gel'fand, S.V. Fomin: *Calculus of Variations* (Prentice-Hall, Englewood Cliffs, NJ 1963)

B8.8 I.M. Gel'fand, G.E. Shilov: *Generalized Functions* (Academic, New York 1964)

B8.9 A.O. Gel'fond: Calcul des Differences Finies (Dunod, Paris 1963)

B8.10 I.I. Gikhman, A.V. Skorokhod: *Theory of Random Processes*, Vol. 1, (Nauka, Moscow 1971) [in Russian]

B8.11 I.S. Gradshtein, I.M. Ryzhik: *Table of Integrals, Series and Products,* 4th ed. (Academic, New York 1965)

B8.12 D. Jackson: *Fourier Series and Orthogonal Polynomials* (The Mathematical Association of America, Oberlin 1941)

B8.13 M.G. Kendall, P.A.P. Moran: *Geometrical Probability* (Hafner, New York 1963)

B8.14 Ia. I. Khurgin, V.P. Iakovlev: *Localized Functions in Physics and Engineering* (Nauka, Moscow 1971) [in Russian]

B8.15 M.G. Krein: Integral equations on semiaxis with kernels depending on the difference of arguments. Usp. Mat. Nauk. *13,* 3–120 (1958)

B8.16 I.A. Kunin: Quantum mechanical formalism in classical wave propagation problems. Int. J. Eng. Sci. *20,* 271–281 (1982)

B8.17 L.D. Landau, E.M. Lifshitz: *Quantum Mechanics: Nonrelativistic Theory,* 3rd ed. (Pergamon, Oxford, New York 1977)

B8.18 P.D. Lax, R.S. Phillips: *Scattering Theory* (Academic, New York, London 1967)

B8.19 B.Iak. Levin: *Distribution of Zeros of Entire Functions* (Am. Math. Soc., Providence, RI 1964)

B8.20 G.Ia. Lubarski: *Groups Theory and its Application in Physics* (Fizmatgiz, Moscow 1958) [in Russian]

B8.21 G. Matheron: *Random Sets and Integral Geometry* (Wiley, New York 1975)

B8.22 M. Morse, H. Feshbach: *Methods of Theoretical Physics* Vol. 1, 2 (McGraw-Hill, New York 1953)

B8.23 A.H. Nayfeh: *Perturbation Methods* (Wiley, New York 1970)

B8.24 R.G. Newton: *Scattering Theory of Waves and Particles* (McGraw-Hill, New York 1966)

B8.25 J.A. Schouten: *Tensor Analysis for Physicists* (Clarendon Press, Oxford 1951)

B8.26 C.E. Shannon: *The Mathematical Theory of Communication* (University of Illinois Press, Urbana 1949)

B8.27 M.M. Vainberg: *Variational Methods for the Study of Non-Linear Operators* (Holden-Day, San Francisco 1964)

B8.28 I.A. Kunin: "Lie group deformations and quantizations", in *Trends in Applications of Pure Mathematics to Mechanics* (Pitman, London 1978) Vol. 2, pp. 171–178

B8.29 I.A. Kunin: "Group-theoretical foundations for interrelations between physical models", in *Continuum Models of Discrete Systems 3,* ed. by E. Kröner, K.-H. Anthony (University of Waterloo Press, Ontario 1980) pp. 43–58

Subject Index

Positrons in Solids

Editor: **P. Hautojärvi**

1979. 66 figures, 25 tables. XIII, 255 pages
(Topics in Current Physics, Volume 12)
ISBN 3-540-09271-4

Contents: P. Hautojärvi, A. Vehanen: Introduction to Positron Annihilation. – P.E. Mijnarends: Electron Momentum Densities in Metals and Alloys. – R.N. West: Positron Studies of Lattice Defects in Metals. – R.M. Nieminen, M.J. Manninen: Positrons in Imperfect Solids: Theory. – A. Dupasquier: Positrons in Ionic Solids.

M. Toda

Theory of Nonlinear Lattices

1981. 38 figures. X, 205 pages. (Springer Series in Solid-State Sciences, Volume 20)
ISBN 3-540-10224-8

Contents: Introduction. – The Lattice with Exponential Interaction. – The Spectrum and Construction of Solutions. – Periodic Systems. – Application of the Hamilton-Jacobi Theory. – Appendices A–J. – Simplified Answers to Main Problems. – References. – Bibliography. – Subject Index. – List of Authors Cited in Text.

D.C. Mattis

The Theory of Magnetism I

Statics and Dynamics

1981. 58 figures. XV, 300 pages
(Springer Series in Solid-State Scienes, Volume 17)
ISBN 3-540-10611-1

"...Mattis's relentless approach will not be to everyone's taste, but his book will be a useful addition to the library of anyone deeply interested in the origins of magnetism and the careful study of mathematical models. The statistical mechanician, or the particle theorist looking for hints on how to solve the lattice gauge theory problem, may, however, prefer to wait for the second volume which will cover thermodynamics and statistical mechanics. Finally, praise must be given for the introductory chapter, 38 pages long, which spells out the history of magnetism from the earliest days to the present, places it in the perspective of the general evolution of physics and the development of Western thought, and is backed up by marvellous quotations and an impressive bibliography. This chapter can be strongly recommended to anyone interested in the history fo science and, almost alone, would justify purchase of the book." *Nature*

The Structure and Properties of Matter

Editor: **T. Matsubara**
With contribution by numerous experts

1982. 229 figures. XI, 446 pages
(Springer Series in Solid-State Sciences, Volume 28)
ISBN 3-540-11098-4

Contents: Atoms as Constituents of Matter. – System of Protons and Electrons. – Helium. – Superfluid Helium 3. – Matals. – Nonmetals. – Localized Electron Approximation. – Magnetism. – Magnetic Properties of Dilute Alloys – the Kondo Effect. – Random Systems. – Coherent Protential Approximation (CPA). – References. – Subject Index.

Springer-Verlag
Berlin
Heidelberg
New York
Tokyo